Creative Regions

Creative Regions examines diverse aspects of the knowledge economy that may influence regional innovation capabilities and generate advantage for regions in global markets.

Uniquely, this book focuses on regional creativity, analysing a variety of factors that affect creativity in regional innovation processes under knowledge economy conditions. Approaching creativity from technological, organisational and regional viewpoints, the book analyses regional creativity from different perspectives and takes account of multi-level interactions in economy and policy.

The variety of papers presented examine:

- how regional actions can be creative and competitive
- how research is outsourced and creative knowledge and innovation are transferred
- types of technology-based cultural activities that are evolving
- sources of creative and innovative knowledge entrepreneurship.

Today, many of these processes are primarily market driven but not completely so, thus *Creative Regions* demonstrates that multi-level policy sectors have significant roles to play in stimulating and promoting creative and innovative developmental actions.

Creative Regions is essential reading on how regional advantage can be constructed for firms and agencies advancing in the global knowledge economy and will be of interest to students, researchers and policy makers working in this and related fields.

Philip Cooke is University Research Professor in Regional Development and founding Director (1993) of the Centre for Advanced Studies, University of Wales, Cardiff. He is a partner in Cardiff's ESRC Centre for Economic and Social Analysis of Genomics (CESAGen) and an Adjunct Professor at the University of Aalborg, Denmark.

Dafna Schwartz is a senior faculty member at the Department of Business Administration at Ben-Gurion University (Israel), head of the area of Entrepreneurship and High-Tech Management and director of the Centre for Entrepreneurship and High-Tech Management. She is an economic consultant in Israel and abroad and a board member of leading Israeli corporations.

Regions and cities

Series editors: Ron Martin, University of Cambridge, UK;
Gernot Grabher, University of Bonn, Germany; Maryann Feldman,
University of Georgia, USA

Regions and Cities is an international, interdisciplinary series that provides authoritative analyses of the new significance of regions and cities for economic, social and cultural development, and public policy experimentation. The series seeks to combine theoretical and empirical insights with constructive policy debate and critically engages with formative processes and policies in regional and urban studies.

Regional Innovation Strategies
The challenge for less-favoured regions
Kevin Morgan and Claire Nauwelaers (eds)

Foreign Direct Investment and the Global Economy
Nicholas A. Phelps and Jeremy Alden (eds)

Restructuring Industry and Territory
The experience of Europe's regions
Anna Giunta, Arnoud Lagendijk and Andy Pike (eds)

Community Economic Development
Graham Haughton (ed)

Out of the Ashes?
The social impact of industrial contraction and regeneration on Britain's mining communities
David Waddington, Chas Critcher, Bella Dicks and David Parry

Creative Regions

Technology, culture and
knowledge entrepreneurship

Edited by Philip Cooke and Dafna Schwartz

Routledge
Taylor & Francis Group

LONDON AND NEW YORK

First published 2007
by Routledge
2 Park Square, Milton Park, Abingdon, Oxon OX14 4RN

Simultaneously published in the USA and Canada
by Routledge
711 Third Avenue, New York, NY 10017

Routledge is an imprint of the Taylor and Francis Group, an informa business

First issued in paperback 2012

Typeset in Bembo by Wearset Ltd, Boldon, Tyne and Wear

British Library Cataloguing in Publication Data
A catalogue record for this book is available from the British Library

Library of Congress Cataloging in Publication Data
  Creative regions: technology, culture and knowledge
entrepreneurship/Philip Cooke and Dafna Schwarz (eds).
    p. cm. – (Regions and cities)
'Simultaneously published in the USA and Canada.'
1. Regional economics. 2. High technology industries—Location.
3. High technology industries—Location—Case studies.
4. Technological innovations—Government policy. 5. Knowledge
management. 6. Industrial clusters. 7. Entrepreneurship. I. Cooke,
Philip (Philip N.) II. Schwartz, Dafna.
HT388.C74 2007
330.9–dc22   2007004530

ISBN13: 978-0-415-43428-7 (hbk)
ISBN13: 978-0-415-54114-5 (pbk)

Contents

Figures

Tables

Contributors

Dietmar Bastian received his PhD in political science from Leipzig University in 2003 and is Associate Professor at the Friedrich Schiller University in Jena, Germany. His main fields of specialisation are political systems in international comparative perspective with a focus on regional, industrial and technology policies, and the political economy of transformation and development processes with a concomitant interest in East–West and North–South perspectives in international relations.

Kean Birch is a Research Fellow in the Centre for Public Policy for Regions (CPPR) at the University of Glasgow and a doctoral student in the Department of Planning at Oxford Brookes University. His thesis focuses on the UK biotechnology industry, while his current research concerns issues of regional development, particularly in relation to less favoured regions in Europe and the role of the social economy in sustainable development.

Frans Boekema is Professor of Economic Geography and Extraordinary Professor of Euroregional Management at Radboud University, Nijmegen, and Associate Professor of Regional Economics at Tilburg University. His research activities include local and regional development, technological development and regional innovation systems, borders, border regions and border crossing activities, clusters, networks and learning regions.

Pieter J. M. de Bruijn is consultant spatial economics at the Netherlands Organisation for Applied Scientific Research, TNO. His research interests concern the role of regional determinants in innovation processes, spatial coherence and networking, and quantitative modelling of regional economic dynamics.

Alex Burfitt is a Lecturer in the Centre for Urban and Regional Studies at the University of Birmingham. He has research interests in the development of regional knowledge economies and the associated restructuring of regional policy. He has also conducted research into the changing nature of housing consumption in the context of the restructuring of regional economies.

Chris Collinge is a Senior Lecturer in the Centre for Urban and Regional Studies at the University of Birmingham. He has research interests in regional and local economic change, particularly the emergence of 'knowledge economies' in various forms, and is currently researching the development of economic knowledge across European regions. He is also interested in the process of urban re/degeneration, and in the significance and conceptualisation of spatial scale.

Philip Cooke is University Research Professor of Regional Development at Cardiff University and Founding Director (1993) of its Centre for Advanced Studies. He is a partner in Cardiff's ESRC Centre for Economic and Social Analysis of Genomics (CESAGen) and an Adjunct Professor at the University of Aalborg, Denmark. His research focuses on Innovation, Bioregions and Policy Platforms. In 2006 he co-edited *Clusters and Regional Development* and *Regional Development in the Knowledge Economy* (both Routledge). In 2007 he co-authored *Regional Knowledge Economies* (Edward Elgar). He is an Academician of the UK Academy of Social Sciences and in 2006 was awarded an honorary doctorate by the University of Lund, Sweden.

Bent Dalum is Associate Professor of Economics at Aalborg University and head of department at the Department of Business Studies. He is a member of IKE, deputy director of DRUID, member of the executive committee of the EU Sixth Framework Network of Excellence DIME, the DIMETIC PhD Programme, a member of the CTIF steering committee and CTIF technical research council, among the organisers of NorCOM and a NOVI A/S board member.

Ulrich Hilpert has been Professor of Comparative Government at the Friedrich Schiller University in Jena, Germany, since 1992. He directs five major international research networks in his main fields of research, which are comparative European–American studies and the role of government policies for techno-industrial innovation, restructuring, environmental technologies, globalisation and regional development.

Hiro Izushi (PhD from the University of California at Berkeley) is a Senior Lecturer at Economics and Strategy Group, Aston Business School, Birmingham. He specialises in technological innovation with an emphasis on innovation management and economic development. He is co-author of the forthcoming book *Competing for Knowledge: The Global Evolution of Economic Growth* (with Robert Huggins; Routledge).

Kati-Jasmin Kosonen is a researcher at the Research Unit for Urban and Regional Development Studies (Sente) at University of Tampere (Finland). She has recently focused on studies concerning the development processes of local and regional innovation environments and new forms of cooperative paces in less favoured Finnish regions. Her expertise is in local and regional development policies and new economic development policy, programme-based policy making, local innovation systems and environments.

Luciana Lazzeretti is Professor of Business Economics at the Faculty of Economics of the University of Florence, Director of Doctoral School in Economics, and Director of the postgraduate programme in 'Economics and Management of Museums'. Her main fields of research are sectoral and local system analysis (cities of art, industrial and cultural districts, urban, creative, and cultural clusters) and the study of strategic behaviour of firms with a multidisciplinary approach.

Jens-Peter Lynov (MSc, PhD) has performed experimental and theoretical research in plasma physics and fluid dynamics in Denmark and as a guest researcher at several foreign universities, including the University of California at Berkeley and the University of New Mexico. Since 1999 he has been Head of the Optics and Plasma Research Department at Risø National Laboratory, and since 2003 he has represented Risø in the Musicon Valley collaboration.

Birgitte Rasmussen (MSc, PhD) is Senior Scientist in the Systems Analysis Department at Risø National Laboratory. Her main areas of work are technology foresight, dialogue processes, knowledge dynamics and practices, and the interaction of science, government and industry. She is engaged in the development of theory and methodology for technology mapping and systems thinking, and the impact of such mapping and thinking on research prioritisation and technological development.

Mika Raunio works as Research Fellow in Research Unit for Urban and Regional Studies (Sente) at the University of Tampere, Finland. In his research he focuses on human resources, the internationalisation of labour markets and the attractiveness of city regions from a regional development point of view.

Roel Rutten is Assistant Professor of Organisation Studies at Tilburg University, the Netherlands. He began his career as a consultant for ERAC (European Regional Affairs Consultants) and specialised in regional innovation policy. At Tilburg University his research focuses on the organisation of innovation networks and the role of proximity. Together with Frans Boekema he is author of several publications on learning regions.

Dagmara Stoerring is presently employed at the European Parliament's Directorate-General for Information. Since 2003 she has been working on a PhD on the cluster emergence process at Department of Business Studies at Aalborg University. She graduated from Aalborg University in 2002 with the degree of Master of Social Science in European Studies. In 1999 she gained the degree of Master of Arts in International Economics from the University of Economics in Poznań, Poland.

Dafna Schwartz, an economist (PhD), is Senior Lecturer in the Department of Business Administration at Ben-Gurion University, Israel. She heads the MBA programme in Entrepreneurship and High-Tech Management and is the director of a Centre for Entrepreneurship and Hi-Tech Management. She is an economic consultant in Israel and elsewhere and has experience as a board member of many leading Israeli corporations.

Frank G. van Oort is Professor of Urban Economics and Spatial Planning at Utrecht University. His research interests include agglomeration economics, urban economics, planning and housing, and the geography of knowledge production. He also works on these subjects at the Netherlands Institute for Spatial Research (RPB) in The Hague.

Rüdiger Wink is Professor for International Economics at Leipzig University of Applied Sciences (HTWK) and Senior Research Fellow at the Ruhr Research Institute on Regional and Innovation Policies (RUFIS) at Ruhr University, Bochum. His areas of research are connected with regional, science and innovation policies and evolutionary and institutional economics.

1 Creative regions: an introduction

Philip Cooke and Dafna Schwartz

Introduction

In this book we bring together for the first time a set of contributions to the knowledge economy literature that combines the innovation and creative dimensions that constitute that designation. The following, and conceivably most important, part of this opening chapter sets out what is intended to be a coherent theoretical framework capable of integrating a moderately diverse group of writings – diverse, but connected, in that each one responded to a conference call that had as its mission 'Enabling knowledge strategies'. The call went as follows:

> This track invites contributions on the role of knowledge institutions in regional development. These may be from science and technology, or culture and the creative arts. Both kinds of work are invited. Perhaps more is written on the former than the latter, and papers on the role of 'memory institutions' such as museums, libraries, galleries and archives as assets for regional growth are particularly welcome. Papers providing 'mapping' for Europe and elsewhere of knowledge centres of either and both kinds, with their associated concentrations of 'talent', are also encouraged. What mechanisms and strategies facilitate the transition from knowledge to business? How do businesses and non-firm organisations manage knowledge? Are there data or cases showing evolution in understanding of the need for organisational change if knowledge sharing is to add significant value? What about territorial knowledge management, boundary-crossing activities or intermediaries where distinct 'communities of practice' (e.g. scientists/entrepreneurs/public servants or bureaucrats/artists/business people) must interact? How does knowledge move from innovator to entrepreneur and what is the role of various intermediaries such as knowledge transfer offices, incubators, investors, collectors, exhibitors?

Notice the emphasis in the call on science *and* culture, mechanisms *and* strategies, with *evolution* and *change*. Each of these categories is core to the interests of many kinds of regional scientist and economic geographer. But they are of special interest to the new school of evolutionary economic geo-graphers (Boschma and Martin, forthcoming). So, the framework to be mapped out in the next section and into which the contributions are placed is rooted in evolutionary economic geography.

The second section of the chapter is devoted to definitional issues, a matter of the highest importance in contemporary academic and policy discourse, since definition of terms seems to be going out of fashion in much contempor-ary theoretical and empirical literature – or, if included, it seems to be stretched almost beyond breaking point. This is a major weakness of otherwise influential concepts like 'clusters', 'regions', development' and the like – a feature of current discourse which is critiqued at some length notably in Asheim *et al.* (2006) regarding 'clusters' and regarding the rest plus a few more in the intro-duction to Cooke and Piccaluga (2006). One result of sloppy or non-existent definition of terms is, naturally, non-communication – a strange ambition for writers – or perhaps it is that the post-modern geographies that have now spawned what might once have been referred to as a 'ginger group' of 'Grumpy Old Geographers' (Rodgers, 2006) now only speak among coteries. To bother to define the code would be all too embarrassing, perchance. So without yet having parted with the subscription fee for the above-mentioned ginger group, we feel no embarrassment in trying to say as clearly as we can what is under discussion in this book – at modest length.

Finally, we will demonstrate how the various chapters relate to the frame-work and to each other. For also included in the call for papers were the following injunctions:

> [W]e are interested in the question of *finance* for innovation and creative activity, ranging from film funds, musical contests/festivals and book-fests, exhibitions and award ceremonies, to business angels and public or private venture capital in stimulating regional growth. Needless to say, monitoring assessments of successful and unsuccessful *policy* strategies, and the forming of new institutions for commercialising innovative creative or technological knowledge, and any other policy-analytic contributions addressing these issues, will be particularly welcome.

Finance *and* policy; perhaps too ambitious, but we thought it necessary to try – and to our delight, papers were forthcoming on all of the following planned elements:

- regional innovation systems
- knowledge transfer and R&D outsourcing
- cultural economy and creative industries
- financing innovation and creativity
- knowledge entrepreneurship.

Each of these is complex enough to warrant a book in its own right, and some, such as regional innovation systems, have already yielded three or four.

So, we had a selection problem, as our publisher also advised. Hence the papers included as chapters in the book are fewer than we initially selected but certainly purer for having survived a rigorous peer review and selection process. Others that are of equal status, but for one reason or another (possibly because they are too close or overlapping with this selection, or because the authors, offered the option, preferred it) are to be published in a forthcoming Special Issue of *European Planning Studies*. Thus, we were keen to stick to the original emphasis on mixing innovation and creativity alongside the support platforms of innovation systems, finance and entrepreneurship.

An evolutionary economic geography approach

Setting a broad evolutionary economic geography framework for this book calls into play a reminder of some key elements broached first by Braczyk *et al.* (1998) and Cooke and Morgan (1998). According to these authors, Witt (1991) points to Schumpeter, Veblen, Marx and Hayek as the godparents of this approach. Schumpeter's great contribution lay in his insight that, contrary to the belief of the neoclassicals in the *equilibrium* tendencies of capitalism, that mode of economic organisation is better understood as a set of *disequilibriating* processes. That is, it is typified by constant movement and change expressed as evolutionary, incremental development *punctuated* by 'gales of creative destruction' brought about by innovations. Such innovations released a *swarming* effect as imitators piled in to seek 'second-comer' advantage from the creativity of the initial innovator. The more radical the innovation, the more likely it was to disequilibriate the economy in the form of the business cycle. From a regional science perspective it is correct to envisage these swarming asymmetries as the origins of, for example, large industry platforms like the coal, steel, engineering and pharmaceuticals industries of the nineteenth-century Ruhr Valley or, earlier, the textile districts of Britain and France.

Nowadays, radical innovation is first associated with *product* or *process* innovation in manufacturing or services, but the work of Klepper (2002) on the economic history of the US automotive industry platform in Detroit shows that *organisational* innovation in the sense of talented managers bringing experience of flow-line assembly from, for example, the armaments, food and tobacco industries and recombining these *routines* in the scaled-up setting of automotive plants was important. To some extent, the advice of Frederick Taylor succeeded these innovations rather than gave birth to them. Moreover, other routines were also carried over from related industries like coach building and rubber manufacture. Akron, Ohio, for example, is not very close to Detroit but in the pre-automobile era it was where rubber was utilised for coach and bicycle wheels (Buenstorf and Klepper, 2005). Local entrepreneurial spin-offs thereafter swarmed around the original producers to meet the huge new market demand from the Detroit automotive industry manufacturers. Boschma and Wenting (2005) tell a similar story of such 'related variety' in accounting for the rise of the UK automotive platform in Birmingham and Coventry.

We shall return to a discussion of the important concept of 'related variety' in due course, but first we briefly note the relevance of the other three godparents of evolutionary economics to our focus, which is evolutionary economic geography. Veblen (1919) was the first to ask: why is economics not an evolutionary science? The answer is that the equilibrium theorists had literally 'marginalised' it. He too was a disequilibrium theorist who developed an *institutional* perspective on economic evolution whereby individual and firm activity were characterised by inertia, custom and habit until punctuated by the impulse to idle curiosity and creativity, seeing economic evolution as a process of *cumulative causation* – a term Veblen invented – but with no predetermined direction towards enrichment or impoverishment. History, he thought, was opaque. From an evolutionary economic geography perspective, there are three clear elements of relevance to spatial processes in Veblen's thought. First, cumulative causation (see also Myrdal, 1957) is a predecessor concept to that of *path dependence*, one of the more widely deployed ideas explaining regional asymmetries along lines explored first as varieties of regional 'lock-in' by Grabher (1993). Second, cumulative causation also has a further useful meaning in the sense of *palimpsest*, or the ways regions are composed of successive historic 'rounds of investment' that evolve distinctive spatial divisions of labour, as memorably put by Massey (1984). But third, Veblen's was a remarkable insight into the creative process and foreshadowed modern *hedonic* approaches in economics with his deduction that 'an evolutionary economics must be a process of *cultural* growth' (Veblen, 1919, p. 77, emphasis added). These aspects of culture and creativity, as it were, effecting regional economic growth are the major insight of research on the nature and impact of the *creative class* on the locations where *talent* settles and, in a knowledge economy, attracts to it or creates through entrepreneurship, new knowledge-driven employment. These, of course, are, all three, core issues studied in this book.

What about Hayek and Marx (strange bedfellows)? Marx was nothing if not a disequilibrium theorist of considerable influence upon Schumpeter, and one of his and Engels' most memorable comments on the ideal creative life was:

> '[I]n in communist society, where nobody has one exclusive sphere of activity but each can become accomplished in any branch he wishes, society regulates the general production and thus makes it possible for me to do one thing today and another tomorrow, to hunt in the morning, fish in the afternoon, rear cattle in the evening, criticize after dinner.
>
> (Marx and Engels, 1970)

But maybe Hayek is more pertinent for his perspective on the primacy of *knowledge* in economic analysis. In Cooke and Morgan (1998) we wrote of

Hayek's notions of markets as settings for 'spontaneous social order [and] sense of economic development as a creative, undetermined, and unpredictable, cumulative process ... of individual, albeit opportunistic actions' (p. 199). To that we wish now to add the relevance of Hayek's observation on the role of knowledge in that spontaneity and creativity. He identified 'the division of knowledge as the really central problem of economics as a social science' (Hayek, 1948, p. 51) and its key question as addressing the puzzle of localised knowledge held by fragmentary firms and individuals nevertheless producing ordered market demand and supply:

> The most significant fact about this system is the economy of knowledge with which it operates, or how little the individual participants need to know in order to be able to take the right action. In abbreviated form, by a kind of symbol, only the most essential information is passed on, and passed on only to those concerned.
>
> (Ibid., p. 86)

In Cooke and Leydesdorff (2006) we noted how Marshall had expressed comparable sentiments thirty years before: 'Capital consists in a great part of knowledge and organisation ... knowledge is our most powerful engine of production ... organisation aids knowledge' (Marshall, 1916, p. 115). It is thus rather remarkable that neoclassical economics did almost nothing to further these insights on the importance of knowledge and a clear opportunity for the evolutionary approach to take up the challenge.

We conclude that the broad lineaments of an evolutionary economic geography of creativity and innovation will be interested in *path dependence* and its varieties of cumulative or superimposed 'lock-in' effects as well, crucially, as creative or innovative processes of escape along new pathways. One chapter that springs instantly to mind on precisely this is that of Rasmussen and Lynov (Chapter 10), which tracks the path of a decommissioned nuclear research institute in Denmark into the creative industries. A comparable process was also cited for Brazil by Cooke (2006a). The evolutionary economic geography framework is also informed by the disequilibrium concept of *punctuated evolution*, driven by radical, or even so-called 'disruptive', innovation (Christensen, 1997), creatively destroying the relative peace of day-to-day cumulative, incremental innovation. The difference between *radical* and *disruptive* innovation is the following: disruptive innovation is revolutionary regarding technological profile, e.g. candle loses out to light bulb. Yet context, i.e. room, house, remains largely unaltered, and its built environment can persist for hundreds of years. However, in the innovation systems tradition, radical innovation changes both technological profile and context or environment, e.g. air travel versus all other historic forms of travel. That is, airports, air traffic control systems, and so on are radical, literally 'root and branch', innovations in historic terms. Evolutionary economic geography's core interest is processes and effects of *agglomeration* and varieties of *proximity* effects and types, and many of the chapters involve such analysis, exploring, in particular, both creative and more prosaic clusters. Lazzeretti's investigation of Florence's 'creative habitats' (Chapter 9) is exemplary of the utilisation of the evolutionary methodology of *population dynamics*. Stoerring and Dalum's chapter on the evolution of the NorCOM mobile telephony and Biomedico clusters in Jutland (Chapter 7) is equally exemplary in its use of innovation systems methodology.

This connects with a fourth interest often prevalent in evolutionary research, which is a strong interest in the role of *innovation* in stimulating economic change. It is fair to say that the evolutionary school has by far the most sophisticated grasp of the nature and variety of categories of innovation, inspired by its neo-Schumpeterian strand that pioneered studies of various types of innovation system. Most of the chapters in this book relate directly to issues concerning innovation and its relationships with knowledge and creativity. The first part of the book is concerned entirely with systems analyses of regional innovation in, for example, Finland, France, Germany, Spain, the United Kingdom and the Netherlands. In these, in line also with the evolutionary interest in *institutions* and the mechanisms by which knowledge is explored and exploited for innovation and learning, including higher education institutions in the training of talent, evidence of evolutionary economic geography processes illuminates the accounts.

Finally, there are dimensions of newer evolutionary economic thinking that has been promoted by geographers that we will highlight as they arise in this book. First, mention has already been made of the powerful evolutionary economic geographical use of the concept of 'related variety'. This solves a debate that has long festered in the orthodox and evolutionary economics literature. It concerns the primacy or otherwise of industry *specialisation* (so-called Marshall–Arrow–Romer, MAR, externalities) versus *diversification* (so-called Jacobs externalities, after Jane Jacobs, 1969) giving the strongest support for innovation, improved productivity and growth. In the language most pertinent to the project of this book we are speaking of *knowledge spillovers* most centrally, but also about the manner in which innovation and creativity relate 'epidemiologically' to each other. Research by Frenken *et al.* (2005) and Boschma (2005) shows that 'related variety' is the optimal industry mix for innovation and growth. By contrast, over-specialisation and over-diversification result in low lateral absorptive capacity whereas *related variety* optimises it. Hence, innovations travel to neighbouring industries more swiftly, receptively because of high lateral absorptive capacity gains, and adaptively owing to the low cognitive distance associated with relational and geographical proximity. Entrepreneurship can be a valuable carrier of related variety innovations across industry boundaries as well as more direct, less intermediated knowledge transfers.

Mapping and defining creative regions

Knowledge is, of course, the core unifying device for the theory and empirical research to be reported on here and in what follows. So, we think of the creative region according to the following conceptual model, a little like Atlas holding up the world. In this book we focus in detail upon all bar the financial and entrepreneurship modules. These are explored in great detail in the Special RSA Aalborg Conference issue of *European Planning Studies*. Possession of a variety of the cultural assets in the visual arts, music, theatre, and so on is a fundamental element of the creative region. However, as we shall elaborate, it is not the same asset as creative industries. The latter may relate to the former and perhaps overlap on occasions, but they are a broader, more diffuse and often profitable market-oriented platform of related activities whereas the 'cultural economy' is often not.

The financing of the creative industries and the cultural economy is therefore fundamental but divergent. Cultural economy activities are subsidised, sometimes free (art galleries) and heavily dependent on public and/or charitable subventions. Creative industry financing takes a variety of market forms ranging from the film industry's typical 'special-purpose vehicles' (SPVs; Fusaro and Miller, 2002) to commissions and retainers from private or public clients. SPVs are risk management tools in the form of separate project (e.g. a film) accounts in which losses (or profits) may be 'hidden' so that losses do not negatively affect overall annual company performance out-turn. SPVs were adopted, with fatal consequences, by Enron, and are also used by venture capital organisations. The financing models for innovative activities that utilise science and technology are well-known forms of risk investment. These range from business start-up 'proof of concept'-type funding, through seed funds, often also public, to business angel investments, venture capital and varieties of public offering and debt financing. Hence, the financing of cultural, creative and innovative activity is a keystone function, the absence of which is problematic for all. Note that they are located, in Figure 1.1's abstract space, outside the regional innovation system to signify that they are not deterministically contained only inside it; they may have presence inside the regional innovation system but they may equally be 'imported'. Knowledge entrepreneurship is crucial to the exploitation capabilities of the creative region whether in the form of new media businesses exploiting internet technologies, or biotechnology and software entrepreneurs – they too may enter the region from outside, possibly as 'serial entrepreneurs'. Mediating these and the innovation system are varieties of knowledge transfer, including talent formation institutions, research laboratories with associated commercialisation institutions, and methods of purchasing research (or R&D) within or, again, beyond the region's innovation system. Because of its relative novelty, developing mainly in the late 1990s and accelerating abroad from advanced economies to knowledgeable regions in Asia, we were interested in and highlighted R&D outsourcing in the conference call. There were intellectual

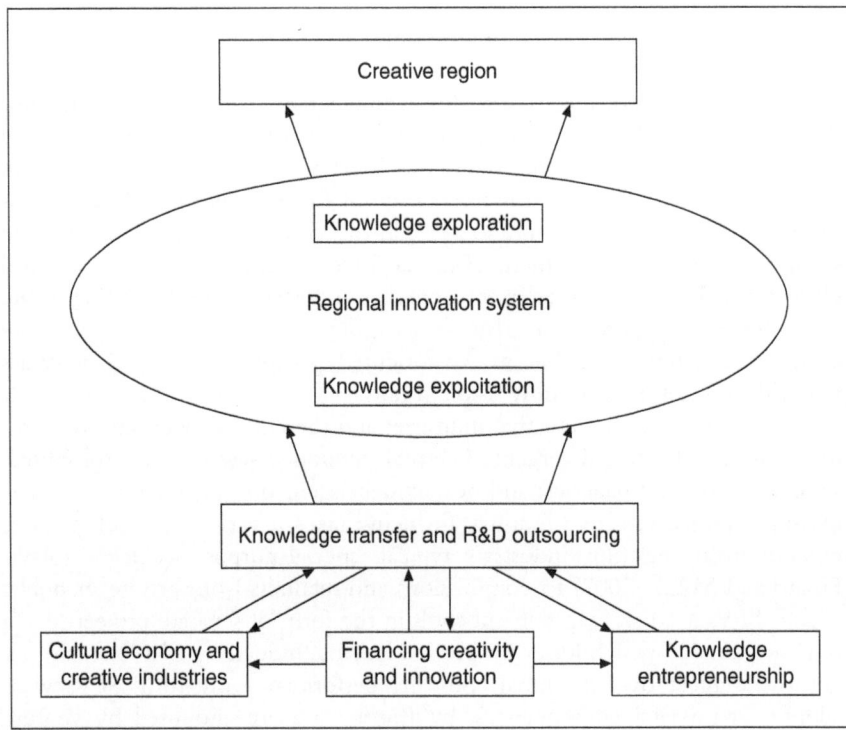

Figure 1.1 Key lineaments of the creative region

reasons too for testing the extent to which 'open science', as practised since time immemorial by academics with a duty both to 'speak truth to power' and repay their public by sharing results openly and freely, was becoming a more marketised 'open innovation' under a regime favouring 'academic entrepreneurship' (Chesbrough, 2003; Shane, 2004).

We come next to the most powerful institutional complex that assists a given region to maximise its creative and innovative potential inside the region, albeit, as we have shown, having institutions and individuals with global search, selection and application capabilities regarding knowledge appreciation and commercialisation. But for the moment let us clarify intellectually and in policy terms what the focus is here.

There is by now a comprehensive literature on regional innovation systems. The concept is precise, involving the idea and practice of interacting sub-systems specialising in knowledge exploration and generation, on the one hand, and knowledge exploitation, on the other. Classic instances of such activities are the discovery of, for example, DNA by Nobel laureates Watson and Crick in 1953 at Cambridge, the discovery in 1975 of monoclonal antibodies (Mabs) by further Nobel laureates Milstein and Köhler at the

Cambridge Molecular Biology Laboratory of the UK Medical Research Council, and the presence of a globally leading spin-out firm, Cambridge Antibody Technologies, independent until its acquisition by pharmaceuticals multinational AstraZeneca for €1.15 billion in 2006 – highly self-contained not even in a region, but a city of 100,000 inhabitants. Admittedly most exploration/exploitation is not as pathbreaking for all humanity as those, nor was there an absence of competing sites, mainly in New York and San Francisco. Indeed, it was the unpatented Milstein-Köhler discovery of Mabs that triggered the dawn of biotechnology in San Francisco when Genentech's founders, Boyer and Cohen, with venture capitalist Swanson, exploited this 'open science', in the process alerting the UK and subsequently other governments to urge scientists to patent and reap financial reward for such effort, ushering in the age of academic entrepreneurship.

But the key point about a regional innovation system is that it may sustain many such innovative business activities. Not infrequently, as in accomplished continental European systems, varieties of engineering excellence underpin regional economic performance, notably in Emilia-Romagna and Baden-Württemberg (Cooke and Morgan, 1998). In the United States, it is well known that northern California hosts ICT and biotechnology, agro-food and financial services in relatively close proximity and that there are technological and knowledge crossovers of some magnitude among them, associated with high *lateral* absorptive capacity (Kenney, 2000). In Greater Boston, including Cambridge and other Massachusetts satellites, biotechnology, biomedical devices, bioinformatics, pharmaceuticals, software and financial services display comparable 'related variety' associated with pronounced inter-industry overlaps and absorptive capacity. This means that in both contexts these innovative industry platforms (IIPs) hasten the recombination of knowledge that underlies rapid and branching innovation from a single originating one. Software can be almost infinitely adapted around its 'kernel' programming; venture capital adapts to distinctive biotechnological, biomedical or bioinformatics financial requirements from what are all fundamentally small and medium-sized enterprise industries (Cooke, 2006b).

It is worth noting, at this point, the important contribution made by regional innovation systems thinking, both specifically to the systems approach to innovation studies, and more broadly to an emergent evolutionary economic geography. With respect to the former, the national innovation systems (NIS) perspective was, and remains, complementary to a *sectoral* analysis – indeed, the concept of 'sectoral innovation systems' (SIS; Breschi and Malerba, 1997) continues as a minority interest in that conceptual domain. Yet sectors are agreed by almost everyone except the statistical offices that curate them to be meaningless fictions, by and large, in the contemporary economy. As with the sister concept of 'technological innovation systems' (TIS; Carlsson, 1995), 'sectors' were a convenient deconstruction of the hugely complex empirical reality confronting NIS analysts. The SIS and TIS approach strengthened NIS analysis by systematically linking *national* phenomena to new thinking about *global* value chains and globalisation more generally. Yet rooting NIS analysis in this non-spatial, *sectoral* perspective weakened the neo-Schumpeterian school's understanding of the *platform-like* nature of emergent and soon to become dominant general-purpose technologies such as ICT, biotechnology and, most recently, nano-technology (Helpman, 1998). Much modern technology – sensors being yet another example – have this pervasive, platform-like nature.

The perspective upon innovation systems that helped constrain neo-Schumpeterians from flirting with a dangerous irrelevancy was that which discovered and sought to analyse the great variety of regional innovation systems (RIS). Founded on an evolutionary, learning and innovation perspective, this had regional science, not industrial economics, lineaments (Braczyk *et al.*, 1998), whereas industrial economists have tended to display a possibly unconsciously vertical perspective that can be innovative, for example Cohen and Levinthal's (1989) original notion of 'absorptive capacity' responding to a crisis theoretically faced by large firms confronted with imperatives to compete through vertical disintegration and outsourcing, particularly of core R&D. Moreover, in policy terms RIS building strategies have been adopted by countries as diverse as Sweden and South Korea and have become a mainstay of development policy advisory work by the likes of UNIDO (Cooke, 2006a), OECD and the European Union (European Commission, 2006).

With respect to evolutionary economic geography, the work of Grabher (1993) had certain evolutionary elements but it was not overtly evolutionary nor particularly strongly spatial. Probably Braczyk *et al.* (1998) provided the first fully fledged theoretical and empirical study of spatial phenomena conducted from an evolutionary perspective. It is worth recalling some of the key concept mapping and defining work conducted in their collection:

> Evolutionary economics ... gives particular emphasis to history, routines, influences of environment and institutions.... In evolutionary economics, firms are differentiated organisations that use differentiated inputs for their production, one of which is knowledge. Knowledge plays a fundamental role: the constitution of a firm is mediated by the knowledge possessed by the founder or 'creative agent' and developed by learning ... they are created, they explore new paths of growth, they discover new routines, develop technological capabilities, capture new opportunities, adapt to new constraints and competition, or cannot respond to this demanding environment and slowly exit from the market.
>
> (Braczyk *et al.*, 1998, p. 8)

Thus, the focus on *knowledge* has been at the forefront of RIS analysis since the earliest formulations – learning, too, although a certain nuancing and possible downgrading of more uncritical, even normative, deployment of the learning notion can be observed. For as James March and others identified as early as March (1991), learning by organisations, whether regional development agencies or firms, is by no means unproblematic. Three vitiating conditions in particular condemn 'learning' almost to futility in such contexts. First, learning has features of a retro-activity in that role models being observed in benchmarking so-called 'best practice' will, in all likelihood, remain superior since they are already experienced at doing, understanding and overcoming institutionally what Deming (1950) observed as the process whereby *variety* is created at every step in a governance process. The causes of 'negative' variety need to be measured and the feedback data should be examined to determine causes of variety, which processes give rise to problems, and focus attention on fixing that subset of processes. This process is unavailable to the learning organisation and it cannot simply be appropriated on a turnkey basis.

A second and related problem of organisational learning was referred to by March (2001) as learning from samples of one or fewer. In other words, there are innumerable examples of good or better practice to benchmark, but resources do not really ever allow even a representative sample to be assessed thoroughly in a manner consistent with Deming's insight regarding negative variety. So, the firm, organisation or development agency is confronted with a high probability of experiencing the 'adverse selection' problem. This is a mainstay of orthodox economics' struggle to master the asymmetric information problem, after Akerlof (1970), who demonstrated that most transactions occur with some degree of asymmetry in the knowledge bundles of transaction partners. When the asymmetries are too great, market transactions fail, at least at the individual, contractual level. This perspective is highly relevant to the study of regional disparities since poor regions tend to have 'thin' knowledge bundles, while the likes of Silicon Valley, as we have seen, contain 'thick' knowledge bundles in *exploration* (research), *examination* (testing and trialling), and *exploitation* (commercialisation) knowledge. Thus, attempts at 'cloning' Silicon Valley through building science parks exemplify this approach; the policy fascination with 'clusters' is yet another instance, and indeed much 'policy formulation' is of this precise kind. It may be rationalised as pursuing a model that peers are also targeting since the appalling prospect of being left out and left behind is at least mitigated by joining the herd. To be fair to actors in what is a highly complex policy context, it is possible that such experiments produce 'positive variety', and many science parks are more or less successful in concentrating high-tech jobs (e.g. Kista, near Stockholm, with some 30,000), but whether truly *generative* growth occurs in such 'clones' is open to question.

The term 'generative growth' has been introduced in preference to the neoclassical 'endogenous growth' because it captures a more evolutionary flavour in the notion of growth in which growth is interactive with its sustenance system, or maybe ecosystem, of more or less sociable, creative, collaborating and competing agents. This is clearly more appropriate to an evolutionary economic geography perspective than the rather disembodied and disembedded notion of endogeneity proposed by new growth theorists such as Romer (1990). Finally, for now, there is a third problem associated with what we might term 'catch-up' institutional learning of the kind under inspection here. All innovative knowledge of utility, by definition meaning new knowledge commercialised, must pass iteratively through similar stages of production from 'discovery' through invention or realisation of the hitherto unrealised experimental form, to a testing process that establishes its reliability and validity, to a state where it has created demand, either market or social. The key thing to note here is that discovery involves not so much 'learning' as 'unlearning' by stepping beyond what can be learned as codified knowledge into the 'unknown' creative space where cognition is tacit but interactive with reality and methodology.

In a trivial sense even 'methodology' may have been 'learned', but it may have to be transcended or even negated in the act of discovery. This often happened in mathematics, where new process methodologies like calculus had to be invented to solve some intractable problem. Similarly in art, Cubism required the unlearning of classic figurative art, even to the extent of appreciation of and experimentation with primitive or child art forms as process methodologies, to escape into 'the new'. These criticisms point in two directions, the one fruitful, the other probably a dead end. The dead-end response is that this is a narrow definition of learning and a broad one can encompass exploration, research, innovation, etc. The response to that was made decades ago by Aaron Wildavsky (1973) in the title of a paper on the perils of over-generalisation, here paraphrased: 'If *learning* is everything, maybe it's nothing.' The more fruitful avenue is that creative regions must work out their own salvation through knowledge accumulation and innovative thinking customised to regional assets and needs. The policy approach is thus an 'envisioning' one but broader than those utilised in economic development thinking hitherto and based on evolutionary methodologies and platform policies.

Finally, we wish to broach an important distinction we introduced at the outset, namely that differentiating a cultural economy from the creative industries. We made some preliminary comments about the tendency for the first to be seen as a public good and subsidised by taxation, foundations or direct sponsorship from private corporations. The creative industries, by contrast, could be conceived of as more like high-tech SMEs, albeit their knowledge sources might not be universities but street fashion, but their funding might overlap with large firm commissions or licences as in high-tech. Consider the following: in the course of research fieldwork in the Marche region, Italy, in 2005, one of us interviewed the manager of and inspected a recently acquired subsidiary of the teenage fashion company Miss Sixty. This former denim workwear company Wash had in one year doubled its turnover to €20 million by entering the 'creative destruction' fashion denim market and supplying Miss Sixty worldwide. Wholesale jeans and jackets were imported from Romania and China for €1 per piece. Each was then systematically hosed with either indigo-consuming enzymes or powdered glass to acquire the appropriate 'distressed' look. The workforce responsible for this part of the destruction process, protected with what appeared to be deep-sea diving suits, was predominantly Muslim, from North Africa and Pakistan. The next stage consisted of sandpapering and generally inflicting various rips and tears on the denim, the workforce here being predominantly South-East Asian and Chinese men. Next, house paint was strategically applied and various graffiti were steamed into the fabric of denim jackets by, predominantly, Italian men and women. Finally, in what was once a Wash building but is now a separate, Chinese-owned sewing plant, thirty Chinese seamstresses were expertly machine-darning repairs to the distressed apparel. The finished product wholesaled to Miss Sixty for €10 and retailed for up to €100 per piece.

In response to a question about where the ideas for such a process originated, the answer was that the company retained the services of a small design consultancy from nearby Rome, staff members of which attended the edgier weekend discos to see what customers were wearing, adapting individual fashion statements for 'manufacturability' by Wash. Clearly, this Schumpeterian industrial process is somewhat different from subsidised opera – although not wholly so, since innovative renditions of standard classics have utilised punk fashion mores on more than one occasion. Probably the key difference comes back to the astonishing profitability of creative destruction in the fashion industry compared to the equally astonishing subventions necessary for mounting high-quality opera in the cultural sphere. So, as Smith and Warfield (2007) put it:

> The roots of the *culture-centric* orientation of policy is conceived of: culture as that which cultivates a citizen to become more civilized, and the '*arts*' as Western, classic, conservative, and traditional (e.g.: opera, ballet, painting, music, museums, etc.).
>
> A major influence behind the growing relevance of the *econo-centric* conception of creativity is the rise of and writings about the knowledge-based 'new economy,' and the importance of the city-region. The econo-centric orientation to creative city discourse also has relations to the rise of urban industrial cluster studies.
>
> (Smith and Warfield, 2007)

However, we will tend to think more of the *econocentric* connotations of the creative industries in what we say about creative activities; thus, we clearly recognise and utilise this *econocentric–culture-centric* distinction in this book, yet we are forced by the chapter by Luciana Lazzeretti (Chapter 9) to recognise also that there exist historic instances when the support services for culture-centric activity can themselves directly produce profitable clusters with global impact, or, perhaps more appropriately in the Italian context, globally competitive industrial districts.

Thus, we are also clear that what unites theoretically and empirically the creative and innovative, the latter defined traditionally as *commercialisation of new knowledge*, is precisely this commercially viable, market-facing but often risk–accommodating economic practice.

Introduction to the chapters

The book consists of fourteen chapters, including this introduction, divided into three parts. The first four chapters focus upon regional innovation systems in European countries. The following five chapters, in Part II, are on cluster policy assessment, cluster evolution in ICT and biotechnology, and cluster variety in the creative industries. The final part is on knowledge transfer, particularly open innovation and knowledge outsourcing to global networks, and knowledge intermediation or boundary spanning. The whole makes for a coherent and integrated set of chapters, written in response to a clear format and sharing a broadly evolutionary perspective and/or more focused systems of innovation, clusters and networks methodology. Thus, think of the structure of this book as represented in Figure 1.2 and note the manner in which the chapters refer to each conceptual element as we proceed to sketch in their content.

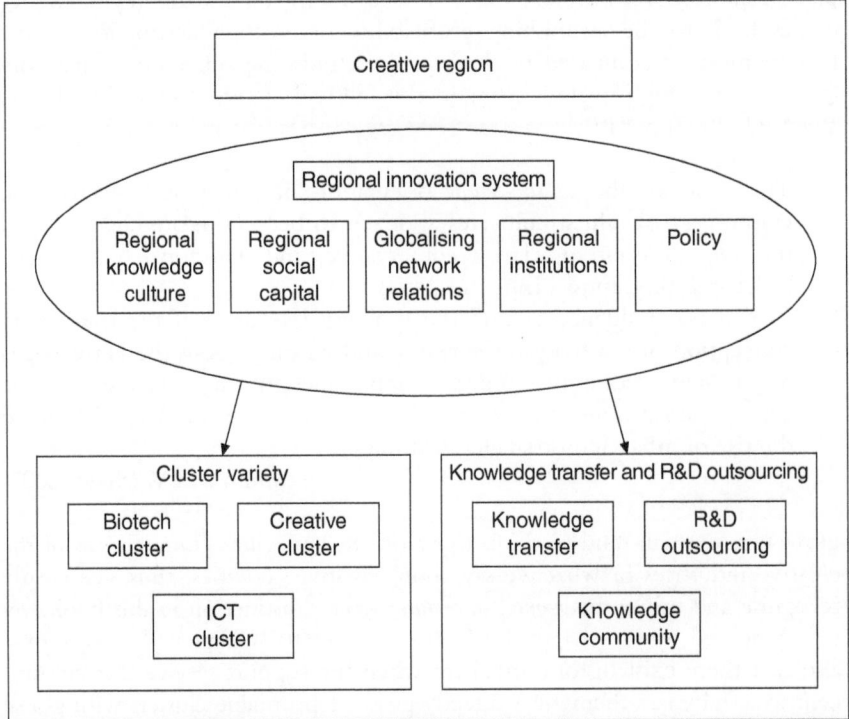

Figure 1.2 Framework of the book

Part I begins with a contribution from Dietmar Bastian and Ulrich Hilpert, who focus on the cultural basis for regional knowledge, including traditional knowledge and competencies, and the prospects for regional development beyond the confines of high-technology industry. The regions of Andalusia, North Rhine-Westphalia, northern England and Poitou-Charentes are compared and contrasted. Regionalised knowledge is shown to evolve, and the chapter shows how different types of region with particular knowledge bases respond in different ways: either by preservation of 'traditional' knowledge (Andalusia and Poitou-Charentes) or by a modernisation of the knowledge base (North Rhine-Westphalia and northern England), each with a prominent role for government policies.

Chapter 3, by Mika Raunio, explores creative regional social capital, which has been a key concept explaining the social relations within innovation environments in recent literature. It tests Florida's (2002) claims that 'creative regions' need open social capital with abundant weak ties. The 'creative region' concept relates to a wider institutional context than is normal in RIS analysis, emphasising such elements as living environment and institutions that steer the social life of the region. The role of social capital as a source of coherence and trust leading to lower transaction costs may lead to cultural 'lock-in' or 'escape', rather as Bastian and Hilpert observe. The core subject here involves exclusion of foreign experts from existing social capital. Regionally and nationally important sources of social capital are brought into contact with foreign experts. This opens up key questions: how and to what extent are social capital and its supporting structures adjusted to *globalisation* to avoid exclusiveness – that is, a mostly unintentional form of discrimination towards foreign expertise? This chapter examines these question by studying the case of foreign ICT professionals working in Finnish innovation environments.

Chapter 4, by Pieter de Bruijn and Frank van Oort, is entitled 'Connectivity and co-location in innovation processes of Dutch firms' and addresses 'related variety' and 'urban field' issues in focusing on the concept of agglomeration economies in relation to firms' competitive advantages, knowledge spillovers and other externalities. It is often argued that these are more easily identified in cities where populations are densely concentrated in a relatively small geographic space. On the contrary, the urban field hypothesis – often applied in the Dutch case – emphasises the absence of differentiated urban contexts in economic accumulation. Despite the large empirical literature on this subject, conclusions are various. In general, on a local scale, agglomeration is usually related to urban density or city size, whereas on a regional scale, agglomeration theories seem to reflect core–periphery patterns. Within these commonly perceived spatial contexts, we describe and model patterns of innovation of firms in relation to environmental assets. The authors make a distinction between firm-internal knowledge assets and assets derived from the regional and global industry context.

Kati-Jasmin Kosonen's Chapter 5 concentrates on less favoured regions and also takes a regional–global perspective in assessing the prospects for growth. In a knowledge-based economy the regional or local *knowledge environment* and *innovation environments* for specific business areas have become more important. The base of knowledge constantly evolves institutionally in regional innovation networks, shaping *technological change* and *transformation.* This chapter highlights how knowledge-intensive high-tech industries in less favoured Finnish regions evolved a changing innovation culture. In Bastian and Hilpert's terms, this meant renewing local knowledge architecture by enhancing locally produced scientific knowledge to become globally competitive. The study also examines what actual efforts were made in these regions to strengthen the *institutional capacity* of the innovation environment, and, more precisely, what efforts were made to bring new knowledge into the town region. The concept of institutional capacity appears in this study as a combination of the local needs for *knowledge resources* and *global partnerships* (coalitions and networks) formed by individual actors (e.g. entrepreneurs, development agencies, university units, municipalities, technology centres) in certain *institutional settings* and certain *knowledge-oriented cooperative spaces referred to as 'shared arenas'.*

Part II of the book is cluster orientated: some innovative clusters, some creative ones. It begins with Chapter 6 by Roel Rutten and Frans Boekema, who evaluate a scheme to evolve sometimes very small 'clusters' in the Dutch region of South-East Brabant, where Eindhoven is located. Cluster firms worked on the development of their new product for two years on average. The chapter begins with a sketch of the Eindhoven region and of the cluster scheme, its objectives and origin. It then discusses the framework that was used to evaluate the scheme and which was developed on the basis of recent insights from theories on regional innovation networks. Next, the results of the evaluation are presented in three subsections: the outcomes of the product development effort in the various clusters, the process of product development within the clusters, and the conditions under which this process took place. The idea behind this is that the process affects the outcomes and that the process, in turn, is affected by several conditions, such as previous relations between the partners. The results show that the cluster scheme has been very helpful in furthering innovation in regional SMEs. Chapter 7, by Dagmara Stoerring and Bent Dalum, has been referred to already. It shows that to support the emergence of new industrial clusters it is important to understand the process behind their evolution. This chapter investigates evolutionary processes by comparing two different cases in the same region, North Jutland, Denmark: (1) a cluster initiative in biomedical technology, Biomedico; and (2) a well-established cluster of wireless telecommunications, NorCOM. The NorCOM cluster was initiated in the 1960s by local companies and illustrates the phenomenon of a high-tech cluster being able to emerge in a peripheral region.

Kean Birch (Chapter 8) also chooses biotechnology as the centrepiece of his chapter, entitled 'The knowledge–space dynamic in the British biotechnology industry: function, relation, and association. He critiques 'cluster' theories inspired by the work of Michael Porter from an implicit 'related variety' perspective. He conducts this critique through an analytical focus on the local linkages between organisations, exploring the *knowledge–space dynamic* in the biotechnology industry. This incorporates the diverse functional, material and associational features of regional theories into a theoretical approach that seeks to understand the strength of organisational and knowledge relationships at different spatial scales. Birch's overall purpose is to show how innovation processes need to be conceptualised in ways that incorporate a variety of approaches and can account for the importance of extra-local linkages. Finally, then, this implies that current policy prescriptions, designed around cluster theories, may 'breed weakness' into regions through the focus on regional specialisation.

Next, Luciana Lazzeretti's Chapter 9, on cultural and creative industries in places of 'high culture', takes part in the debate on modern economies and the relationship between culture, creativity and models of local development, paying special attention to Florence, city of art and culture. She briefly recalls the cultural cluster/district approach and the economic enhancement of high-culture cities and places, and the creative economy approach, in particular Florida's models of the creative class, and highlight a few points of connection between them. A first notion of 'creativity for high culture places' is thus introduced, especially with a cross-fertilisation of the concepts of creative district atmosphere and the 'creative habitat' described by the creative economy. This notion finds its place in the guiding principle which looks upon the possible evolutions of cultural clusters and districts into creative ones and the evolution of culture from a production factor into a source of innovation that can set off or redevelop a variety of filières of production and professions. While Florida mainly concentrates on the creative class, this chapter focuses on creative firms, with the aim of assessing by empirical evidence the consistency of a phenomenon of economic enhancement of culture. Finally, Chapter 10, by Birgitte Rasmussen and Jens-Peter Lynov, already mentioned, is about the development of the Musicon Valley in Roskilde, Denmark. This initiative is discussed from the perspective of Risø National Laboratory, a governmental research institution. Risø has to face at least four main challenges that can also be seen as four incentives for identifying new research areas together with new commercial and scientific constellations, networks and partners. These are: (1) science's new role in society, which requires science to demonstrate the societal benefits of research investments; (2) the demand for successful innovation and product development in a globalised economy; (3) the economic and social transformation in the United States and Europe from an industrial to a creative economy; and (4) the focus on research institutes as dynamos in regional knowledge-based economies.

Part III, which is on knowledge transfer, with a particular interest in R&D outsourcing to global networks, has four chapters. Chapter 11, by Philip Cooke, reports on the elaboration and testing of a new theoretical approach to understanding economic development. The focus is on micro-economic geography, informed by theoretical insights on spatial monopoly, on the knowledge capabilities of firms and their networks, and on 'open innovation'. Sectors in focus are ICT and biotechnology. During the 1990s it became clear that a significant shift in location of R&D by large firms was occurring through 'open innovation' or outsourcing to small, specialist research-intensive firms and public research organisations.

Chapter 12, by Alex Burfitt and Chris Collinge, traces knowledge outsourcing through the evolution of a knowledge-intensive R&D services sector within a traditional manufacturing region, and uses this to probe *path dependence* as a way of conceptualising regional economic change. It then examines how the case can be accounted for in terms of path dependence. While strongly influenced by the economic trajectory established by traditional manufacturing industries, these firms illustrate the possibility of 'switching' within a regional economy based on a reorientation in the exploitation of knowledge resources by firms.

Chapter 13, by Rüdiger Wink, is also on outsourcing in knowledge-intensive services. It also brings the impact of internationalisation to the associated processes. Based on the example of the aircraft sector in the metropolitan region of Hamburg, Wink's chapter investigates the effect of increased use of modular sourcing and outsourcing of knowledge-intensive engineering services on the generation and diffusion of new knowledge for an aircraft cluster. Engineering and design services have become the main technological drivers within the aircraft sector. Wink shows how the engineering and design sector is integrated into the urbanised structure and culture of Hamburg, connecting location with urbanisation effects. The empirical investigation is based on a set of interviews with representatives of companies, research and cluster organisations and authorities, done in 2005. The main part, however, is the development of a theoretical argument on the interplay between creativity, openness and supporting institutional conditions.

In the final chapter, Hiro Izushi notes that in the literature on innovation, boundary spanning is considered to be a key to successful management of innovation. Communication and collaboration between individuals or organisational units have a tendency to take place within boundaries at different levels. Such boundaries often stifle innovation by deterring coordination, exchange and combination of different sets of resources (and knowledge in particular), and thus preventing novel ideas from emerging. Boundary spanning refers to collaboration and communication both across internal boundaries formed by teams, departments and divisions within firms, and across external boundaries formed by organisations, sectors, regions and nations. Against the background, this chapter introduces a new boundary formed by R&D workers researching into the same area of knowledge: the 'knowledge community'. The idea of the 'knowledge community' derives from a contradiction in the growth performance of advanced economies. In spite of a phenomenal growth in the number of workers devoted to creation of new ideas, advanced economies have exhibited constant average productivity growth rates during the past fifty years. The chapter describes a model that resolves the contradiction.

References

Akerlof, G. (1970). The market for 'lemons': quality uncertainty and the market mechanism, *Quarterly Journal of Economics*, 84: 488–500.

Asheim, B., Cooke, P. and Martin, R. (2006) (eds) *Clusters and Regional Development*, London: Routledge.

Boschma, R. (2005) Proximity and innovation: a critical assessment, *Regional Studies*, 39: 61–74.

Boschma, R. and Martin, R. (eds) (forthcoming) *Evolutionary Economic Geography*, Cheltenham: Edward Elgar.

Boschma, R. and Wenting, R. (2005) The spatial evolution of the British automobile industry, *Papers in Evolutionary Economic Geography* 05.04, Utrecht: Utrecht University.

Braczyk, H., Cooke, P. and Heidenreich, M. (eds) (1998) *Regional Innovation Systems*, London: UCL Press.

Breschi, S. and Malerba, F. (1997) Sectoral innovation systems: technological regimes, Schumpeterian dynamics and spatial boundaries, in C. Edquist (ed.) *Systems of Innovation*, London: Pinter.

Buenstorf, G. and Klepper, S. (2005) Heritage and agglomeration: the Akron tyre cluster revisited, *Papers on Economics and Evolution 0508*, Jena: Max Planck Institute.

Carlsson, B. (ed.) (1995) *Technological Systems and Economic Performance*, Dordrecht: Kluwer.

Chesbrough, H. (2003) *Open Innovation*, Boston: Harvard Business School Press.

Christensen, C. (1997) *The Innovator's Dilemma*, Boston: Harvard Business School Press.

Cohen, W. and Levinthal, D. (1989) Innovation and learning: the two faces of R&D, *Economic Journal*, 99: 569–596.

Cooke, P. (2006a) *Regional Innovation Systems as Public Goods*, Vienna: UNIDO.

Cooke, P. (2006b) Global bioregional networks: a new economic geography of bio-scientific knowledge, *European Planning Studies*, 14: 1267–1287.

Cooke, P. and Leydesdorff, L. (2006) Regional development in the knowledge-based economy: the construction of advantage, *Journal of Technology Transfer*, 31: 5–15.

Cooke, P. and Morgan, K. (1998) *The Associational Economy*, Oxford: Oxford University Press.

Cooke, P. and Piccaluga, A. (eds) (2006) *Regional Development in the Knowledge Economy*, London: Routledge.

Deming, E. (1950) *Elementary Principles of Statistical Control Quality*, Tokyo: Union of Japanese Scientists and Engineers.

European Commission (2006) *Constructing Regional Advantage*, Brussels: DG Research.

Florida, R. (2002) *The Rise of the Creative Class, and How It's Transforming Work, Leisure, Community, and Everyday Life*, New York: Basic Books.

Frenken, K., Van Oort, F. G., Verburg, T. and Boschma, R. A. (2005) Variety and regional economic growth in the Netherlands, *Papers in Evolutionary Economic Geography* 05.02, Utrecht: Utrecht University.

Fusaro, P. and Miller, R. (2002) *What Went Wrong at Enron*, New York: Wiley.

Grabher, G. (ed.) (1993) *The Embedded Firm: On the Socio-economics of Industrial Networks*, London: Routledge.

Hayek, F. (1948) Economics and knowledge, in *Individualism and Economic Order*, Chicago: University of Chicago Press.

Helpman, E. (ed.) (1998) *General Purpose Technologies and Economic Growth*, Cambridge, MA: MIT Press.

Jacobs, J. (1969) *The Economy of Cities*, New York: Vintage.

Kenney, M. (2000) *The Silicon Valley Edge*, Stanford, CA: Stanford University Press.

Klepper, S. (2002) The capabilities of new firms and the evolution of the US automobile industry, *Industrial and Corporate Change*, 11: 645–666.

March, J. (1991) Exploration and exploitation in organisational learning, *Organisation Science*, 2: 71–87.

March, J., Sproull, L. and Tamuz, M. (1991) Learning from samples of one or fewer, *Organisation Science*, 2: 1–13.

Marshall, A. (1916) *Principles of Economics*, London: Macmillan.

Marx, K. and Engels, F. (1970) *The German Ideology*, London: Lawrence & Wishart.

Massey, D. (1984) *Spatial Divisions of Labour*, London: Macmillan.

Myrdal, G. (1957) *Economic Theory and Underdeveloped Regions*, London: Duckworth.

Romer, P. (1990) Endogenous technical change, *Journal of Political Economy*, 98: 338–354.

Shane, S. (2004) *Academic Entrepreneurship*, Cheltenham, UK: Edward Elgar.

Smith, R. and Warfield, K. (2007) The creative city: a matter of values, in P. Cooke and L. Lazzeretti (eds) *Creative Cities*, Cheltenham, UK: Edward Elgar.

Rodgers, M. (2006) Grumpies survey field's slag heaps, *The Times Higher Education Supplement*, 1 September, p. 1.

Veblen, T. (1919) Why is economics not an evolutionary science?, *Quarterly journal of Economics*, 12: 373–397.

Wildavsky, A. (1973) If planning is everything, maybe it's nothing, *Policy Sciences*, 4: 127–153.

Witt, U. (1991) Reflections on the present state of evolutionary economics, in G. Hodgson and E. Screpanti (eds) *Rethinking Economics: Markets, Technology and Economic Evolution*, Cheltenham, UK: Edward Elgar.

Part I

Regional innovation systems

2 The regionalisation of knowledge: the territorial basis of development

Dietmar Bastian and Ulrich Hilpert

Introduction

While the variety of locations for manufacturing has been growing in the global economy, there is the question of why the rationality of low costs, deregulation, and cheap labour does not always lead to industrial relocation. There are certain locations that manage to continue both economic development and a remarkable level of employment quite successfully while keeping to rather traditional products. Some locations manage to maintain their position in global processes of development even though there is an ongoing race to innovate, as well as increasing competition.

So, there are at least two traditions that can both lead to rather dynamic processes: (1) the tradition of enterprises able to make products of outstanding quality; and (2) the tradition of locations able to manage change. Both enterprises (through their employees) and entire regions sometimes establish an expertise in particular economic activities; these regions' universities are known to be at the leading edge of research, and governments have introduced public policies that serve appropriately to foster economic development and innovation. Such competences in the different areas of activities need both to match and to meet market opportunities. But, even more important, these arrangements cannot be introduced quickly even though they might become the subject of public intervention or appropriate policies. These arrangements cannot be explained merely in terms of costs, and competitive advantages based on costs; clearly, competences, knowledge and places are involved, and these must be appropriately arranged. In order to transcend the mere economic analysis of regional development, such arrangements demand broader understanding of regional identity that also includes the cultural dimension – culture, however, conceived not in terms of cultural entertainment, but in terms of adequate attitudes towards innovation at the level of both individuals and institutions.

This approach first of all draws attention to the increasing role of knowledge in regional development. Here, the discussion is mainly focused on technology transfer, the recruitment of both highly trained labour and gifted academics, as well as on the idea of fostering enterprises by low taxation or specific support programmes. Nevertheless, the knowledge is created under specific conditions and it is used according to the regional opportunities to be met. First, the skills of the labour force inhibit specific attitudes towards both the role of labour in an enterprise and the way employees may participate in it. Second, a region's stock of knowledge is to be found in enterprises, the labour force and the products produced.

A region's tradition can provide both the knowledge and the basis for development. Its capability both to apply new technological opportunities and to market particular products is crucial for its participation in economic development. Economic development is increasingly related to markets, demands and changes that are located outside the region. Regional development indicates the way a region is prepared to initiate such relationships and how it may regionalise such opportunities. Policies that are effective in a region can provide for such arrangements as to take advantage of the new and additional knowledge required. So, tradition and innovation refer to both the region's societal basis and the policies that are effective in a region. The study thus confirms approaches combining evolutionary economics and economic geography (Boschma and Lambooy 1999), and contributes to an emerging knowledge theory of the region (Lawson 1999).

On the basis of findings from four case studies (Hilpert 2006a), this chapter will argue that regions develop distinct bodies of knowledge. This 'regionalised' knowledge is shaped by contributions from economic, cultural and institutional factors. The specific character of this body of knowledge is, we believe, strongly influenced by the nature, structure and traditions of the regional economic base. Regionalised knowledge is also subject to change, and the chapter argues that different types of region with particular knowledge bases will respond in different ways: either by preservation of 'traditional' knowledge or by a modernisation of the knowledge base. In either case there is a prominent role for government policies. By focusing on the role of cultural knowledge and regional culture within these processes of regional change, two cases (Poitou-Charentes and Andalusia) illustrate the manner in which regions with 'tradition-based competences' have achieved divergent economic outcomes based on differing cultural traditions. These findings are contrasted by the cases of North Rhine-Westphalia and northern England, both regions that have attempted to modernise their knowledge bases, once again with differing outcomes based on regional cultural features.

The regionalisation of knowledge based on cultural identity

Levels of economic development vary considerably among the most advanced industrialised countries, and even within these countries. Locations generate particular competences in relation to their economic profile, and they continue to maintain such competences over long periods of time. Examples include the continuing competence in microelectronics in Silicon Valley, the tradition of internationally outstanding research at Massachusetts' universities, the revival of optical industries and precision engineering in Jena or of shoe industries in Hungary, the success of mechanical engineering in so-called Third Italy (Terza Italia), or the production of cognac in Poitou-Charentes, etc. Like many others, these locations continue their development on the basis of existing tradition and knowledge, and the capability to make use of new opportunities resulting from the changes in products, markets or technologies.

There is a particular pattern of regionalisation of knowledge that emerges and that also reintroduces the relationship with a particular territory where the knowledge can be found. It is important to see that such knowledge is not necessarily based on new technologies or new research findings. It can also be based on a particular industrial knowledge inherent in enterprises and their employees, who make use of it while manufacturing, adjusting to consumer demands and needs or increasing the quality of the products. Regional or local networks reflect such knowledge arrangements as well as particular milieux, or modes of manufacturing. Anthropocentric production systems (Lehner 1991; Wobbe 1992) can be understood clearly as examples where particular knowledge opportunities relate to a particular territory. These refer to a tradition that is different from that of Taylorist industries; and industrial labour gives rise to attitudes of craftsmanship and provides the competences for industries and products that are to be found at a limited number of locations and that rarely relocate. In addition, it is interesting that such regionalisation of economic opportunities is also often to be found in products based on pre-industrial knowledge, as in the case of food production (Calafati 2006) or the cognac industry.

The regionalisation of knowledge, as a consequence, can be found in different situations. It may refer to new research that is constantly being generated, to a particular kind of product that constantly adds competence to the already existing stock of knowledge in the region, or it can also be induced because of climate, tradition or nature. But there is also the knowledge that is built up in regional societies and government administrations of how to cope with change and to continue elements of a region's stock of knowledge when facing new situations. Such changes may result from new opportunities, but they are often caused by markets and demands that call for an adjustment to products, enterprises and industrial opportunities. Where markets cause problems to existing enterprises and industries, governments are expected to induce an adjustment to change that may continue the economic development of the region. Existing knowledge is simultaneously questioned, and it is uncertain how to continue such competences under different conditions. A region's cultural identity may offer helpful references for government policies if this identity comprises knowledge of previous adjustment efforts. The regionalisation of such politico-administrative knowledge is clearly bound up with the problems that are to be solved in a region, and to the attitudes of regional societies when dealing with such problems: either oriented in investment for structural modernisation or oriented in following the rationality of costs. The experience in conducting policies builds up a stock of knowledge that is linked to a particular region and is empirically bound to the situation in this territory.

Here, it is important to see that the regions' economic basis is different, and is based on divergent industries and patterns of organisation of enterprises. Large enterprises that are oriented to mass production (e.g. the tobacco industry, mass steel production or textiles for clothing) will not require the kind of knowledge that is demanded in smaller and highly specialised enterprises (e.g. in mechanical engineering, precision engineering or research in biotechnology). The pattern of organisation of enterprises in smaller and more flexible units (Piore and Sabel 1984) demands different kinds of knowledge and a particular kind of culture that is effective in the enterprises and in the region. Regional societies can provide different grounds for regional industries, regional economic development and for the regionalisation of such developments to take place at such locations. Clearly, the knowledge body in Terza Italia, Massachusetts or Baden-Württemberg is different from the ones at Nord-Pas-de-Calais, northern England or Pittsburgh. Industrial history has provided for divergent opportunities that have then accumulated different stocks of knowledge (e.g. Hickie 2006). Now these knowledge bodies of enterprises turn out to be fundamental to a region's future. But in food and beverages too, as indicated by the globalisation of consumers and tastes, such knowledge, which is inherent in the regions and closely related to the territory, makes a difference.

Furthermore, governmental structures and traditions of public intervention show important differences. The fact that German *Länder* or American states can focus on the problems in their regions, govern change and indicate new opportunities reflects more than just a particular polity structure. While from a national perspective it may not be very important where a particular development takes place as long as it does take place, for regional governments it is fundamentally important that it takes place within their territory. A particular form of politico-administrative knowledge adds up to a stock of knowledge that can deal more appropriately and often also more effectively with the development of a region. So, some knowledge that is available, and extraordinarily important, in one region may even not be needed, or of any interest, in a different region. Knowledge of particular situations and the competence to deal with individual problems to be solved are constantly developed in relation to the existing socio-economic situation, and reflect these specific regional arrangements of industrial structures. So, governmental structures inhibit specific and different knowledge bodies.

The different kinds of knowledge indicate patterns of regionalisation, no matter whether it is industrial knowledge – application of new findings from science and research – or administrative knowledge built up within governmental structures. Given that the knowledge provides the basis for a continuation of socio-economic development, then enterprises will not relocate; they prefer to stay in the region that provides the basis for their competence and success. So, the regionalised knowledge that increasingly relates a particular stock of knowledge with a particular territory is continuously changing. The tradition is reused and adjusted to new situations; there is a change of knowledge.

But not all regions are prepared to induce such a change of knowledge. Some sources of a region's knowledge may not be ready to continue, or only in particular niches (e.g. shipbuilding, textiles, design, steel, or individual food processing). Some regions manage to keep their traditional knowledge, whereas others do not manage to do so, even though the regional situations may be rather similar. In some situations new opportunities are added to the regional body of knowledge to provide for a modernised tradition in the territory, but similar initial conditions may end up in clearly less advantageous processes. So, it is obvious that it is not just tradition, costs and markets that provide for the use of knowledge. It is also a question for the role of government and public policies with regard to regions managing the change; and this activity will be necessarily bound to the territory, because regional societies, regional authorities and the institutional structure, including enterprises as knowledge institutions, are defined by the territory. On the basis of the cultural identity of a given territory, regional decision-makers in politics and industry try to adapt with new opportunities and modernise traditional competences; they aim at a renewed body of knowledge and attempt to make efficient use of the existing tradition-based knowledge in order to participate successfully in globalisation and regionalisation of development.

Culture, change and economic development: the role of government for culture-based advancement

In the light of different cultural arrangements that can be identified in conjunction with paths of socio-economic development, regional development demands more than just competitive production costs. Tradition and cultural settings play an important role in particular development processes at different locations (Hilpert 2006a; Calafati 2006). In addition, there is variety even among the industrial sectors in a particular region: while some industries may pass through economic crisis, others may be at the forefront of innovation, and finally there are some that may be able to progress by using certain aspects of their industrial tradition in conjunction with new technological opportunities. So, there are varied industries, divergent processes of socio-economic development, and variety in the processes of innovation that may be found simultaneously at particular regions and locations (Singh and Allen 2006; Goldberg 2006).

Regional opportunities, as a consequence, are not related just to one of the various cultural arrangements required for socio-economic development, but are characterised by both the dominant cultural setting and the role of tradition in the regional processes in question. So, industrial restructuring, socio-economic development and cultural change are closely interrelated. Industrial restructuring points to a change of cultural settings that provides the basis for development at different periods in time. Regional processes are related to both cultural stability and cultural change. The cultural settings that provide for socio-economic development need to continue to be effective, while those in turbulence or crisis need to be changed. Regional identities and their underlying cultural settings are in permanent change and adjustment (Hilpert 2006b). These parallel challenges correspond on the one hand to capabilities emerging from tradition and modernisation, and on the other hand to economic opportunities due to industrial settings or markets.

In order to meet market opportunities, cultural settings need to be appropriate to provide the required products. Inappropriate cultural settings, embodying traditions that have not undergone necessary adjustments or modernisation, cannot continue to contribute to regional development (Bastian 2006) and will disappear as a result of the economic crisis within the industries concerned. So, there is demand either for a permanent, continuing adjustment of the regional identity, or for restructuring and cultural change. Since advanced socio-economic development requires necessary initial conditions that cannot be arranged through enterprises and industries alone, but demand government activities, a question arises about the relationship between such culture-based development and public policies.

The role of culture in economic development: tradition-based competences, industrial capabilities and production costs

Opportunities for regional development vary greatly according to industrial structures and their respective cultural arrangements. So, it is the product and the ability to bring this product to potentially changing markets that are important if advantage is to be taken of traditions and particular competences. Depending on the processes of specialisation occurring within the overarching process of globalisation, there may be a greater variety of such opportunities. Open market economies provide larger markets for products that respond to a very specific situation or tradition. The region of Poitou-Charentes indicates the unique and outstanding competences in producing cognac. The development of this product has grown from a long tradition. There is a great deal of tacit knowledge that is incremental to wine growing, picking grapes and treatment of the distillery. This traditional knowledge is merged with modern methods regarding yields and productivity, and a modern understanding of the biological processes that take place during the making of cognac. The commercialisation of cognac, which is produced in Charente, is controlled by Martel, Hennessy, Courvoisier and Rémy Martin. However, when the massive exports of the 1960s and the extension of vineyards soon led to a decrease in quality, the cognac industry faced the need to react by intensifying research on diversification, imposing quotas and putting premiums on sustainable vintage methods. This reaction may well be illustrative of an underlying readiness and ability to defend existing (world) market niches. Based on luxury production, like cognac, with export shares of more than 80 per cent – one-third of overall regional exports – the primary sector triggers strong demand in down- and upstream sectors such as packaging, printing or glass manufacturing, thus stimulating growth in the respective regional branches. Including all related industries, the cognac industry employs some 80,000 persons, or nearly a quarter of the active population in the two Charentes *départements* (Bastian 2003a). The uniqueness of the product and its quality have given it a dominant position in all markets. Because of the tradition of cognac making, this regional industry and its producers can flourish economically even though the process of production is not yet industrial, but relates to pre-industrial competences and a long-established tradition. The knowledge accumulated on this basis is mostly experience, rather than systematically generated by scientific research or product design. The pre-industrial source of the product's success is still of primary importance. The region has developed an identity and a set of cultural arrangements that is clearly based on these pre-industrial competences.

Situations that are similarly deep-rooted in tradition can be identified in Andalusia. There is also a long tradition of wine making and food processing, with plenty of experience-based knowledge about production in relation to regional specialities. But in contrast to the situation in Poitou-Charentes, this pre-industrial knowledge in Andalusia cannot provide for attractive economic development, because there are no products that can take full advantage of such a regional identity. With a share of almost 10 per cent, the contribution of the primary sector in the make-up of Andalusia's GDP is outstanding, and twice the overall Spanish figure. Furthermore, the agro-food subsector is the one carrying the biggest relative weight in Andalusian industry, which makes up close to 40 per cent of total business volume in the sector and accounts for nearly 30 per cent of the working population (Junta de Andalucía 1999). Yet only the cultivation of fruit and vegetables (which now contributes nearly one-third of total agricultural production) constitutes a promising pole; the rest of the agricultural sector faces a more complicated situation, be it due to climatic factors in dry regions, high production costs (irrigation areas, for example, along the Guadalquivir), a lack of adequate technologies or the structure of demand favouring single cropping.

The role of regional government in processes of socio–economic change

The differences also indicate a different role for government policies in the future of the two regions' development. In Poitou-Charentes, or similar regions with a dominant industry doing well, government policies do not need to focus on equalising regional disparities. Instead, there is the idea of establishing a broader variety of industries to provide a range of different jobs. It is interesting that such jobs are predominantly in Taylorist industries (e.g. the automobile or rubber industries). The pre-industrial arrangements did not require strong institutions in the fields of research and education because existing attitudes in the region were adequate and appropriate for such industries. They are, however, not sufficient for more modern or advanced industries. This is not the case with regard to the industries additionally brought to the region in the course of the central government's decentralisation policies. So, there is a clear cultural identity related to cognac production. Since this provides for important elements of regional economic development, careful attention has to be paid to the modernisation of cultural arrangements in the territory; the cultural change or cultural arrangements that emerge as a result of new industries do not change the broader regional identity.

These additional industries in Poitou-Charentes cannot develop a particular regional dynamic in the way it can be identified in relation to cognac production. So, in this situation Taylorist industries are not nearly as attractive to the region's development as is the pre-industrial setting. Poitou-Charentes has been facing a difficult economic situation since it was set up as a new administrative entity, owing to a structural inheritance that is mainly characterised by traditional specialisation in primary-sector activities and Taylorist industries implemented in the region under the technocratic decentralisation strategies that have been implemented since the 1960s. Likewise, in political terms the process of region building seems to be still in the making (INSEE 1997), and appropriate support of innovation can be guaranteed only to a minor extent by regional resources. While traditionally the University of Poitiers had been the focal point for research and education in the region, these capacities have become more extended and dispersed in recent years. However, university education, techno-scientific research, and application in industry are not sufficiently interrelated to meet the requirements for a sustainable regionalisation of modern skill-based socio-economic development.

As no tradition of cultural trust in the ability to realise product or process innovation has arisen in the region, neither the public nor the industrial research landscape system can be expected to provide major innovative inputs for the regional economy. Although regions' capacities have been extended over the past decade, research and innovation so far have remained unrelated elements of regional identity, given the incomplete coherence of education, research and application in industry. As a further important aspect of limitation of culture, it is important to note that an extension of public research facilities and an emergence of the regional system of technology transfer institutions have to be understood in the framework of centrally developing the respective national structures and thus express little regional originality.

The fixed nature of the French system can likewise be noticed in attitudes towards regional policy making. Limited in financial means and competence (Préfecture de la région Poitou-Charentes 1999), the regional government is not able to pursue large-scale enabling approaches beyond the existing economic specialisations and clusters. Attitudes in regional government reveal a culture of using the opportunities of the prefabricated national system of public support instruments more efficiently and of using learning effects under this system for an adaptation to European funding opportunities (Poitou-Charentes Entreprises 1998).

Major innovative injections tend to come from outside, as the Futuroscope project near Poitiers has demonstrated in recent years. Accordingly, a further typical attitude under a culture of limitation is to expect stimulation to come primarily from outside the region. Similar orientations are not just politically induced by the French system of relations between the centralist state and its regions, but can be found in research and education as well as in industry, where decision-making often takes place outside the region.

Although not among Europe's vanguard regions in terms of regional identity, Poitou-Charentes thus gives an illustrative example of combining a comparatively slow pace of socio-economic development with only gradual changes in attitudes – a combination, though, that allows some characteristics of the region not related to modernisation to be conserved. As a contrasting experience to Europe's innovative centres, Poitou-Charentes may, then, show how innovation can be assimilated under a culture of provincialism.

In Andalusia and similar regions, in contrast, the pre-industrial cultural settings were unable to generate a similar situation giving rise to an internationally successful product or industry. Here, in contrast, government policies are expected to balance regional disparities. Because of the lack of capabilities in research and education, there is a need to make the region attractive for Taylorist industries. Such regional settings cannot generate economic dynamism and cannot take advantage of local cultural arrangements because these lack both marketability and the capacity to adjust to change.

A further structural characteristic results from the enterprise structure: in the light of Andalusia's dual economy of some few larger enterprises with a large share of extra-regional commerce and a large number of smaller units engaged in diversified activities and basically oriented towards regional or national markets, the corresponding segments of the regional industry are not equally exposed to innovative impulses from outside. Such injections, which moreover are usually limited to changes in business culture (adaptation of internationalised marketing strategies), have little impact on the domestic economy, owing to the absence of cross-sectoral linkages, which have been shown to be functional carriers for the spread of changes in industry.

With no significant tradition in industry, but positive growth rates in the second half of the 1980s and again since 1995, Andalusia – usually thought of in terms of high unemployment rates, sunny beaches and traditional folklore – has been able to profit from a certain economic dynamism. A first glance at broad economic indicators furthermore suggests the presence of a relatively modern structure with a strong service sector, often regarded as favourable for advanced forms of socio-economic development (Castells and Hall 1992). Likewise, in political terms, dynamic tendencies have been apparent during the past two decades: since Andalusia gained a measure of autonomy in the aftermath of Spain's transition towards democracy, regional policy making has been experiencing emancipation from Madrid and, at the same time, has been able to benefit from the distribution of European funds (CTPS 1996). Thus, changes in regional politics and economics have contributed to a certain degree of modernisation, yet the potential for a self-centred regional identity in terms of cultural attitudes has so far not been fully addressed.

The impression of an apparent dynamism in Andalusia soon fades away if one takes a closer look at economic indicators, however. Even worse, the ongoing changes have had hardly any major cultural effects on the mix of attitudes to be met in the region. Growth rates in industry have been largely due to foreign investments which did not yield major extensions of productive capacities or enterprise organisation. Accordingly, the production system continues to be characterised by 'Taylorist' orientations, while adjustments have materialised above all in internationalised marketing strategies. As a consequence, the full scale of new concepts in enterprise organisation is rarely addressed. Anthropocentric production systems that emphasise horizontal information flows between manufacturing, marketing and R&D departments, decentralised decision-making and a learning approach by staff at all levels simply lack the necessary social traditions. Likewise, the impact on industrial R&D activities has remained marginal. An extension of public capacities in education has not been paralleled by a sufficient intensification and specialisation of research that could then become functional for wider application in regional industry (Román 1999). The Andalusian government has profited from a transfer in competence in both legal and financial terms but so far has proved unable to elaborate a clear-cut strategy for catching up with more advanced forms of socio-economic developments (Junta de Andalucía 1998). Accordingly, made-to-measure policies supporting innovative streams in research and industry are in their infancy, if not completely absent.

The problem of regions like Andalusia is that they develop a regional identity that is strongly dependent on decisions taken outside the region and does not develop from within a range of regional opportunities from which to choose (Bastian 2003b). Since Taylorist industries are not positively culture related and do not relate to tradition-based development, this situation puts them into immediate competition with other regions that can provide low production costs and cheaper industrial labour. So, similar regional traditions and cultural settings need not necessarily generate similar opportunities for socio-economic development. Regional identities may vary quite widely, because they are characterised by the dominant relationship between culture and socio-economic development. So, in fact, similar cultural settings may provide for divergent socio-economic tendencies, while non-cultural industrial paths of development (related to Taylorist industries and low production costs) may appear even in culturally highly diverse regions.

As indicated by the example of Poitou-Charentes, it is the uniqueness of the product that can provide for such attractive development processes, similar in their effect to having a position at the forefront of techno-industrial innovation. To provide for such attractive development, culture needs to be directed towards bringing about a successful industrial set-up within the region. The result may be a culture-related territorial immobility of the industry in question, because it relates to particular opportunities that are to be found and are set up only in that particular region. So, the existing regional identity to a great extent defines socio-economic success, the path of development and the ability to participate in global markets. Beyond the regional cultural arrangements and traditions as such, these identities define the areas and opportunities for government policies.

In the light of open economies and globalised processes, continuing socio-economic development requires a region's industrial sectors to be advanced. Structural change, as a consequence, is characterised by the continuation of some industries and the growth or demise of others. So, structural change, in particular, is characterised by a blend of stability and change in the cultural settings of the region. Regional identity represents a typification of the dominant processes, but, simultaneously, there are different processes going on. Most significantly, this is indicated by the differences between processes of techno-industrial innovation and sectoral crisis that can take place in the same region. These are related not merely to different industrial sectors, but also to the different cultural settings associated with it.

A strong traditional and an innovative part to the regional economy can coexist separately. The phenomenon of a region having these two cultural arrangements existing simultaneously can become an element of the regional identity, provided they are both clearly characteristic of the situation. As long as both cultural arrangements can provide economic development, there should be no problems. Indeed, this varied cultural basis of socio-economic development provides an enhanced variety of opportunities for taking advantage of the regional identity, provided such regional arrangements can accommodate continuous transition, adaptation and change.

Different cultures in regional development

The lack of a sufficient variety of cultural settings is fundamental to the problems of old industrial regions (Pichieri 1992). Industrial mono-structures give few opportunities for parallel cultures, and Taylorist-type industries hardly provide for a tradition that could be the basis for emerging new enterprises. So, while the arrangements discussed above relate to regional identities that can generate both multiple cultural arrangements and an additional knowledge body autonomously, the situation of old industrial regions becomes by contrast all the more critical. Here, regional arrangements ought to be changed, and the new production systems that are to be established demand particular sets of competences that hardly exist in the region. In such a situation, positive regional socio-economic development can be induced only by establishing a parallel culture; the new culture cannot emerge autonomously out of the industrial structure or the labour force itself.

Illustrative of this mechanism is the case of North Rhine-Westphalia, which has at its heart the old traditional industrial region of the Ruhr area. The Ruhr has been characterised by old industries such as mining and steel, and to a certain extent it still is. Nevertheless, although such traditional sectors continue to be an important part of the *Land*'s industrial structure, they have been modernised and restructured. Because of these industries, with their inhibiting attitudes towards economic activity, there was the question of how to realise restructuring and build up a new culture of innovation after the region had been so deeply immersed in a traditional Taylorist orientation over a long period of industrial history. More important for the change to culture-based developments than the Taylorist industries that moved newly to the region may have been the emphasis on strengthening engineering capabilities related to innovation in mining equipment or on changes towards introducing specialised steel or new materials. The challenges posed by environmental problems, in addition, have provided opportunities for developing modern solutions close to the area where they occur.

So, a changing industrial structure emerged, related to high-quality services and new more technology-oriented sectors. The large enterprises have also undergone diversification of their activities. Modern telecommunications are now to be found as well as computing, new media and modern biotechnology. This has led to a change in the industrial structure of North Rhine-Westphalia, a change that to some extent has taken place in the Ruhr area and has had a significant impact on industrialisation through links with modern industrial sectors outside this old industrial region.

New kinds of manufacturing organisation and of modern enterprises in general led to a situation in which there was a significant demand for personnel with different attitudes towards their places of employment. The flexibility of these new or reorganised and innovative enterprises, the openness towards collaboration, and the broad participation in decision making and the enterprises' development changed the kind of employees they need.

This structural change and the satisfaction of the demand for this new type of workforce took place over a period of 30 years and is still under way. Cultural changes based on the outlook of the participating individuals obviously have much to do with the generational change in the labour force over time, as younger people entered the labour market. In order to realise this change, working conditions have had to be rearranged as well. The new attitudes towards employment were restricted to people who had been well educated, and who had received higher education. The educational system and the foundation of new universities in the Ruhr area in particular had a strong enabling impact on these changes.

They were established with strong orientations in the areas of relevance to the region's industrial sectors or addressed to the new industries that were emerging in the process of structural change. So, in this regional research arrangement, from the very beginning an attitude could be found of being addressed to the region's needs as well as to the state of the art as defined in the international scientific community. In this process of transforming from one cultural model of socio-economic development based on outdated knowledge into another, more modern one related to modern industries and

state-of-the-art knowledge, the design and establishing of government programmes has played an important role. The *Land*'s science and technology
policies have established schools and universities for upgrading skill levels.
These activities provided the basis for the new outlook that is fundamental
for a modernised labour force. Investment policies were attracting new plants
and enterprises to the region, and economic policies made it easier for the
region's enterprises to realise the change. Environmental policies again had an
impact on creating demand for new and modern products developed and
manufactured in the region. And, finally, the training and labour market policies were fundamental for providing the modern labour required as well as
avoiding more severe unemployment problems.

These government policies followed a distinct idea of enabling industries,
research and labour to bring about the changes that structural modernisation
requires. When compared with earlier government activities, an entirely different politico-administrative culture provided the basis for this. There was
the orientation in the processes of socio-economic development instead of
dealing with the effects of change, and there was the orientation of actively
arranging the situation in the direction of innovation and modernisation. The
regional government was taking an active role in this process and was now
rearranging the cultural setting.

This provided the basis for a realisation of a merger between cultures of new
industrial sectors with the culture of the traditional industrial sectors. The attitudes of openness in industry and of actively governing the change in administration have provided for a situation in which traditional sectors and modern
sectors can develop simultaneously and sometimes even in interrelation.

In contrast, and in addition to cultural change associated with restructuring, government policies do not merely provide for change in order to create
a more dynamic development process. Government policies here induce elements required for additional cultural arrangements. This parallel culture is
needed to transform an old industrial region in crisis into a region under
restructuring. If governments do not even attempt to do so, or fail, as can be
seen in northern England, a new cultural arrangement cannot develop and
the situation will continue to be highly critical.

In term of West and South Yorkshire, there are clear differences between
the two sub-regions (Yorkshire Forward 1999). West Yorkshire has experienced a longer period of decline, running through from the Second World
War. Its staple industries (textiles and engineering) were faced by growing
foreign competitors. The long, slow decline saw the development of some
innovative patterns in the remaining industries, with the development of
niche markets. Growth has been in financial services and a range of supporting activities, high-technology development in electronics and a modernisation of some key activities (e.g. packaging and pharmaceuticals). This has
been centred in the regional capital, Leeds, which in many ways has come to
represent the classic post-industrial city, with two very large universities
undertaking significant research activity, a pivotal communications role both

within the region and in the wider UK framework, and a reconstructed labour force in which trade unionism has been largely eradicated and which is based on youth, gender and ethnicity, and distinguished by its flexibility in terms of employment rights, contracts and length of employment.

On the other hand, the position in South Yorkshire is markedly different. In essence, the cultural milieu of South Yorkshire is such that it is much less likely to be involved with economic innovation. Up to the late 1970s the region was significantly richer than West Yorkshire, and its staple industries – coal mining, steel production and metal manufacture – remained powerful entities. This diminution in the labour force has been largely related to productivity increases and the utilisation of new technology (Wren 1990). Over the past two decades the South Yorkshire economy has been devastated from a position where it was the largest coal mining region in the United Kingdom, to one in which there are fewer than ten mines. A similar story has occurred in steel and metal manufacture.

The consequence has been that the sub-region is now the poorest industrial region in the United Kingdom and suffers from a series of potentially difficult features that are likely to inhibit economic development. These problems are deeply rooted and will continue for a long time. There is a regional culture in which education and training do not command a high status. Previous industries were not knowledge driven, so did not recognise high levels of educational attainment, and training was 'on the job'. Today this is manifested in low scores in school results, low staying-on rates and low numbers progressing into higher education. An aversion to change also continued in the regional culture, because the dominating enterprises were firms with high levels of paternalism, with features such as miners' welfare, sports facilities, continuation of support after retirement, and the like.

This situation did not include an orientation towards change with an openness regarding new business, dynamic small enterprises or a search for South Yorkshire's place in an increasingly globalised economy. Universities, on the other hand, were not put into a position to generate research findings as a basis for new enterprises that may emerge over time. So, there was no new and economically valuable knowledge body based in the territory, partly because the regional society did not change and continued to be orientated towards a now globalised economy. This situation showed communications to be a major problem. The pit villages were separate entities in which virtually the whole male population worked at the local pit, and a coherent communications infrastructure never developed. In this situation, neither the regional society nor the regional economy was prepared to initiate major changes. Policies that would be peculiar to this region would have been required.

It is important to be aware that the development of northern England had experienced an important change when the Conservative government of Margaret Thatcher came to office. Following the ideas of competitiveness as against subsidising industrial change, attempts to change industries and modernise regional structures were widely scaled down. So, new knowledge did not enter the region to the extent needed to build new structures and provide a basis for vital new and modern industries. The further development of northern England was characterised by decline and unemployment simultaneously with a devaluation of competences. The territory was still linked with knowledge, but this body of knowledge had already lost its economic value. New knowledge following government policies to build up modern industrial structures was never acquired. A disadvantageous relation of the territory to outdated knowledge was continued.

In old industrial regions in general, attempts to manage the crisis and introduce structural change are often associated with a certain alienation towards the cultural settings already in existence (Schulze 1991). A strong focus on the education provided for younger people or the attractiveness for international enterprises to locate in the restructuring regions were both political targets and provided for a more modern basis for economic development in the decades following the 1960s and 1970s. While the cultural arrangements needed to be developed in these regions, in other regions international economic tendencies were providing for opportunities to make use of regional identities and traditions. Together with modern technologies and advanced research, such traditions can provide opportunities for attractive economic specialisation. Varieties of industrial competences create an increasing number of economic and industrial opportunities for favoured regions. While they were under the dominance of Taylorist industries, these regions could not arrange attractive locational settings; now they can do, so provided they build again on their cultural strengths. These economic opportunities are fostered by appropriate government policies that create parallel cultures.

Cultural changes required in the regions of industrialised countries, in order that they can participate in the international division of labour, are generally induced by governments. Different types of development relate to parallel cultures that provide for future or additional socio-economic opportunities. Since economic forces do not create such arrangements themselves, government policies become more and more important for industrial advancement. Government policies can provide for a simultaneity of different cultural arrangements, or they can provide for the emergence of additional cultural settings by supporting the bringing of new industries to a region; or new cultures may emerge in a region along with new industries. So, tendencies towards culture-based socio-economic development relate to proactive policies and the success of such policies.

Conclusions

Different paths of development of regions indicate a role for traditions, knowledge and a regional basis for processes of socio-economic development. Traditional and even pre-industrial knowledge, as in Poitou-Charentes or Andalusia, can provide a basis of development – given that markets for products exist and can be served. If such market-oriented attitudes do not exist, even similar situations may not provide for such processes. The role of both willingness to engage in development and the fostering of regional opportunities is indicated by examples such as the Ruhr area or northern England. While the one, based on the regional government of North Rhine-Westphalia, is in a situation to modernise the knowledge body of the region by means of education policies, new universities, research institutes and science-based enterprises, these factors are hardly to be met in northern England. The political attempt of the Ruhr's regional government did continue a tradition of cooperation among the social partners and political actors; in addition, programmes fostered opportunities for new competence building; whereas the era of Margaret Thatcher focused on costs and efficiency, and stopped similar activities in this old industrial region at a very early stage. Maintaining regional identity, as was attempted in the Ruhr area, was not considered an important factor in the case of northern England.

The four case studies illustrate that tradition and culture play an important role and they provide for the initiatives based on industrial culture promulgated by enterprises and their employees or by governmental systems and their administrations. In different ways, culture and tradition can be supportive of socio-economic development; regions can be linked to global development even beyond high technology and production costs, provided they focus on their competences regarding particular products and markets. The linkage with a regional society as a basis of a region's development indicates the relationship with tradition and the role it is expected to play in industrial change. In effect, knowledge is widely bound to a particular territory and loses its economic value when transferred elsewhere to a different situation. So, the region becomes the focus of interest and is a necessary basis for particular paths of development. This situation demands appropriate policies that keep knowledge economically effective and requires political systems that are both ready to intervene and fundamentally related to the regional society and the regional economy.

References

Bastian, Dietmar (2003a) *Regional Identity and Limitation in Culture: The Case of Poitou-Charentes*, Working Paper on Regional Identity, Friedrich-Schiller-Universität, Jena.

Bastian, Dietmar (2003b) *Approaching the Thresholds of Cultural Change: The Case of Andalucia.* Working Paper on Regional Identity, Friedrich-Schiller-Universität, Jena.

Bastian, Dietmar (2006) Modes of knowledge migration, regional assimilation of knowledge and the politics of bringing knowledge into the region, in Hilpert, Ulrich (ed.) *Knowledge in the Region*, European Planning Studies – Special Issue, London: Routledge.

Boschma, R. and Lambooy, J. (1999), Evolutionary economics and economic geography, *Journal of Evolutionary Economics*, 9: 411–429.

Calafati, Antonio G. (2006) 'Traditional knowledge' and local development trajectories, in Hilpert, Ulrich (ed.) *Knowledge in the Region*, European Planning Studies – Special Issue, London: Routledge.

Castells, Manuel and Hall, Peter (1992) *Andalucía: innovación tecnológica y desarrollo económico*, 2 vols, Seville: Espasa Calpe.

Center for Technology and Policy Studies (CTPS) (1996) *Innovación y transferencia de tecnología en Andalucía*, Instituto de Fomento de Andalucía; Instituto Andaluz de Tecnología, Proyecto RITTS 037, Andalusia.

Goldberg, Michael A. (2006) Knowledge creation, use and innovation: the role of urban and regional innovation strategies and policies, in Hilpert, Ulrich (ed.) *Knowledge in the Region*, European Planning Studies – Special Issue, London: Routledge.

Hickie, Desmond (2006) Knowledge and competitiveness in the aerospace industry: the cases of Toulouse, Seattle and north west England, in Hilpert, Ulrich (ed.) *Knowledge in the Region*, European Planning Studies – Special Issue, London: Routledge.

Hilpert, Ulrich (2006a) *Government and the Culture of Economy*, London, Routledge.

Hilpert, Ulrich (2006b) Knowledge in the region: development based on tradition, culture and change, in Hilpert, Ulrich (ed.) *Knowledge in the Region*, European Planning Studies – Special Issue, London: Routledge.

Institut National de la Statistique et des Etudes Economiques (INSEE) (1997): *La France et ses regions*, Paris: INSEE.

Junta de Andalucía (1998) *Programa industrial para Andalucía 1998–2001*, Seville: Consejería de Trabajo e Industria.

Junta de Andalucía (1999) *Andalusia Basic Data*, Seville: Instituto de Estadística.

Lawson, C. (1999) Towards a competence theory of the region, *Cambridge Journal of Economics*, 23: 151–166.

Lehner, Franz (1991) *Anthropocentric Production Systems: The European Response to Advanced Manufacturing and Globalization*, Synthesis Report, Gelsenkirchen.

Pichieri, A. (1992) Regionale Strukturkrisen und ihre politische Bewältigung, in Häußermann, H. (ed.) *Ökonomie und Politik in alten Industrieregionen Europas*, Basel, Boston and Berlin: Birkhäuser Verlag.

Piore, M. J. and Sabel, C. (1984) *The Second Industrial Divide*, New York: Basic Books.

Poitou-Charente Entreprises (1998) *La Dynamique des projets. Dispositif régional en faveur des entreprises*, Poitiers.

Préfecture de la région Poitou-Charente and Région Poitou-Charente (1999) *Poitou-Charente. Contrat de Plan Etat-Région 1994–1998*, Poitiers.

Román, Carlos (1999) *Learning How to Innovate*, Seville: Institute for Regional Development.

Schulze, Rainer (ed.) (1991) *Industrieregionen im Umbruch: historische Voraussetzungen und Verlaufsmuster des regionalen Strukturwandels im europäischen Vergleich*, Essen: Klartext.

Singh, Vijai P. and Allen, Thomas (2006) Institutional contexts for scientific innovation and economic transformation, in Hilpert, Ulrich (ed.) *Knowledge in the Region*, European Planning Studies – Special Issue, London: Routledge.

Wobbe, W. (1992) *What are Anthropocentric Production Systems? Why are They a Strategic Issue for Europe?*, Brussels: Commission of the European Communities.

Wren, Colin (1990) Regional policy in the 1980s, in National Westminster Bank (ed.) *National Westminster Bank Quarterly Review*, pp. 52–64.

Yorkshire Forward (1999) *Yorkshire and the Humber Region*, Draft Regional Economic Strategy, Leeds.

3 Creative regions and globalizing social capital: connecting foreign ICT experts to Finnish innovation environments

Mika Raunio

Introduction

Since the beginning of civilization, intellectual capital has moved in response to demand. During the Renaissance, Italian seafarers were hired in Spain, while Italian architects and painters were much in demand in France and Northern Europe. In seventeenth-century England, Dutch expertise in instrument making and weights and measures was actively sought after as a way to help England establish a lead in this area. In the twenty-first century, demand for IT specialists was driving industrial demand for foreigners with such skills, especially in the United States and in European OECD countries (Guellec and Cervantes 2002, p. 79).

Thus, foreign experts have always been an important asset for knowledge-based regional development. Their role has become even more important with the emergence of the global knowledge economy.[1] At the heart of the debate on the knowledge economy is the argument that innovation, which results from the application of technical and other knowledge, is a major contributor to industrial competitiveness. The knowledge economy is thus an innovation-driven economy (Boden and Miles 2000; Schienstock and Hämäläinen 2001), and the key sources of innovation are the knowledge and creativity of individual knowledge holders. As a result, the concepts of regional innovation system (RIS) and its broader version, innovation environment, have been used both to explain and to scrutinize regional economic development around the world. Also, these approaches clearly recognize that different innovative regions utilize different spatial levels – local/regional, national and global – in order to renew their knowledge base (Asheim and Isaksen 2002; Cooke *et al.* 2004; Cooke 2004a, b).

In this chapter the "human aspect" of global knowledge flows is examined. The possibilities for regional innovation systems to connect with the global pool of intellectual capital through expert foreign labour moving in have improved recently. Immigration legislation is increasingly based on skills and qualifications in most advanced economies, and restrictions to the mobility of labour have been lowered by international agreements. In addition, transnational companies provide pathways for crossing borders and multicultural work environments for individuals, who have often acquired at least the basic skills of multicultural interaction (Ghosh 2000; Findlay 1998; Stalker 2000). These processes are strengthened by the diminishing labour force in most advanced nations and the growing demand for immigrant labour.

Social capital has been the key concept explaining the social relations of individual within the innovation environments in recent literature (Maskell 2000; Doloreux 2002). Richard Florida (2002) claims that "creative regions" need open social capital with abundant weak ties. The phrase "creative region" refers to a wider regional context than RIS, emphasizing human aspects such as living environment and institutions that steer the social life of the region. According to Formhold-Eisebith (2004), innovation environment and social capital may be regarded as somewhat controversial concepts by their nature. "Innovative milieu" refers to the change-oriented cooperative actions that induce innovation and industrial change, whereas "social capital" refers to stabilizing relationships that provide stability and support for agents (ibid., p. 761). Thus, the role of social capital as a source of coherence and trust which lower transaction costs might, at worst, turn into an obstacle for renewal. This includes exclusion of foreign experts from existing social capital.

In this chapter, regionally and nationally important sources of social capital are brought into contact with foreign experts. This opens up interesting questions concerning regional innovation environments, or rather creative regions, that are about to utilize the global labour market of experts in order to renew and complement their knowledge base. How are social capital and its supporting structures to be adjusted to the "global age" in order to avoid exclusiveness – that is, a mostly unintentional form of discrimination towards foreign experts? This chapter examines this question by studying the case of foreign ICT professionals working in Finnish innovation environments. How do they interpret their career advancement opportunities and income level in Finland?

Theoretical framework and the data

Global social capital

In analysing the relationships of individual knowledge holders in the context of regional innovation environments, the concept of social capital has turned out to be extremely useful. To take one example, Maskell (2000) argues that social capital may constitute a competitive advantage for a company, because it is not abundant in all communities and cannot be bought or acquired. Most significantly, it is impossible to imitate, replicate or substitute for three different but interrelated reasons:

- First, asset mass efficiencies are present since communities that already possess a large stock of social capital are often in a better position to accumulate additional social capital than communities with a limited initial stock. To become a late starter on the right track and to match the first movers' rate of social capital accumulation might require more than luck and blind reliance on the possible beneficial but anticipated consequences of doing something else.
- Second, the accumulation of social capital is at least in part the unintended and unanticipated outcome of activities performed to achieve another purpose. This gives rise to all sorts of causal ambiguity, the disentangling or unravelling of which might prove impossible. Unlike certainty, ambiguity cannot be reduced by the collection of more facts. A community's stock of social capital represents a complex web of relationships and linkages woven over time while leaving the precise nature of the means–ends relationship blurred. Initially significant institutions might even over the years mutate or interact in the organization of derivatives, with profound influence on the specific qualities of the social capital of the community. Many aspects are "in the air" but not easily associated with any specific institution, formal or informal.
- Third, social capital accumulation always requires time-consuming reiteration and habituation, and generally no short cuts are available. Attempts to catch up with first movers, already in possession of large stocks of social capital, are faced with time compression diseconomies.

(Maskell 2000: 118)

It is precisely the same reasons that make global social capital valuable in the global knowledge economy. The opportunity to create global social capital may be hindered by national institutions and cultures if they cause difficulties for newcomers in integrating with social capital as equal partners.

Regional innovation environments with strong traditions of attracting foreign talent and with strong global connections, Silicon Valley being a prime example, may overcome this problem with the help of existing transnational ties and diverse ethnic communities. Such regions have become strong transnational hubs for business and technology experts around the world. They have the potential to exchange both explicit and tacit knowledge through mobile individuals or "global nomads", who move transnationally at will. Similarly, regions with a substantial return-migration of experts and the highly skilled have succeeded in establishing strong global networks and in strengthening the local knowledge base with extra-regional expertise. The latter has been the case in Ireland and in parts of China, for example. These existing connections both attract and retain newcomers from the global field (Saxenian 2002; Luo and Wang 2002; Grimes and White 2005).

These are "natural habitats" for the emergence of global social capital. In the case of Silicon Valley, ethnic business communities had been established as early as the 1960s by successful immigrants. In the case of China, numerous highly educated return-migrants have recently found that the growing Chinese economy provides competitive business and living opportunities that make returning worthwhile. Also, migration literature points out the important role of transnational social capital as cause and effect for international migration and formation of transnational communities (Findlay 1998; Faist 2000; Sassen 1999; Massey and Taylor 2004).

However, most innovation environments do not have strong multicultural environments and global networks, or significant numbers of individuals living in a "scientific diaspora" who could return home. Thus, their situation is very different. In the case of Finland, multicultural and international (expert) communities within the country are extremely small, and potential return migrants have not shown an interest in massive return-migration to Finland[2] (Table 3.1). The endogenous emergence of global social capital is not likely to occur in this case.

The need for foreign experts arises – in addition to the general trend of a diminishing labour force in all advanced economies – from specific reasons associated with the nature of innovation-based competition in the global field. A statement by Hallstein Moerk, Vice President and head of human resources at the Nokia group, summarizes the issue:

> Diversity is extremely important for the following reasons. Firstly, it is important to understand the needs of our global customers. So, we have to have an organization that more or less represents the cultures of our customers. Secondly, diversity increases the creativity of the work community. Working in diverse teams is not always easy, but it frequently leads to more innovative outcomes, because different approaches are used. Thirdly, we want to hire the best experts, so our recruiting has to be global.
>
> (Raunio 2005, p. 36)

Table 3.1 Percentage share of non-nationals in selected high-skill occupational groups

Nation	Austria	Belgium	Germany	Denmark	Spain	Finland	France	Greece	Italy	Holland	Portugal	Sweden	UK	EU
Managers[a]	5.0	10.1	7.7	2.7	1.5	0.4	4.7	0.9	1.1	2.5	1.4	2.3	4.2	4.3
Experts[b]	6.0	5.0	4.0	2.4	1.0	0.6	2.6	0.9	0.7	2.6	1.7	3.2	4.5	3.1

Source: Auriol and Sexton (2002).

Notes
a ISCO 1: Production and operation department managers. Other (including computing services) department managers. General managers.
b ISCO 2: Physical, mathematical and engineering science professionals. Life science and health professionals. Teaching professionals. Other professionals. ISCO 3: Physical and engineering science associate professionals. Life science and health associate professionals. Teaching associate professionals. Other associate professionals.

Essentially, this is the key message, which is also found in literature on diversity management and creativity management literature (see Fleming and Sorenson 2004; Parkhe 1991).

In the literature on innovation environments, the concept of global pipelines refers to the need for external knowledge sources in order to revitalize and renew the local knowledge base. Problems with extra-regional or international pipelines have arisen as a result of the "different languages" that partners from very divergent cultures may have. Knowledge that is not part of the firm's repertoire may also be too different from the present mental representations and thus ignored or not taken seriously enough. The concept of absorptive capacity emphasizes the role that diversity of expertise and its distribution within the company play in creating new mental maps that integrate new knowledge coming from external sources. If all individuals share precisely the same specialized language and symbolic representations, they will not be able to tap into diverse external knowledge sources, even if the relevant pipelines are in place (Bathelt *et al.* 2002, 18–19).

Although excessively strong external pipelines may threaten the long-term existence of a cluster by reducing its coherence, there are clear advantages that cannot be ignored. Foreign experts may support the absorptive capacity and diversity of expertise and, rather than reducing local coherence, they may increase the global reach of the organization. Having an equal share in local social capital increases global reach without reducing the local coherence. If newcomers are excluded from local networks and contributions which sharing in local social capital could bring, it is likely that their networks will favour other directions and so reduce local coherence.

In order to use the rather fuzzy concept of social capital as a tool for understanding the different ways to integrate a community, three key dimensions of the concept should be clarified. First, from an individual's point of view, the depth of integration to social capital may be evaluated at three levels:

- *form of the network* (nature, depth and structural aspects of social ties, the scope of the network)
- *shared norms* (which may result in feelings of trust and obligation and actions of reciprocity)
- *access to resources* (additional social networks, relationships, information, language, money, physical goods, etc.).

(see Monkman *et al.* 2005)

The most interesting measure for assessing whether or not an individual is *included* in a social network is access to resources. A foreign expert may formally belong to a network but be excluded from resources in terms of receiving the same information as other members of the network or having an equal chance of promotion, etc. This chapter focuses on the *resources* potentially delivered by social capital, instead of form or norms.

The second dimension to clarify concerns the idea of openness. Openness refers to the capability to absorb newcomers and integrate them as part of the community. Florida suggests that social capital based on strong ties may even hinder the economic development of regions, whereas it is fostered by a wide set of weak ties and quasi-anonymity of individuals (2002, pp. 267–268). Argument concerning the significant role of weak ties refers to Mark Granovetter's study (1973), according to which weak ties were highly important in terms of careers and job opportunities for highly educated professionals and experts. However, more precise distinctions of social capital related to its openness are needed for analysis. In this chapter the following distinctions are used:

- *bonding* social capital, characterized by strong bonds – or "social glue" – such as between family members or among an ethnic group
- *bridging* social capital, characterized by weaker, less close but more cross-cutting ties or "social oil", such as between business associates, acquaintances, friends from different ethnic groups and friends of friends
- *linking* social capital, characterized by connections between those with different levels of power or social status, such as links between the political elite and the general public or between individuals from different social classes.

(Huber and Skidmore 2003, p. 66; Woolcock 1998)

These concepts are useful when considering the purpose of social capital – with whom do we share the resources?

Third, the relation of social capital to institutions and structures supporting it, should be acknowledged. Work by Granovetter (1985) and Bourdieu (1986) explained how social capital is related to other forms of capital, by introducing the concept of embeddedness and cultural capital. Without going any deeper into these studies, it is enough to say that they also revealed the processes through which social relations and power structures among them may create institutional forms based on local interactive culture and values. Indeed, Harrison states that "culture is the mother and the institutions are her children" (2001, p. 121). Manuel Castells (1997) provides detailed analysis of how social relations and "collective identity" that challenge the dominating structures and institutions of society often radically renew the very basic structures of society. In short, it is important to note that values, cultural traits and forms of social relations are often institutionalized as regulations, legislation and stable modes of operations. If cultural traits and social relations are about to change, institutional structures have to respond to this change, and vice versa.

According to Katz (2004, p. 228), homogeneity is a common feature among groups that work together for longer periods of time. Individuals stabilize their work settings and patterns of communication, which may lead to insulation from other, more heterogeneous groups. Similarity between the interacting individuals' values, beliefs, etc. prevails.

Florida (2002, pp. 323–324) states that when the nature of the economy has changed, old institutions stop functioning. People and social groups can no longer relate to each other as they once did because their economic roles are different. He even claims that communities with strong (bonding) social capital will not pass muster because people work differently today and desire very different kinds of lives. The key is to create new mechanisms for building social cohesion. Strong communities, instead of institutions within them, are the key to social cohesion (ibid., pp. 323–324). This chapter aims to recognize how Finnish creative regions are seeking new social cohesion through collisions of different forms of social capital.

Research data

The research data include empirical data on foreign professionals who have worked or still work in Finland. The empirical data in this chapter are based on data collected during the "Should I Stay or Should I Go?" project in 2001–2002. The research process included 30 background interviews, including:

- Interviews with persons responsible for the recruiting of foreign professionals and other persons with a significant role in the process (relocation consultants, managers and HR personnel).
- Personnel dealing with these issues in public-sector organizations in urban regions.
- An internet questionnaire for foreign professionals working in Finland (mostly in the ICT field; 2 per cent of the respondents work in the field of biotechnology). The aim was to find out whether foreign professionals are satisfied or dissatisfied with their situation and what factors affect the prevailing level of satisfaction (556 responses).
- Personal interviews with foreign professionals (59 individuals) and their (foreign) spouses (33 individuals). Spouses were interviewed in order to enhance understanding of the pros and cons experienced in the everyday lives of foreign families or couples.

The study was conducted between September 2001 and May 2002 in Finland. The study regions were the four biggest city-regions of Finland: Helsinki (Espoo), Tampere, Turku (Salo) and Oulu (Figure 3.1).

The respondents' views are mostly based on a fairly long time spent in Finland. Almost half of the respondents had lived in Finland for at least three years. A typical respondent profile was a 34-year-old European male with higher education, working in Finland on a local contract and living with his family or spouse. However, the backgrounds of the respondents were quite diverse. These data have now been reinterpreted and analysed in order to serve the research question presented in this chapter which aims at a more refined theoretical understanding of the issue. Characteristics of respondent to questionnaire ($n = 556$)

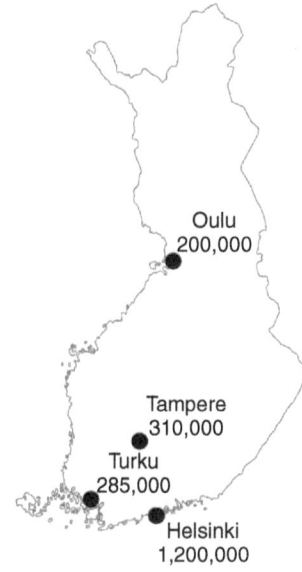

Oulu
200,000

Tampere
310,000

Turku
285,000

Helsinki
1,200,000

Figure 3.1 Location of Finland in Europe, regions studied in Finland and their populations

- Response rate: 41 per cent (questionnaire was sent to 1,365 foreign experts)
- Educational level: doctoral 7 per cent, Master's 54 per cent, bachelor's 31 per cent, other 7 per cent
- Place of birth: Northern Europe 38 per cent, Eastern and Central Europe 19 per cent, Southern Europe 8 per cent, India and Southeast Asia 12 per cent, China 7 per cent, North America and Australia 11 per cent, the Middle East and Africa 5 per cent
- Gender: male 80 per cent, female 20 per cent
- Household type: single 28 per cent, couple 38 per cent, family 31 per cent, other 3 per cent
- Type of contract: permanent local contract 85 per cent, expatriate/temporary assignment 10 per cent, other 6 per cent.

Regarding the type of contract, it should be noticed that "local contract" refers to local salary level and other local conditions in labour markets. "Expatriate assignments" may include substantial financial benefits including higher salaries, housing benefits and fees for international schools, etc. Thus, our data focus on foreign professionals working in Finland with local contracts.

Access to resources bridging corporate cultures and linking national institutions

In our project "Should I Stay or Should I Go?" we found out that the main reasons pushing foreign professionals away from Finland were the lack of career opportunities for non-Finnish professionals, low income level and social relations. Often people do not really need push factors for moving away, because they have come to Finland for a temporary stay to begin with. However, from the viewpoint of social capital these results offer quite an interesting starting point for our analysis. Of the total of 556 respondents, 25 per cent intended to stay in Finland permanently.

It seems that individuals' interpreted satisfaction with rewards, such as career advancements and income, increase the likelihood of their staying. Sixty per cent of those who intended to stay in Finland agree or mostly agree with the statement that "career prospects in my current home region in Finland are good", in comparison to only 52 per cent of those who were about to move to a third country and 45 per cent of those who were going to return home. Differences between those who were going to stay and those who were about to leave were even higher as regards income-related factors (salary, tax, pension) and social factors (communication, atmosphere). Especially in the case of tax rate, those who were going to stay generally found it far more tolerable (43 per cent) than those who were going to continue their career in a third country (12 per cent) (Table 3.2).

Exclusion from the social network and dissatisfaction with rewards (professional and economic) refer to a social capital that is not genuinely open to foreigners or ready to share its resources with them. The following chapters seek to explain why this dissatisfaction and increased urge to leave are at least partly due to locally and nationally created social capitals and their, in some cases unintentional, exclusiveness.

Career advancement supported by bridging social capital

The first aspect to be discussed is the career opportunities and bridging efforts in multinational corporations (MNCs). According to Earley and Mosakowski (2004), knowing what makes groups and individuals tick is important in a world where crossing boundaries is routine. Cultural intelligence becomes a vitally important aptitude and skill; learning to cope with different national, corporate and vocational cultures becomes a necessity (ibid., p. 139). This is especially important in expert organizations, where communication and the flow of information are essential parts of the successful work process. In order to make global alliances and multicultural work environments effective, companies create and apply corporate cultures, global management styles, diversity management and respect for individual approaches, all of which are designed for coping in diverse and turbulent environments (Larkey 1996; Parkhe 1991).

Table 3.2 Satisfaction with certain environmental factors and plans for the future (*n* = 556)

Strongly agree or mostly agree ("strongly agree in brackets")	I will stay in Finland permanently	I will move back to my home country	I will move on to a third country	Total (n = 556)[a]
My salary and other financial incentives are sufficient	51% (7%)	32% (7%)	26% (6%)	34% (6%)
My current tax rate is tolerable	43% (2%)	24% (6%)	12% (2%)	24% (3%)
In Finland my pension benefits are working all right	43% (12%)	33% (4%)	27% (8%)	34% (8%)
Career prospects in my current home region in Finland are good	60% (17%)	45% (11%)	52% (16%)	55% (15%)
The working culture in my unit is communicative	80% (28%)	66% (21%)	56% (18%)	66% (21%)
I feel I have been treated equally when dealing with Finnish authorities	77% (37%)	68% (20%)	60% (24%)	70% (20%)
The atmosphere of my current home region is open and people are approachable	67% (20%)	42% (6%)	36% (8%)	49% (11%)

Source: Raunio (2003).

Note

a Includes those who had not yet considered the matter (28% of the respondents).

In order to overcome communication problems at the basic level, English is often used as the "official language" of global business and as the working language in the offices of MNCs regardless of their location. A shared language and diversity management in corporate cultures aim to create bridging social capital among employees, in order to avoid a situation where differences between individuals would hinder the key business processes. The aim of diversity management is not so much to increase global equity among professionals as to increase the competitiveness of the company. However, in practice there have been cases where the human resources (HR) unit of a MNC has instructed the local management that they should recruit more different people, and globally, for the sake of diversity and in order to get the best brains. The local management naturally have the last word in the selection of their employees, but the global employer clearly introduces the forms of bridging social capital to local innovation environments.

> In our selection we aim at diversity: we do not recruit people who are just like ourselves or whom we like. We recruit those who think differently from us. Different opinions create more synthesis than like-mindedness. Divergence always creates more than "yes-men" do. Nice guys are not an asset for the company.
>
> (Finnish HR Manager, ICT)

According to this logic, individuals should be valued and rewarded for their contributions for the key business processes of the company. However, this is not always the case. Foreign workers frequently perceived their chances of career advancement within the company as slightly poorer than those of their Finnish colleagues. In some cases the respondents even took it for granted that when the choice for promotion lies between a Finn and a foreigner, it is the Finn who will be preferred almost without exception. The reason for this was considered to be the existing social networks, in which people who are already acquainted with one another select each other as subordinates. The "Finn-boy bias" was seen to be clear in some top and middle management selections (vertical career path), but in the management of technical projects the phenomenon did not emerge (horizontal career path). In most cases the bias was not felt to be extremely strong, but some highly critical views were also voiced.

> I think it definitely helps if you are Finnish, in my opinion it is a fact. I've seen it again and again that a Finn-boy gets promoted before someone else. It is a matter of network, of who you know. When you spent some time studying with someone or something, it helps. If I think of all my immediate managers and look all the way up the management line, I don't see any foreigners there. Technical groups are more open to foreigners but the management circle is quite closed.
>
> (ICT professional, Canada)

It thus seems fairly evident that bonding social capital based on rather homogeneous groups (Finns) really detracts from the opportunities of newcomers to join in and enjoy the resources of social capital in terms of career advancement and job opportunities. The "Finn-boy bias" has been created over a long period of time and has only recently started to open up, thanks to such methods as diversity management and increasing the numbers of foreign professionals and students. The latter are especially important for the emergence of global social capital, because without genuine social networks and a true understanding of the modes of operation of the enterprise as a part of local culture, the chances of both promotion and inclusion in social capital are probably weakened.

In general, many of the respondents had slightly more negative views about broader career prospects in the region where they were living than about the opportunities of career advancement within their employer's organization. The Finnish job markets do not have a great deal to offer to those without a command of Finnish. In addition to the language problem, the choices available were seen to be limited, at least to some extent, by the assumed incapability of Finnish enterprises of acculturating into the work community an expert with no command of Finnish. The job markets in Finnish urban regions are quite small, since the only region with a population of more than 500,000 is the capital region, with 1.3 million people (Figure 3.1). Indeed, hunting for a new job within Finland had received very little serious consideration, as a change of countries was more likely if a change of jobs was in the offing and there were no personal ties to Finland.

This makes it extremely significant that regions willing to attract foreign professionals should support the openness of the local culture, and especially the openness of local employers. The innovation regions in Finland are not alone in facing the problem, but they lack experience of multicultural interaction and endogenous bridging social capital. In particular, the further respondents were, ethnically, from the Finns, the more they felt that promotion opportunities were not equal. In general, it is confirmed by several studies, including this one, that ethnically more different individuals encounter more negative feedback and xenophobia than those who look and behave more like natives (Pitkänen 2006; Raunio 2003). If ethnic difference judged on the basis of the individual's appearance plays a significant role in the kind of rewards he or she will receive in Finnish creative regions, it is not likely that such a region will attract the world's best brains, since most of the world's population is already excluded by this attitude.

However, it seems that the creation of bridging social capital is much more a matter of will than of skill. Units where managers were willing to put some effort into developing their management styles or other means of improving the social and cultural qualities of their group were able to overcome the problem within a fairly short period of time. In addition, many foreigners had advanced in their careers and perceived their chances of advancement within the company as good. Although the respondents saw the exclusion as fairly natural and real in many other countries as well, it is likely that the monocultural Finnish society slows down the foreigners' advancement and the recognition of their expertise more than would be the case in countries where cultural or ethnic diversity is commonplace. Moreover, the equality and reliability characteristic of Finnish culture and the emphasis placed on these as essential features of society served to exaggerate a contradictory situation in which the cultural and ethnic difference of an individual prevents the enjoyment of the rewards of social capital. An innovation environment's culture emerging from local management seems to be less eager to connect with global labour markets than the global strategic level of MNCs seems to be. Thus, MNCs and other global employers seem to provide globally bridging social capital to the region.

Taxes and income as linking social capital

The second view on supportive structures of exclusive social capital may be seen in institutions of the nation state, where the aim has been to offer benefits and equality for all groups of people within the nation state by creating linking social capital. According to Straubhaar (2000), the twentieth century began worldwide with the (partly artificial) forming of nation states. Nation states have been very efficient as institutions in minimizing transaction costs in the era of industrialization.[3] In a world of increasing globalization of more and more business activities, however, the politically defined territoriality of nation states faces growing economic pressure. Several nation states have disintegrated (the Soviet Union, Yugoslavia, etc.). Others have started to congregate in regional integration blocs (NAFTA, the EU, ASEAN, Mercosur). Many national borders have been abolished and some integrated areas without national borders have arisen (the EU). There is a growing body of international organizations which deal with "international public goods" like the United Nations for security, the World Bank and IMF for financial matters or the World Trade Organization for trade matters. These are intended to optimize the benefits from the interplay of national actors in a global game. The global world needs global rules, because matters of national policy or market failures have now become topics of international concern, with their repercussions easily crossing national territories, such as in the case of macroeconomic instabilities causing migration issues (Straubhaar 2000, pp. 110–111).

In the making of nation states, one crucial factor was the aim to create social capital by producing shared cultures that separate "us" from "others". The bridging, bonding and linking capitals were created through language, schooling system, history, politics, etc. The ideology has created work permit practices to safeguard local labour markets that have now been opened up by the emergence of a global labour market along with the global economy. The emergence of national cultures was supported by several strong institutions of which the taxation system is the most fundamental, since it forms the finan-cial base for the existence of the nation state, and its construction reveals the quality of the capitalism supported by each state.

The taxation system in Finland was created in the spirit of the coordinated market economy (CME) and of the *universal* welfare state to diminish income differences between different societal groups. Thus, it is an institution that supports the emergence of linking social capital, and has in fact been remarkably effective. According to the OECD, income disparities in Finland are among the lowest in the OECD countries, even though the average income level is reasonably high (Föster and Pearson 2002; Föster and d'Ercole 2005, p. 10). The small income disparities in Finland, resulting from a progressive taxation policy and small wage differentials, reflect the value base of the welfare state. The aim is to ensure that everyone has at least an adequate basic standard of living through income transfers, whereas in a performance-oriented society such as the United States or United Kingdom, poverty is seen to be largely a consequence of the individual's own choices and it is considered that society should not subsidize lifestyles not regarded as mainstream (Hofstede 1991, p. 97). Partly as result of intentional policy choices, in Finland the potential economic benefit provided by higher education and professional skill is among the lowest in Europe (Figure 3.2).

The nation state has assumed a major role as the provider of linking social capital between individuals by attempting to reduce differences between the socio-economic classes, and has done that quite effectively. If one were to exaggerate a little, one could say that Finland has become "a country of cheap labour for research and development", which is partly a consequence of the small income disparities between different educational levels (Castells and Himanen 2001).

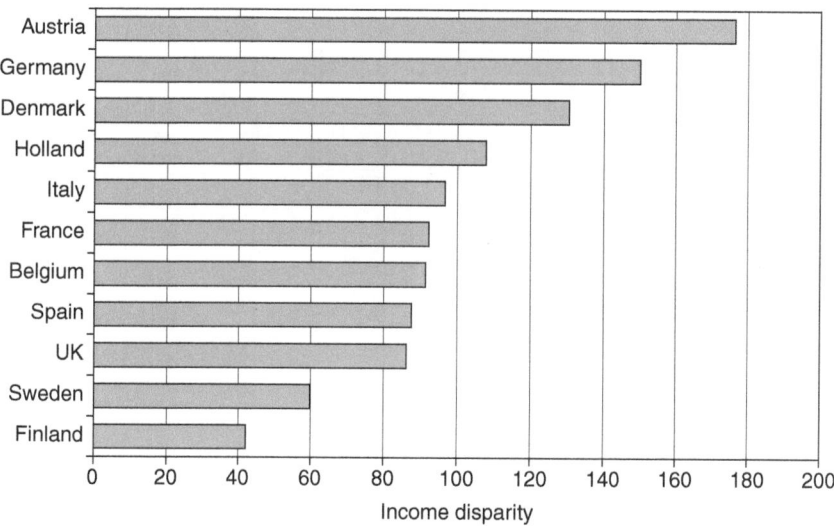

Figure 3.2 Increase in income compared between employees with lower-secondary-level and university-level education (%). Source: Eurostat 2005

When I was looking for another job, Finnish employers weren't able to pay the "market price". This doesn't concern only my line of business; look at nurses' salaries, for example – no wonder they go to work abroad. If Finland can't keep its own nurses, how could it attract foreigners to Finland?"

(ICT professional, UK)

In fact, taxation in Finland is among the highest in the world, especially for those on high incomes (Figure 3.3).

The respondents frequently compared income level to that in high-income countries such as the United Kingdom or the United States, because these are the most likely destinations for ICT experts moving globally. In both the United States and the United Kingdom the income differences are substantial and the income level for ICT experts is high. Comparisons between different countries are extremely difficult and should be approached with caution and treated as rough approximations to reality. In contrast, assessments made by individuals are in most cases based on their own experiences and are treated as more or less proven by them. Thus, statistical national comparisons provide fairly limited information in the sense of what individuals think of as attractive. However, Table 3.3 shows a comparison of engineers' salaries in three countries. The comparison is based on the annual salaries of engineers between the ages of 30 and 35. This is a relevant age group since in the data for this study, most professionals moving internationally were quite young. The average age of our 556 respondents was 34 years.

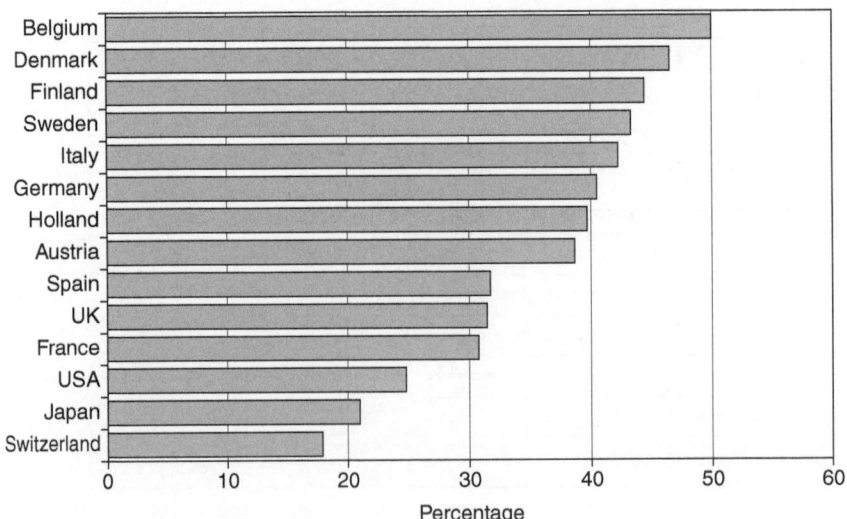

Figure 3.3 Combined share of income taxes in some OECD countries, for households with two salaries and two children aged under 18 years (income €45,000–€96,500). Source: Kurjenoja (2003)

Table 3.3 Income comparison between Finland, Germany and the United Kingdom, 2000 (€)

	Finland (2000)	Germany (2000)	UK (2001)
Annual salary of graduate engineer[a] (university degree)	37,000	50,000	53,000

Note
a Aged 30–35 years.

For most respondents, however, their income level had not fallen very dramatically on their coming to Finland, but it is nevertheless quite often necessary to accept a drop in economic well-being, or at least in net income, when opting for Finland. In only 13 per cent of responses ($n = 556$) had economic prosperity improved in consequence of a respondent's moving to Finland. On the other hand, this is indicative of a great preference for work and career prospects over economic factors in the choices of global experts, but it may also be seen as a competitive handicap both to attracting experts and to securing their commitment (Raunio 2003).

The key dilemma is that in a universal welfare state individuals are seen as citizens whose lifespan comprises three main phases: childhood and youth, when free schooling and health care are provided; working life, when the citizen works and pays taxes as a contribution to the community; and old age, when the community returns the contribution in the form of pensions and health care. This line of thinking fits quite poorly with foreign experts, who spend two to ten years in Finland and then move on.

The role of progressive taxation should be considered from a broader perspective: should it be related to services used rather than just income earned? Education, for example, may have been extremely expensive to acquire for an individual moving to Finland. It is obvious that the perspective on taxation differs between those only temporarily resident in the country and those whose residence is relatively permanent.

> The taxes I pay would be justified if I was born or if I was going to retire here, but I am not and so I am not a burden for the state in the time that I am here. So why the hell should I be paying and treated the same as those who have been or will be that burden? My tax rate is 48 per cent and I pay 22 per cent VAT. If I add up all the taxes I believe I have 25 per cent left in my back pocket, I believe that my tax rate is 75 per cent all in all.
>
> (ICT professional, South Africa)

The high taxation is not necessarily mirrored in the high quality of the public services used by the target group and, on the other hand, single persons and childless couples use only a few services. In Finland the level of salaries and taxation means that a couple with a family "must" both work, and the general way of life is partly dependent on the public sector, as far as child care outside the family is concerned. This is partly due to the principle that the welfare state pays for public services which elsewhere are paid for through private funds or insurance. An individual arriving from a different kind of culture is prompted to ask why one should pay the public sector for health care services, child care and pensions, when one could pay the private sector to the extent deemed necessary, or even save for one's old age out of a better net income.

However, there already exist some approaches to the creation of linking social capital among all individuals who will contribute to the local economy regardless of how long they have lived in the region. As an example, progressive taxation based on length of residence, which is applied by some countries *vis-à-vis* foreigners, is seen as a fair practice. In this case the tax rate increases year by year and reaches the local level during the sixth year of residence, for example. Progressive taxation of this type is regarded as justified because a person who only lives in the country for a few years does not feel that he or she benefits from free education prior to entering working life or from the pension benefits after retirement or other services for the elderly. A progressive system would reduce the amount of tax paid "without benefit" during a short sojourn, and after the person has lived in the country for a sufficiently long period, the realization of the benefits produced by society in an individual's life would "justify" raising the taxation to the local level.

In Finland, inequality has been decreased by a special tax arrangement for foreign experts working in Finland. The prerequisite for eligibility is that the individual is either an expert earning more than €5,800 per month or working in an academic position as teacher, professor or researcher. In these cases he or she is eligible for a special income tax rate of 35 per cent, which will be applied for two years. This is, however, inadequate, since the pay level of experts does not often exceed the limit, and the two-year period is too short for many R&D-related projects. At the time of the study there were only a few dozen foreign specialists enjoying this benefit. However, the arrangement does indicate that there are means of improving the institutional base in response to the requirements of the globalizing environment, by creating supporting institutional structures for the linking of social capital at the global level.

Conclusions

In this chapter the social relations in regional innovation environments have been scrutinized through the concept of social capital. Richard Florida brought up the need for "openness" towards newcomers if regions are to succeed in the global knowledge economy or "creative economy", as he puts it. In previous work the qualities of social capital in Finnish innovation environments, or creative regions (the four biggest city regions), were examined from an openness point of view. Openness of social capital was seen as an access to *resources* that shared social capital could provide: career opportunities within the region and income level.

Career opportunities were seen as slightly poorer for foreigners without bonding social capital. Foreigners often lack access to the "right" networks, which are formed over time, and since local managers often prefer to hire people they are familiar with, the situation is not likely to change soon. Income level is rather low, because the universal welfare state lowers income differences through progressive taxation, and public goods financed through high taxation are paid back in the form of free education and care for the elderly. Both services often neglect foreign experts, since their education is in most cases obtained in countries other than Finland and often they also leave Finland before needing care for the elderly. These features clearly make foreign experts different from other groups of foreigners, whose stay in the country is often more permanent.

Evidently, there are three overlapping forms of social capital emerging in Finnish creative regions that distribute resources according to quite different logics. The roles and key advocates of various forms of social capital may be defined as follows:

- The global level is represented by global MNCs and other global actors, which would like to create social capital around their core business. The key aim is to get the best brains and individuals who are capable of working and doing successful business in the global economy. In this case, the method utilized by the MNCs was diversity management, and the aim was to decrease inequality among the company employees. *The MNCs support the emergence of bridging social capital at the global level.*
- The national level is represented by the state and its institutions. Although immigration laws and tax arrangements are modified in order to use the global talent pool and labour in general, the basic goal is to ensure the coherence of the nation state. Thus, the decreasing of income disparities and inequality between individuals is still the key goal of the universal welfare state. *National institutions provide linking social capital at the national level.*
- At the local level, it seems that recruiting managers are not very willing to recruit people who are *too* different, in order to make the work unit function smoothly. Their key motive is to create a *functional work unit by increasing bonding social capital and coherence within the work community.*

It might be argued that the globalization of labour markets requires fundamental societal and institutional changes. However, at least in this case it seems that rather minor modifications to individual, organizational and institutional levels may be very effective if the aim is to connect the Finnish regional innovation environment with the global pool of talent. It is not just a question of providing the weak ties and bridging social capital; also, the bonding social capital and linking social capital should be given new forms and become more inclusive than they are at present. Linking and bridging social capital are routes to more flexible and diverse bonding social capital that is the aim of an open society with inclusive communities and high absorptive capacity. Examples of how basic institutions of the nation state like the taxation system may proceed towards more flexible linking capital were given in this chapter. Also, a role for bridging social capital was described. Supporting structures for methods such as diversity management can be used to strengthen the emergence of the bridging social capital within the region. As a part of the "creative region's infrastructure", knowledge-intensive business services or training organizations may provide the content for these services. Of course, local ethnic and/or multicultural associations and organizations, etc. may participate in the process.

However, the bottom line is that there are three different actors from three different spatial levels, all building coherence within the same creative region by providing the social capital that is most useful from their perspective. The change seems to be incremental rather than radical, but it takes place at all the levels of society. The birth of the nation state saw national institutions create linking social capital, and nationwide organizations create bridging social capital. between the existing communities of the time. Now some globally oriented MNCs and other global agents are building global bridging capital upon these nationally based communities. It depends on the national institutions and local communities how they respond to this challenge: are local and national actors able and willing to add some new elements to their prevailing social capital, and even slightly redefine their identity to be more open – a bit more global?

According to Forsander (2004), the Nordic countries should redefine themselves as nation states that distinguish between ethnicity and nation state. European nation states are built on the myth of national boundaries which are identical politically, culturally and ethnically. Rewriting the national myth to make it more open to diversity might help in developing more of the bridging Nordic social capital while becoming more inclusive to all members of these societies (ibid., p. 227). Bridging social capital is a start, but also the very essence of bonding and linking social capitals should change somewhat. This is a precondition for providing genuinely open social capital with global networks where instead of forms and norms the issue is to share the resources of the network equitably.

Notes

1 This article is part of the "Technology, Talent and Tolerance in European Cities: A Comparative Analysis" project, funded by the Academy of Finland/European Science Foundation (no. 203524).
2 The globalizing knowledge economy is an economy that is directly based on the production, distribution and use of knowledge and information. In an increasing number of economic fields, flows of goods have been or will soon be replaced by flows of information and knowledge. The fastest-growing countries and regions are those that manage to generate and diffuse knowledge most rapidly.

3 Unlike Finland, most West European countries suffered from labour shortages following the Second World War and recruited labour from abroad, and rather large ethnic minorities were created in these countries (Joronen 2002, p. 135).
4 Transaction costs consist of all additional costs of market activities not included in the production costs.

References

Asheim, B. T. and Isaksen, A. (2002) Regional innovation systems: the integration of local "sticky" and global "ubiquitous" knowledge. *Journal of Technology Transfer*, 27: 77–86.

Auriol, L. and Sexton, J. (2002) Human resources in science and technology: measurement issues and international mobility. In OECD, *International Mobility of the Highly Skilled*. Paris: OECD.

Bathelt, H., Malmberg, A. and Maskell, P. (2002) Clusters and knowledge: local buzz, global pipelines and the process of knowledge creation. DRUID Working Paper no. 01-12.

Boden, M. and Miles, I. (2000) Conclusions: beyond the service economy. In Boden, M. and Miles, I. (eds) *Services and the Knowledge Based Economy*. Continuum, London.

Bourdieu, P. (1986) The forms of capital. In Richardson, G. B. (ed.) *Handbook of Theory and Research for the Sociology of Education*. Greenwood, New York.

Castells, M. (1997) *The Information Age: Economy, Society and Culture*. Vol. 2, *The Power of Identity*. Blackwell, Oxford.

Castells, M. and Himanen, P. (2001) *The Finnish Model of the Information Society*. Wsoy, Helsinki.

Cooke. P. (2004a) The role of research in regional innovation systems: new models meeting knowledge economy demands. *International Journal of Technology Management*, 28, pp. 507–533.

Cooke, P. (2004b) Introduction: regional innovation systems – an evolutionary approach. In Cooke, P., Heidenreich, M. and Braczyk, H.-J. (eds) *Regional Innovation Systems: The Role of Governance in a Globalized World* (2nd edition). Routledge, London.

Cooke, P., Heidenreich, M. and Braczyk, H.-J. (eds) (2004) *Regional Innovation Systems: The Role of Governance in a Globalized World* (2nd edition). Routledge, London.

Doloreux, D. (2002) What we should know about regional systems of innovation. *Technology in Society*, 24, pp. 243–263.

Early, P. E. and Mosakowski, E. (2004) Cultural intelligence. *Harvard Business Review*, October, pp. 139–146.

Eurostat (2005) Average annual earnings varied significantly across the EU25. News release STAT/05/68.

Faist, T. (2000) *The Volume and Dynamics of International Migration and Transnational Social Spaces*. Oxford University Press, Oxford.

Findlay, A. M. (1998) A migration channels approach to the study of professionals moving to and from Hong Kong. *International Migration Review*, 32, (3), pp. 682–703.

Fleming, L. and Sorenson, O. (2004) Science as a map in technological search. *Strategic Management Journal*, 25, pp. 909–928.

Florida, R. (2002) *The Rise of the Creative Class, and How It's Transforming Work, Leisure, Community, and Everyday Life*. Basic Books, New York.

Forsander, A. (2004) Social capital in the context of immigration and diversity; economic participation in the Nordic Welfare States. *JIMI/RIMI*, 5 (2), pp. 207–227.

Formhold-Eisebith, M. (2004) Innovative milieu and social capital: complementary or redundant concepts of collaboration-based regional development. *European Planning Studies*, 12 (6), pp. 747–765.

Föster, M. and d'Ercole, M. M. (2005) Income distribution and poverty in OECD countries in the second half of the 1990s. OECD Social, Employment and Migration Papers 22. OECD, Paris.

Föster, M. and Pearson, M. (2002) Income distribution and poverty in the OECD area: trends and driving forces. *OECD Economic Studies*, 34.

Ghosh, B. (2000) Towards a new international regime of orderly movements of people. In Ghosh, B. (ed.) *Managing Migration: Time for a New International Regime?* Oxford University Press, Oxford.

Granovetter, M. (1973) The strength of the weak ties. *American Journal of Sociology*, 78, pp. 1360–1380.

Granovetter, M. (1985) Economic action and social structure: the problem of embeddedness. *American Journal of Sociology*, 91, pp. 481–510.

Guellec, D. and Cervantes, M. (2002) International mobility of highly skilled workers: from statistical analysis to policy formulation. In OECD, *International Mobility of the Highly Skilled*. OECD, Paris.

Harrison, L. E. (2001) Why culture matters. In Harrison, L. E. and Huntington, S. P. (eds) *Culture Matters: How Values Shape Human Progress*. Basic Books, New York.

Hofstede, G. (1997) *Cultures and Organizations: Software of the Mind*. McGraw-Hill, New York.

Huber, J. and Skidmore, P. (2003) *The New Old: Why the Baby Boomers Won't Be Pensioned Off*. Demos, London.

Joronen, T. (2002) Immigrant entrepreneurship in Finland in the 1990s. In Forsander, A. (ed.) *Immigration and Economy in the Globalisation Process: The Case of Finland*. Sitra Report Series 20, Helsinki.

Katz, R. (2004) Managing professional careers: the influence of job longevity and group age. In Tushman, M. L. and Anderson, P. (eds) *Managing Strategic Innovation and Change*. Oxford University Press, Oxford.

Kurjenoja, J. (2003) *Kansainvälinen palkkaverovertailu 2003*. Finnish Taxpayers' Association, Helsinki.

Larkey, L. K (1996) Toward a theory of communicative interactions in culturally diverse work-groups. *The Academy of Management Review*, 21 (2), pp. 463–491.

Luo, Y.-L. and Wang, W.-J. (2002) High-skill migration and Chinese Taipei's industrial development. In OECD, *International Mobility of the Highly Skilled*. OECD, Paris.

Maskell, P. (2000) Social capital, innovation and competitiveness. In Baron, S., Field, J. and Schuller, T. (eds) *Social Capital: Critical Perspectives*. Oxford University Press, Oxford.

Massey, D. S. and Taylor, J. E. (2004) Back to the future: immigration research, immigration policy, and globalization in the twenty-first century. In Massey, D. S. and Taylor, J. E. (eds) *International Migration: Prospect and Policies in a Global Market*. Oxford University Press, Oxford.

Monkman, K., Ronald, M. and Théramène, F. D. (2005) Social and cultural capital in an urban Latino school community. *Urban Education*, 40 (1), pp. 4–33.

Parkhe, A. (1991) Interfirm diversity, organisational learning and longevity in strategic global alliances. *Journal of International Business Studies*, 22 (4), pp. 579–601.

Pitkänen, P. (2006) *Etninen ja kulttuurinen monimuotoisuus viranomaistyössä*. Edita, Helsinki.

Raunio, M. (2003) Should I Stay or Should I Go? The images and realities of foreign ICT professionals in Finnish work and living environments – English summary. University of Tampere, Sente Working Papers 7.

Raunio, M. (2005) *Aivovuodosta aivokiertoon. Huippuosaajat talouden voimavarana*. Finnish Business and Policy Forum (EVA), EVA report, Taloustieto Oy, Helsinki.

Sassen, S. (1999) *Guests and Aliens*. The New Press, New York.

Saxenian, A. (2002) *Local and Global Networks of Immigrant Professionals in Silicon Valley*. PPIC, San Francisco.

Schienstock, G. and Hämäläinen, T. (2001) *Transformation of the Finnish Innovation system. A Network Approach*. Sitra Reports 7, Helsinki.

Stalker, P. (2000) *Workers without Frontiers: The Impact of Globalization on International Migration*. Lynne Rienner, Boulder, CO.

Straubhaar, T. (2000) Why do we need a general agreement on movements of people (GAMP)? In Ghosh, B. (ed.) *Managing Migration: Time for a New International Regime?* Oxford University Press, Oxford.

TEK (2002) Teknisen alan palkkavertailu eri maissa. (www.tek.fi.)

Woolcock, M. (1998) Social capital and economic development: toward theoretical synthesis and policy framework. *Theory and Society*, 27 (2), pp. 151–208.

4 Connectivity and co-location in innovation processes of Dutch firms

Pieter J. M. de Bruijn and Frank G. van Oort

Introduction

This chapter focuses on the concept of agglomeration economies in relation to firms' competitive advantages, knowledge spillovers and knowledge externalities. It is often argued that these are more easily identified in cities, where many people are concentrated into a relatively small geographic space. By contrast, the urban field hypothesis – often applied in the Dutch case – emphasizes the absence of differentiated urban contexts in economic accumulation. Despite the large empirical literature on this subject, conclusions are manifold. In general, on a local scale agglomeration is usually related to urban density or city size, whereas on a regional scale agglomeration theories seem to reflect core–periphery patterns. Within these commonly perceived spatial contexts, we describe and model patterns of innovation of firms in relation to assets of the environment. We make a distinction in firm-internal knowledge assets and assets derived from the production environment in terms of the regional and international environment. Building on Porter's cluster concept, we conclude on commonalities and complementarities in spatial production structures by applying a subdivision between network links (partnerships in innovation projects) and co-location in innovation processes of firms. Our empirical evidence supports the thesis that the regional fabric has to be linked to international networks of innovative firms. To leave out the international network dimension means missing out potentially important externality sources. Further, the urban field hypothesis regarding innovative behavior of firms in the Netherlands is rejected. The chapter is further structured as follows. The next section summarizes briefly the debate in the literature on the role of the regional production environment in innovation studies. A further section explains the necessity to extend the regional production environment of innovative firms with network-based externalities in studies on spatial externalities. The fourth section describes how innovation in the Netherlands is measured, and how subsequently geographical patterns are visualized. Spatial econometric models on municipal innovation intensity are presented in the final main section, followed by a conclusion.

The regional production environment in innovation studies

This contribution deals with the role of the production environment in the innovation processes of Dutch firms. As knowledge, learning processes and innovation are regarded as main driving factors for sustainable comparative advantage, the past twenty years have witnessed increasing attention being paid to the importance of the regional production environment in creating competitive strength (Scott and Storper, 1987; Aydalot and Keeble, 1988; Becattini, 1989; Florida, 1995; Cooke *et al.*, 1997; Porter, 1998). Although contributions to this debate are embedded in different research traditions, they share a common interest in synergies in processes of learning and innovation derived from the regional production environment (Moulaert and Sekia, 2003).

Two aspects of the regional production environment are of crucial importance in gaining innovative strength (Porter, 2000, pp. 261–262). First, the division of labor in firms within regional concentrations of innovative activities potentially provides complementarities to other firms. This leads to economies of scale and scope in production networks of intermediate deliveries and supplies. Facilitated by ongoing relationships with other cluster entities and possibilities for face-to-face contacts, firms embedded in clusters perceive new technological opportunities and changing buyer needs early compared to their more isolated counterparts. Second, proximity to other innovative companies provides innovating firms with indirect synergies, commonalities, through better access to employees, specialized information, research infrastructure and other facilities. According to Porter (1998, p. 80) 'a cluster allows each member to benefit as if it had greater scale or as if it had joined with others formally – without requirements to sacrifice its flexibility'. On the basis of these regional synergies, territorial innovation models claim a geographical distinction in types of interaction (Lagendijk, 2001). It is argued that within the region, processes of accumulation of tacit knowledge are present. This is also characterized by collective learning and the growth of associational structures on the basis of untraded interdependencies. Coordination then occurs through trust, reciprocity and long-term strategic agreements, upheld by regionally embedded structures. Outside the region is the global marketplace, driven by shortening product life cycles to which regional clusters have to respond.

Though a large literature on this subject has emerged, the empirical validation of these models is questionable. First of all, in neoclassical economic contributions it is stressed that the interaction of companies with their production environment is not a necessity for innovation (Freeman and Soete, 1997). Approximately three-quarters of all innovating companies innovate through their own efforts (Poot, 2004). Additionally, innovative networks cannot be put on a par with spatial clusters of innovative activities. On the one hand, case studies of successful examples of innovative clusters like

Silicon Valley (Saxenian, 1994) point to the importance of proximity and regional clustering in innovation trajectories. On the other hand, in more generally designed surveys, the importance of regional cooperation, as opposed to international network links, appears to be less convincing (Curran and Blackburn, 1994). Although in some cases relational factors seem to lead to geometrically concentrated patterns of learning and production among firms (Scott, 1988), these patterns cannot be regarded as a general phenomenon in industrial organization (Markusen, 1998; Martin and Sunley, 2003). International collaboration is barely theorized in territorial innovation models (Humphrey and Schmitz, 2002). Dicken *et al.* (2001, p. 90) state that 'too often a particular or a bifurcated geographical scale of analysis is used in ways that, in effect, preclude alternatives and that obscure the subtle variations within, and interconnections between, different scales'. This conception of space can be related to the proximity debate in economic geography (Boschma, 2005). Whereas theory automatically takes proximity to mean spatial proximity, proximity does not necessarily need to be spatial. Alternative conceptions range from economic space (Perroux, 1950), organizational space (Hudson, 1999) to emotional space (Taylor, 2005). In territorial innovation models, the concept of proximity is not always used in an explicitly geometrical sense (Oinas and Malecki, 2002). Rather, cultural and institutional dimensions of space are *supposed* to be related to geometrical distances, which can be expressed in terms of geometrical units such as kilometers or miles. Rigid divisions between 'in here' and 'out there' are to be displaced by relational conceptions of space. In this relational or network perspective on space and places, the focus lies on the interconnectedness between mutually overlapping spatial scales (Amin, 2002).

Extending the regional to network production environments

In our contribution we aim to extend the production environment through the inclusion of the international production environment in territorial innovation models. In this light, Foray (2004) makes a useful distinction between the functional and the physical production environment. Functional networks are organized around a specific technology and are often international in their nature (Jaffe and Trajtenberg, 2002). The production environment in terms of physical proximity often refers to urban regions with a highly educated labor force and highly sophisticated supply companies, customers and supporting services (Acs, 2002). This interpretation of the regional production environment touches upon the spatial context under which cluster synergies express themselves. In this respect the literature on agglomeration economies is of relevance (Lambooy, 1998; Van Oort, 2004). The concept of agglomeration economies stresses the role of the urban environment in firms' competitive advantage. Knowledge spillovers and knowledge externalities are more easily identified in cities, where many people are concentrated into a relatively small geographic space. On the contrary, the urban field hypothesis emphasizes the absence of any effect of differentiated urban contexts in economic accumulation.[1] Despite the large empirical literature on this subject, research results are not robust on the spatial scale of analysis that should be central when researching the agglomeration and cluster circumstances of knowledge and the innovative economy. On a local scale, agglomeration can be related to urban density or city size, whereas on a regional scale agglomeration theories seem to reflect core–periphery patterns. Within these spatial contexts we describe and model patterns of innovation in relation to assets of the environment. In our conceptual framework a distinction is made between firm-internal knowledge assets and assets derived from the production environment in terms of the regional or even wider environment. Following Porter's (1998) distinction between commonalities and complementarities, a subdivision is made between network links and co-location in the regional production environment.

Questions on the spatial dimension of innovation are especially interesting in the context of the Netherlands. First, the Netherlands is a relatively small country. In less than four hours it is possible to travel from any location in the Netherlands to any other part of the country. Second, it is a highly urbanized country. Unlike many other European countries such as Spain or France, which show a more centralized urban structure, core-periphery patterns in the Netherlands are less pronounced (see Figure 4.1A). Still, a spatial distinction can be drawn between the Randstad in the western part of the country, which encompasses the four largest cities (Amsterdam, Rotterdam, The Hague and Utrecht), an intermediate zone closely linked to the Randstad, and a peripheral zone that encompasses the southwestern, the northern

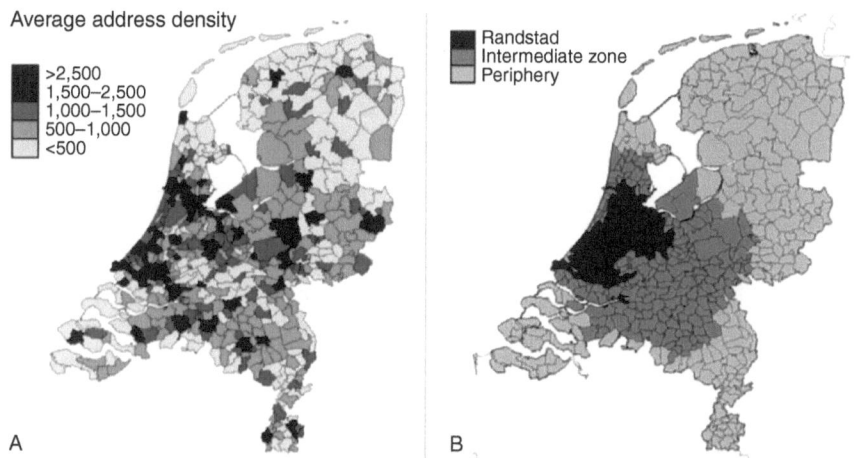

Figure 4.1 Urban structures in the Netherlands. A: Urban density. Source: Statistics Netherlands, Statline. B: Core and periphery. Source: Van Oort (2002)

and the southeastern part of the country (Figure 4.1B). This zoning distinction is hypothesized to be important by many studies on economic growth in the Netherlands, in the sense that the Randstad region traditionally has better economic potential for development than other regions (Van Oort 2004; Manshanden 1996). Especially in densely populated countries, such as the Netherlands, agglomeration effects might not be exclusively constrained to specific urbanized areas, but are more evenly spread across wider areas. In this sense the urban field can be seen as the spatial counterpart of the term 'foot-loose' (Wever, 1987). Following the above reasoning, two sets of spatial regimes are distinguished, each indicating aspects of urban structures at different spatial scales. Urban density is based on the average address density of the surrounding area of 500 by 500 grid squares (Dulk *et al.*, 1992). The surrounding area is the area within a radius of one kilometer. On the macro level, three national zoning regimes have been distinguished to form the core–periphery taxonomy: the Randstad core region, the so-called intermediate zone and the national periphery. The distinction between macroeconomic zones in the Netherlands is based on a gravity model of total employment concerning data from 1996.

As Table 4.1 reveals, despite the presumed urban field character of the Netherlands, location patterns of different companies clearly discriminate between different parts of the country. Agriculture and manufacturing are strongly oriented to less urbanized areas and peripheral zones, whereas service sectors are strongly oriented to densely populated cities and the western part of the Netherlands.

Recently, many empirical contributions have been presented on the character of the regional production environment, focusing on the questions of whether localization economies (single-sector concentration), Jacobs-externalities (multisector co-location) or urbanization economies (location

Table 4.1 Specialization indexes in agriculture, manufacture and services, by urban density and center-periphery location, 2002

	Agriculture	Manufacturing	Commercial services	Non-commercial services
Urban density				
>2,500	0.4	11.5	52.6	35.5
1,500–2,500	0.9	19.5	46.8	32.8
1,000–1,500	3.4	23.3	46.5	26.8
500–1,000	8.5	27.1	42.4	22.1
<500	15.5	24.2	40.2	20.0
Core-periphery				
Core	1.9	14.1	53.9	30.1
Intermediate zone	3.7	22.4	45.7	28.2
Periphery	6.9	24.9	39.7	28.5

Source: TNO, on the basis of LISA and Statistics Netherlands.

advantages due to urban amenities and infrastructure) are related to externalities that benefit innovation and growth potentials of firms (for an overview see Van Oort, 2004). Network advantages of firms are not much researched directly, mainly because of lack of appropriate data. It is especially the network source of externalities that we make central to our models. By distinguishing between national and international linkages, we extend the potential relevant production environment of firm's innovation performances.

Measurement and visualization of local innovation intensity of firms

Because of its rich variable content and high response figures, we chose the Community Innovation Survey (CIS) as the empirical basis for our analyses. The third innovation survey covers a broad range of aspects of innovative companies in the Netherlands during the period 1998–2000. The Innovation Survey is based on innovation processes from a system perspective (Statistics Netherlands, 1998). Innovation is not considered to be solely dependent on 'linear' determinants of learning (for example, expenditure on in-house research and development), but is analysed within a framework in which interaction and knowledge diffusion play an important role. Substitution of the linear model, upon which the bulk of information collection and data methodology for R&D and innovative activities is still heavily dependent (Evangelista *et al.*, 2002, p. 175), with the chain-linked model (Kline and Rosenberg, 1986) is more capable of empirically capturing the arguments upon which territorial innovation models are based. In the innovation survey, innovating companies are identified on the basis of whether they produce technologically new and improved products. The evaluation of CIS indicators is summarized by Kleinknecht *et al.* (2002). Although some interpretation problems can arise out of the use of CIS, especially in inter-country comparisons, there is wide agreement that CIS indicators provide a broader understanding of innovation processes than can be obtained by relying solely on traditional indicators like expenditures on R&D and patenting.

In identifying geometrical patterns of innovation, the use of the Innovation Survey is limited by the choice of geographical units in the analysis. Due to the complication that the administrative location of companies' headquarters can lead to biased results, it is only possible to analyse regionalized innovation figures in the Netherlands on the level of 12 NUTS-II regions (provinces), which may not necessarily reflect homogeneous innovation systems, the central focus of the territorial innovation models. Furthermore, the interpretation of the location of large companies is somewhat questionable given the fact that often those companies consist of more establishments than merely the head office. This might lead to some difficulties in the spatial interpretation of the survey results (Statistics Netherlands, 1999). To obtain spatial insights on detailed geometrical levels, we combined NUTS-II-level sector aggregates of input, throughput and output indicators of innovative performances in the Innovation Survey with employment figures of the

Netherlands Information System on Labor (LISA) on the municipal level. In a shift-and-share analysis we estimated innovation figures of municipalities within each NUTS-II region on the basis of sector differentiation between those municipalities, following the algorithm below in which I_M represents the number of employees in innovating companies as a share of the total number of employees in municipality M situated in province P, $E_{S,M}$ the number of employees in sector[2] S in municipality M, E_M the total number of employees in municipality M, $I_{S,P}$ the number of employees in innovative companies in sector S in province P, and $E_{S,P}$ the total number of employees in sector S in province P:

$$I_M = \sum_S \left(\left(\frac{E_{S,M}}{E_M} \right) \cdot \left(\frac{I_{S,P}}{E_{S,P}} \right) \right)$$

It is important to notice that our estimates must be regarded as indications of spatial patterns in innovation and must not be considered as definite figures.

As output indicator of innovation we estimated the number of employees working in innovative firms that experience significant positive effects of innovation on their market share, as a share of the total number of employees working in innovative firms. By relating innovation to economic effects of technologically new or improved products and processes on market share, we explicitly take into account Schumpeter's (1949) distinction between innovation and invention. As input measure for innovation the number of academically educated employees and R&D personnel, as a share of the total number of employees, is used. The number of employees in companies that innovate in partnership, at both national and international scale, as a share of the total number of employees, is used as throughput measure of innovation relating to the network orientation of innovating companies.[3]

Figures 4.2 to 4.4 depict spatial patterns in innovation processes, as defined above.[4] The effect of innovation on market share is particularly strong in parts of North Holland (the Amsterdam region and Het Gooi, a region strongly engaged in multimedia services) and Overijssel. Peripheral regions (the northern provinces and Zeeland) score relatively weakly. All municipalities that house a technical university find themselves in the top segment of municipalities according to employment in innovative companies that estimate the effect of innovation on market share at least as significant. Given the patterns of input dimensions of innovation, depicted in Figure 3.3, innovation intensity in the core of the Netherlands, the Randstad region, appears to be driven by academically educated employees, whereas innovation in the intermediate zone can be related to R&D. These differences reflect the sector composition of both parts of the country (Bruijn, 2004). In the Randstad region, knowledge-intensive services take up a disproportionate share in total economic activity. In the so-called intermediate zone (regions adjacent to the Randstad region) manufacturing takes up a dominant position. While innovation in manufacturing companies is predominantly driven by R&D

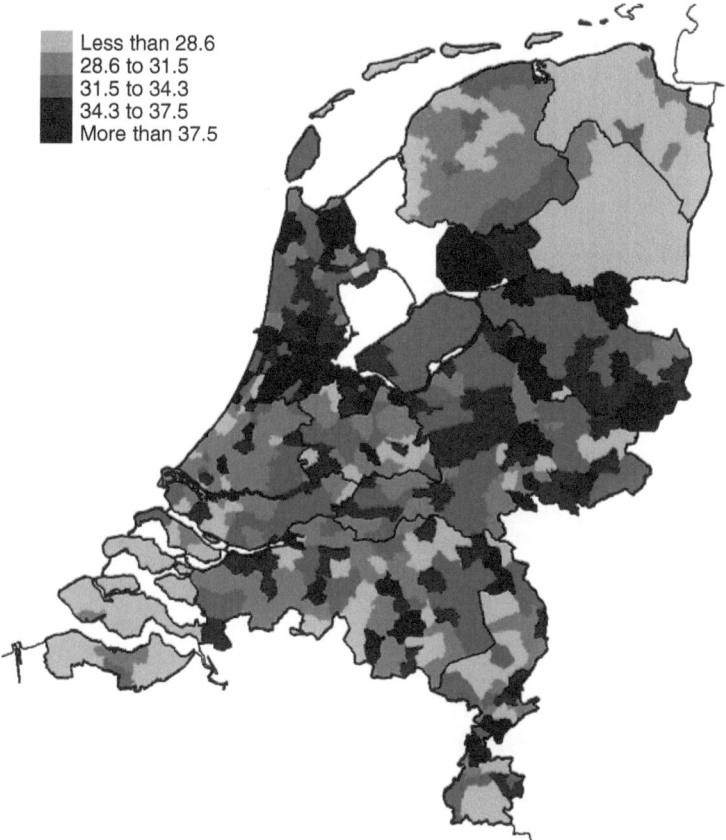

Less than 28.6
28.6 to 31.5
31.5 to 34.3
34.3 to 37.5
More than 37.5

Figure 4.2 Effect of innovation on market share (employees working in innovative firms that experience significant effects of innovation on their market share, as share in the total number of employees) by municipality in the Netherlands, 1998–2000. Source: TNO, on the basis of CBS/LISA

innovation in service industries is characterized by its input from human resources more generally.

Figure 4.4 shows maps in which spatial patterns of employees working in firms that innovate in partnership are depicted. These figures are relatively high in the Randstad region. The southern part of the Randstad is strongly oriented to domestically based partners, whereas companies in the northern part of the region are more oriented toward international partners in their innovation trajectories. Partnership, on both national and international scale is also strongly prevalent in the southern part of Limburg, in the southeastern fringes of the Netherlands.

The figures suggest strong patterns of spatial co-location. To test this formally, Table 4.2 contains Moran *I* coefficients on input, throughput and

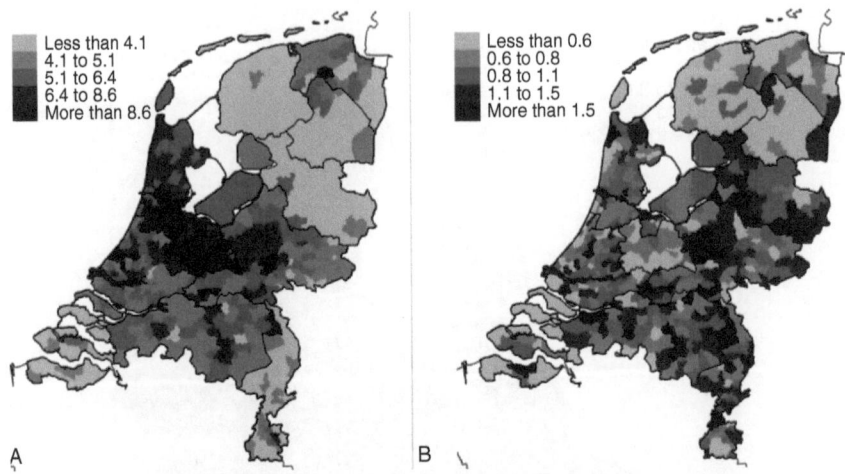

Figure 4.3 Human knowledge capital A: academically educated employees; and B: employees working in R&D functions, both as a share in the total number of employees) by municipality in the Netherlands, 1998–2000. Source: TNO, on the basis of CBS/LISA

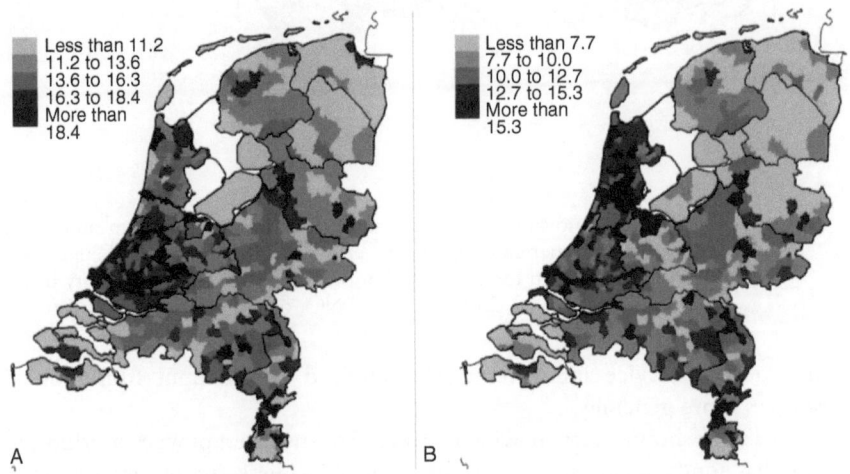

Figure 4.4 Partnership in innovation processes (employees working in innovative firms which innovate in partnership with (A) national and (B) international partners, both as a share in the total number of employees) by municipality in the Netherlands, 1998–2000. Source: TNO, on the basis of CBS/LISA

Table 4.2 Spatial clustering of effect of innovation, human knowledge capital and partnership in innovation processes (Moran I coefficients)

	I	z
Effect of innovation	0.06	19.17
Academically educated employees	0.14	42.17
Research and development	0.03	8.52
National partnership	0.08	24.42
International partnership	0.08	24.24

Source: TNO, on the basis of CBS/LISA.

output aspects of innovation. The Moran I coefficient provides an indication of spatial correlation of municipal scores with their surrounding areas[5] and ranges from minus 1 (extreme negative spatial autocorrelation – comparable to a checkerboard pattern) to plus 1 (extreme spatial clustering). For large populations the expected value tends to zero. The standard error depends on the assumptions made. The z-values relate to the hypothesis that spatial patterns under consideration are an expression of a random configuration in space under the assumption of normality of the underlying variables.[6] The positive values of Moran's I are partly a result of the shift-share estimation technique, and therefore these values must be interpreted with caution. The patterns for all variables differ significantly from a random spatial distribution. The spatial pattern of the share of academically educated employees reveals the strongest extent of spatial co-location. A relatively strong correlation also exists for partnership on both the national and the international level.

Tables 4.3 and 4.4 depict aggregated patterns of input, throughput and output measures of innovation. The definition of the variables followed the definitions given in Figures 4.2 to 4.4. Innovation is positively correlated to agglomeration, in terms of both national core–periphery patterns and urban density. Employment in firms that estimate the effect of innovation to be significant is highest in the Randstad region and in mid-sized urban settlements. The highest differentiation between national core–periphery regions and levels of urban density can be found for the share of academically educated employees in total employment. R&D intensity is highest in the

Table 4.3 Effect of innovation, human knowledge capital and partnership in innovation processes by national core–periphery zones

	Randstad	Intermediate zone	Periphery
Effect of innovation	34.8	33.4	31.8
Academically educated employees	9.9	6.8	4.7
Research and development	0.9	1.1	1.2
National partnership	16.8	14.9	14.6
International partnership	12.8	11.0	10.9

Source: TNO, on the basis of CBS/LISA.

Table 4.4 Effect of innovation, human knowledge capital and partnership in innovation processes by urban density

	Average address density in surrounding areas				
	>2,500	1,500–2,500	1,000–1500	500–1,000	<500
Effect of innovation	34.1	34.2	34.4	32.4	29.0
Academically educated employees	10.1	7.5	7.1	5.3	4.3
Research and development	0.9	1.2	1.1	1.0	0.8
National partnership	17.2	16.5	15.2	13.9	12.1
International partnership	13.2	12.1	11.7	10.5	8.5

Source: TNO, on the basis of CBS/LISA.

intermediate zone and the national periphery. In terms of urban density, R&D intensity is relatively high in moderately urbanized regions. Like employment of academically educated employees, partnership relations in innovation are characterized by a pattern strongly focused on highly urbanized areas.

Econometric modeling of localized innovation intensities

In this section, innovative output is econometrically modeled using a set of explaining variables relating to internal knowledge assets, the network environment in terms of connectivity and the physical environment in terms of co-location. The number of employees in innovative companies that estimate the effect of innovation on their market share at least as significant, as a share in total employment, serves as dependent variable.[7] The number of academically educated employees and the number of employees in R&D functions, both as a share of total employment, relate to internal knowledge assets. As indicators of the network environment we use the number of employees working in companies that innovate in partnership with national partners, and the number of employees in companies that innovate in partnership with international partners, both as a share of the total number of employees. To implement the physical environment in our analyses, we made use of spatial econometrics (Anselin, 1988), in which both the explained variable and explaining variables are spatially weighted by a distance decay function on the basis of travel time distances by car. Because of our estimation technique, the provincial environment accounts for a disproportionate effect on both the explained and the explaining variables. Therefore, we controlled the municipal scores for the provincial aggregates in which they are situated. In this sense, over- and under-achievement by municipalities relative to their provincial surroundings are the central focus of the model. Hence, the model only explains intra-provincial differences in innovative strength, a fact that must be kept in mind for the interpretation of the results. As we have seen, differences between municipal estimates situated in the same province are fully determined by sector composition. Although we acknowledge the fact that limits to data availability constrain us to impose the somewhat unrealistic assumption that innovative patterns are fully determined by industry structure, theoretical contributions (Malerba, 2002) and empirical analyses (Pavitt, 1984; Klepper, 1996) do emphasize important structuring effects of sectors with respect to innovation.

Apart from an estimation of the model for the population of 496 municipalities in the Netherlands in general, we differentiated our analysis by spatial regimes to account for spatial heterogeneity on the basis of agglomeration effects. On a national scale, spatial regimes are arranged on the basis of a core–periphery distinction. Agglomeration effects on a local scale are incorporated through regimes based on urban density. Spatial heterogeneity is modeled by spatial regimes, involving change-of-slope regression estimation over various types of locations that theoretically 'perform' differently.

Table 4.5 summarizes the results of the analysis of the general model. From the ordinary least squares (OLS) estimators it follows that both R&D and partnership at an international level are decisively important for local innovative output in terms of the estimated impact of innovation on market share. Apart from the standard OLS estimators a number of indicators relating to spatial dependence and spatial heterogeneity are given. The Lagrange multipliers LM (ρ) and LM (λ) refer to potential improvements of an estimation that takes into account spatial lag dependence or spatial error dependence. In spatial lag modeling, local scores are related to scores in municipalities nearby. To account for spatial lag dependence, the dependent variable, spatially 'lagged' by a distance decay function is incorporated in the model as independent variable, next to the variables related to the input and through-put factors in the municipality under consideration. The spatial error model is not reported, since in our case the model cannot be accepted.[8]

Table 4.5 Ordinary least squares and spatial lag estimates (*t*-value in parentheses) on the basis of linear distance decay with dependent effect of innovation on market share (*n* = 496)

	OLS model	Spatial lag model
Constant	0.62 (29.12)★★	−0.03 (−0.19)
Academically educated employees	0.03 (1.38)	0.03 (1.40)
Employees in R&D-functions	0.09 (6.76)★★	0.09 (6.86)★★
National partnership	0.04 (1.40)	0.04 (1.53)
International partnership	0.20 (8.63)★★	0.20 (8.55)★★
P	–	0.69 (3.84)★★
R^2	0.54	0.54
ML/AIC	483.6/−957.2	486.5/−961.1
LM (ρ)	6.8★★	–
LM (λ)	24.9★★	32.6★★
BP (KB)	(12.0★)	18.6★★

Source: TNO, on the basis of CBS/LISA.

Note
Confidence levels: ★★99%; ★95%.

Corrected for spatial dependence, both R&D intensity and international partnership still prove to be significant for the effect of innovation. The correction factor ρ can be interpreted as the impact of the effect of innovation in surrounding municipalities on the effect of innovation in the municipality under consideration. In this sense, the significance of the estimator can be related to externalities based on spatial co-location of innovative companies. Since estimations depend to a large extent on the distance decay function on which the spatially weighted scores are based, we also estimated the spatial lag model on the basis of squared distance decay. Contrary to the estimations based on linear distance decay, the factor ρ does not prove significant at a level of α of 5 percent. Therefore, estimations based on quadratic distance decay are not reported. Externalities thus play a significant role on the *regional* level. On a local scale, externalities do not prove to have a significant impact on the effect of innovation on market share.

Tables 4.6 and 4.7 present analyses that differentiate by spatial regimes in terms of urban density classes and the core–periphery taxonomy. Considering the significant values of the Chow–Wald (CW) criterion,[9] the inclusion of both regimes provides an improvement compared to the general model. Significant effects, however, do not differ much. A remarkable outcome is that international partnership (used as a network externality indicator) is not significant in the Randstad region. Instead, national partnership has a significant effect on the effect of innovation. We do not have a clear-cut theoretical explanation for these results. These effects of national and international partnership might be related to the fact that small and medium-sized enterprises and services, which are overrepresented in the Randstad region, are

Table 4.6 Spatial lag estimates (*t*-value in parentheses) on the basis of linear distance decay with dependent effect of innovation on market share, by national center–periphery zone: summary of results

	Randstad	Intermediate zone	Periphery
Constant	−0.06 (−0.37)	−0.20 (−1.16)	−0.02 (−0.11)
Academically educated employees)	0.07 (1.46)	0.10 (3.05)**	−0.04 (−1.36
Employees in R&D functions	0.14 (5.65)**	0.06 (2.11)*	0.09 (4.47)**
National partnership	0.17 (2.50)*	0.08 (1.44)	0.03 (0.86)
International partnership	−0.01 (−0.21)	0.26 (5.81)**	0.23 (7.69)**
P	–	0.73 (4.30)**	–
N	90	175	231
R^2	–	0.57	–
ML/AIC	–	504.3/−976.7	–
CW	–	36.9**	–
BP	–	9.6**	–

Source: TNO, on the basis of CBS/LISA.

Note
Confidence levels: **99%; *95%.

Table 4.7 Spatial lag estimates (*t*-value in parentheses) on the basis of linear distance decay with dependent effect of innovation on market share, by urban density regime: summary of results

Average address density in surrounding areas	1,000 and over	500–1,000	lower than 500
Constant	−0.01 (−0.08)	0.04 (0.21)	0.07 (0.34)
Academically educated employees	0.11 (3.22)★★	−0.02 (−0.45)	−0.01 (−0.30)
Employees in R&D functions	0.07 (2.89)★★	0.10 (4.98)★★	0.10 (4.16)★★
National partnership	0.04 (0.85)	0.13 (2.79)★★	−0.09 (−1.74)
International partnership	0.18 (3.95)★★	0.15 (4.29)★★	0.31 (7.14)★★
ρ	–	0.63 (3.06)★★	–
N	90	175	231
R^2	–	0.56	–
ML/AIC	–	498.3/−964.5	–
CW	–	24.0★★	–
BP	–	7.7★	–

Source: TNO, on the basis of CBS/LISA.

Note
Confidence levels: ★★99%; ★95%.

more dependent on the regional production environment. Still, the descriptive analyses initially do not endorse this explanation, and the causal structure is also not expected from the theoretical point of view, which states that core regions are embedded in international innovation networks to a higher extent than peripheral regions. Further, in contrast to the general model, the share of academically educated employees in total employment relates significantly to the dependent variable for both the intermediate zone and highly urbanized areas. One thing is explicitly clear from the analyses, though. Since the spatial regimes clearly discriminate over the model estimates, on the basis of our analyses the urban field hypothesis is rejected for the Netherlands.

Conclusions

From the spatial patterns of input (academically educated employees and employees in R&D functions), throughput (partnership at national and international scale) and output measures of innovation (the impact of innovation on market share), as presented in this chapter, we learn that innovation tends to be clustered in space in the Netherlands. Agglomeration proves to be highly significant in spatial patterns of innovation. Cities inhibit above-average scores on innovation. In a national context, employment in firms that estimate the effect of innovation to be significant is highest in the Randstad region. R&D intensity is relatively high in Dutch peripheral zones, which can be regarded as a reflection of sector composition in the sense that manufacturing firms, which are overrepresented in the periphery, are the main actors in R&D. Spatial correlation proves to be high for the number of academically educated employees as a share of total employment as well. This can be explained by highly specific location preferences on the part of knowledge-intensive business services in urban agglomerations in the northern part of the Randstad region.

The effect of innovation is modeled against a set of explaining variables relating to internal knowledge assets (academically educated employees and employees in R&D functions), the network environment (partnership on a national and international scale) and the physical environment (the spatially lagged dependent variable in surrounding municipalities). We corrected the municipal scores for the provincial score in which they are situated. Both R&D and international partnerships prove to be significantly related to the impact of innovation on market share. The effect of national partnership does not exercise a significant effect on regional innovative output. These results contradict the main arguments in territorial innovation models that are largely based on case study analysis of regional success stories. Most case studies focus on specific sectors, target groups (like small and medium-sized enterprises or high-technology firms) and regions. However, the arguments that territorial innovation models reveal in specific situations do not always apply for regional development in general. The results, however, are in line with research outcomes from more generally designed surveys. Manshanden (1996) relates innovation to certainty and uncertainty to proximity, and concludes that proximity does not necessarily lead to more exchange of information in innovative networks. From our model exercises, regional co-location appears to be more decisive for innovative output than local clustering, which is in line with most theoretical contributions that focus on the regional level rather than on a local level.

Regional innovation policies across Europe are to a large degree aimed at creating regional synergies and enhancement of regional networking. Our results show that innovation policies should not exclusively focus on the regional level. International links are also, or even more, important as potential spillover sources. We conclude that our empirical evidence supports the thesis that the regional fabric has to be linked to international networks of innovative firms. Leaving out the international network dimension means omitting potentially important externality sources. The relation of this international externalities indicator with innovative output does not differ substantially over regimes defined by agglomeration degree. Despite the fact that the outcomes of our analyses do not exclusively limit themselves to clear-cut theoretically comprehensible differences, we reject the urban field hypothesis for innovative activities in the Netherlands. The Netherlands cannot be regarded as an urban field for innovative activities since our descriptive analyses show remarkable differences in innovative strength over different region types based on urban density and the national center–periphery zoning regimes.

Notes

The empirical part of this chapter was carried out at the Centre for Research of Economic Microdata (Cerem) at Statistics Netherlands. The opinions expressed are solely those of the authors. The authors thank Statistics Netherlands (CBS) for the opportunity to use microeconomic data and the Cerem team for their support on-site. The chapter benefited from comments of attendants at the Statistics Netherlands Cerem seminar on micro-data research (Voorburg, December 2004) and the Regional Studies Association's international conference on regional growth agendas (Aalborg, May 2005) and two anonymous referees. All errors remain ours.

1 This interpretation of the urban field concept is typical for applications on the Dutch case (Manshanden, 1988). Originally the concept was introduced by Friedman and Miller (1965), who focused on interdependencies within a wide area (approximately 100 miles, 160 kilometers) around urban centers in the United States.

2 Sectors are incorporated in the analysis in terms of two-digit codes in the Standard Industrial Classification. Since SIC 74 (other business services) comprises a broad range of different business activities, activities within other business services are incorporated in the analysis at a more detailed level of three-digit classification codes.

3 Unfortunately, the third CIS does not make a distinction between regional and extra-regional partnerships. However, from the second CIS it follows that the vast majority of national partnerships are held between regional partners located from each other at a distance of not more than 50 kilometers.

4 Classification of regional scores on innovation is based on a cumulative percentile distribution of 0 to 25, 25 to 50, 50 to 75, 75 to 90 and 90 to 100.

5 The Moran I coefficient for a variable x in a population of n regions can be given by:

$$I = \frac{n}{S} \frac{\sum_i \sum_j w_{ij}(x_i-\mu)(x_j-\mu)}{\sum_i (x_i-\mu)^2}$$

in which x_i is the observation in region i and x_j is the spatially weighted observation in surrounding areas obtained through a distance decay function on the basis of a weight matrix w. In our analyses spatial weights are based on travel time by car. S is a scaling factor based on the weight matrix. For an extensive elaboration, see Anselin (1988).

6 Another assumption relates to randomization in the sense that all regional configurations are equally likely. Apart from these assumptions, the hypotheses can also be tested through simulation. The results obtained under the assumption of randomization and through simulation do not differ significantly from the results under the assumption of normalization and are not presented in this chapter. For an extensive elaboration, see Anselin (1988).

7 By this estimator we measure localized employment connected to innovative firms, as a share of total employment. In the case of innovative small firms or firms engaged in productivity-enhancing innovations (industries), employment might not be a suitable indicator. Because of lack of productivity data, this could not be tested in our models. Employment indicators do link to international comparative research, however (Van Oort, 2004).

8 Spatial error models correct for non-normality of error terms through a correction factor λ. Non-normality of error terms can only be attributed to a *spatial* dimension of error terms when λ and the correction factor in the spatially lagged model λ_w do not differ from each other significantly. This demand is quantified through the common factor hypothesis. In our case, the test on the common factor hypothesis proves to be highly significant. Therefore, the spatial error model cannot be accepted. For a more extensive elaboration, see Anselin (1988).

9 The spatial Chow–Wald test is distributed as an F variate and tests for structural instability of the regression coefficients over regimes (Anselin, 1995, p. 32).

References

Acs, Z. J. (2002) *Innovation and the growth of cities*, Northampton, MA: Edward Elgar.

Amin, A. (2002) Spatialities of globalization, *Environment and Planning A*, 34, pp. 385–399.

Anselin, L. (1988) *Spatial econometrics: methods and models*, Dordrecht: Kluwer Academic Publishers.

Anselin, L. (1995) *SpaceStat: a software program for the analysis of spatial data (version 1.80)*. Morgantown: Regional Research Institute, West Virginia University.

Aydalot, P. and D. Keeble (1988) High technology industries and innovative environments in Europe: an overview, in P. Aydalot and D. Keeble (eds) *High technology industries and innovative environments: the European experience*, London: Routledge.

Becattini, G. (1989) Sectors and/or districts: some remarks on the conceptual foundations of industrial economics, in E. Goodman, J. Bamford and P. Saynor (eds) *Small firms and industrial districts in Italy*, London: Routledge.

Boschma, R. A. (2005) The role of proximity in interaction and performance: conceptual and empirical challenges, *Regional Studies* 39 (1), pp. 41–45.

Bruijn, P. J. M. de (2004) Mapping innovation: regional dimensions of innovation and networking in the Netherlands, *Tijdschrift voor Economische en Sociale Geografie* 95 (4), pp. 433–440.

Cooke, P., M. G. Uranga and G. Etxebarria (1997) Regional innovation systems: institutional and organisational dimensions, *Research Policy* 26 (4–5), pp. 475–491.

Curran, J. and R. Blackburn (1994) *Small firms and local economic networks: the death of the local economy?* London: Paul Chapman Publishing.

Dicken, P., P. F. Kelly, K. Olds and H. W. C. Yeung (2001) Chains and networks, territories and scales: towards a relational framework for analysing the global economy, *Global Networks* 1 (2), pp. 89–112.

Dulk, C., H. den, H. van de Stadt and J. M. Vliegen (1992) A new measure for urban density: average address density, *Maandstatistiek van de bevolking* 1992 (July), pp. 14–27.

Evangelista, R., S. Iammarino, V. Mastrostefano and A. Silvani (2002) Looking for regional systems of innovation: evidence from the Italian Innovation Survey, *Regional Studies* 36 (2), pp. 173–186.

Florida, R. (1995) Towards the learning region, *Futures* 27, pp. 527–536.

Foray, D. (2004) *The economics of knowledge*, Cambridge, MA: MIT Press.

Freeman, C. and L. Soete (1997) *The economics of industrial innovation*, London: Pinter.

Friedman, J. and J. Miller (1965) The urban field, *Journal of the American Institute of Planners*, 31 (4), pp. 312–320.

Hudson, R. (1999) The learning economy, the learning firm and the learning region: a sympathetic critique of the limits to learning, *European Urban and Regional Studies* 6 (1), pp. 59–72.

Humphrey, J. and H. Schmitz (2002) How does insertion in global value chains affect upgrading in industrial clusters? *Regional Studies* 36 (9), pp. 1017–1027.

Jaffe, A. B. and M. Trajtenberg (2002) *Patents, citations and innovations: a window on the knowledge economy*, Cambridge, MA: MIT Press.

Kleinknecht, A., K. van Montfort and E. Brouwer (2002) The non-trivial choice between innovation indicators, *Economics of Innovation and New Technology* 11 (2), pp. 109–122.

Klepper, S. (1996) Entry, exit, growth and innovation over the product life cycle, *American Economic Review* 86 (3), pp. 562–583.

Kline, G. J. and N. Rosenberg (1986) An overview of innovation, in R. Landau and N. Rosenberg (eds) *The positive sum strategy: harnessing technology for economic growth*, Washington, DC: National Academy Press.

Lagendijk, A. (2001) Three stories about regional salience: 'regional worlds', 'political mobilisation' and 'performativity', *Zeitschrift für Wirtschaftsgeographie*, 45 (3–4), pp. 139–158.

Lambooy, J. G. (1998) Agglomeration advantages and spatial development: cities in the era of the knowledge economy. Inaugural speech [Agglomeratievoordelen en ruimtelijke ontwikkeling. Steden in het tijdperk van de kenniseconomie. Oratie]. Utrecht, Universiteit Utrecht, Faculteit Ruimtelijke Wetenschappen.

Malerba, F. (2002) Sectoral systems of innovation and production, *Research Policy* 31, pp. 247–264.

Manshanden, W. J. J. (1988) Urban field in the Netherlands: what's in a name? [in Dutch], *Planning Practice and Research* 3 (4), pp. 43–45.

Manshanden, W. J. J. (1996) *Business services and regional economic development; Economics of proximity* [in Dutch]. Amsterdam: Koninklijk Nederlands Aardrijkskundig Genootschap.

Markusen, A. (1998) Fuzzy concepts, scanty evidence, policy distance: the case for rigour and policy relevance in critical regional studies, *Regional Studies* 33 (9), pp. 869–884.

Martin, R. and P. Sunley (2003) Deconstructing clusters: chaotic concept or policy panacea? *Journal of Economic Geography* 3 (1), pp. 5–35.

Moulaert, F. and F. Sekia (2003) Territorial innovation models: a critical survey, *Regional Studies* 37 (3), pp. 289–302.

Oinas, P. and E. J. Malecki (2002) The evolution of technologies in time and space: from national and regional to spatial innovation systems, *International Regional Science Review* 25 (1), pp. 102–131.

Pavitt, K. (1984) Sectoral patterns of technological change: towards a taxonomy and a theory, *Research Policy* 13 (6), pp. 343–373.

Perroux, F. (1950) Economic space: theory and applications, *Quarterly Journal of Economics* 64, pp. 89–104.

Poot, A. P. (2004) *Determinants of knowledge-intensive R&D cooperation* [in Dutch]. Delft: Delft University Press.

Porter, M. E. (1998) Clusters and the new economics of competition, *Harvard Business Review*, November–December, pp. 77–90.

Porter, M. E. (2000) Locations, clusters, company strategy, in G. L. Clark, M. P. Feldman and M. S. Gertler (eds) *The Oxford handbook of economic geography*, Oxford: Oxford University Press.

Saxenian, A. L. (1994) *Regional advantage: Culture and competition in Silicon Valley and Route 128*, Cambridge, MA: Harvard University Press.

Schumpeter, J. A. (1949) *The theory of economic development: an inquiry into profits, capital, credit interest and the business cycle*, New York: Oxford University Press.

Scott, A. J. (1988) *New industrial spaces: flexible production organization and regional development in North America and Western Europe*, London: Pion.

Scott, A. J. and M. Storper (1987) High technology industry and regional development: a theoretical critique and reconstruction, *International Social Science Journal* 112, pp. 215–232.

Statistics Netherlands (1998) *Knowledge and economy 1998. Research and innovation in the Netherlands* [in Dutch]. Voorburg: Centraal Bureau voor de Statistiek.

Statistics Netherlands (1999) *Innovation and provinces 1999. Regional innovation profiles of SMEs in the Netherlands* [in Dutch]. Voorburg: Centraal Bureau voor de Statistiek.

Taylor, M. (2005) Clusters: the mesmerising mantra. Paper presented at the Regional Studies Association Conference on Regional Growth Agendas, Aalborg, 28–31 May.

Van Oort, F. G. (2002) Innovation and agglomeration in the Netherlands, *Tijdschrift voor Economische en Sociale Geografie* 93 (3), pp. 344–360.

Van Oort, F. G. (2004) *Urban growth and innovation: spatially bounded externalities in the Netherlands*, Aldershot, UK: Ashgate.

Wever, E. (1987) The spatial pattern of high-growth activities in the Netherlands, in B. Van der Knaap and E. Wever (eds) *Technology and regional development*, Wolfebore: Croom Helm.

5 On strengthening the knowledge base of knowledge-intensive SMEs in less favoured regions in Finland

Kati-Jasmin Kosonen

Introduction

In the era of knowledge-based economy, the regional or local knowledge environment and innovation environments for specific business areas have become more important. In the knowledge-based economy, the base of knowledge constantly evolves institutionally. New institutions are taking part in the local innovation networks, shaping the technological *change* and trans-formation in the region for the benefits of all parties, local businesses, universities or other higher education institutions and local or national development authorities. The institutional evolution comes about by linking different kinds of knowledge creation institutions to the knowledge exploitation organizations and sub-systems through new kinds of knowledge-enhancing mechanisms, and mainly from the R&D conducted in relation to regional capabilities (see Cooke and Leydesdorff, 2006).

Changes in the world economy have had major implications for economic development strategies and territorial governance in securing or boosting regional economic success in the twin processes of globalization and localization (Goddard and Chatterton 1999). While it is announced widely that innovation is an interactive process between firms and research institutions, between the different functions in the firm, between producers and users at the inter-firm level and between firms and the wider institutional milieu, in the institutional approach[1] the argumentation goes even further by stating that public organizations and institutions can have a significant role in promoting innovations. If the institutional base is 'thin' in the specific region, firms in the emerging sector do not get the appropriate assistance in their growth and internationalization processes, and actors widely find it more difficult to trans-form information (resources) into new knowledge and innovations. These kinds of regions are called here 'less favoured regions'. In this study, the Seinäjoki and Pori town regions are examples of the less favoured regions of Finland that are building an institutional base for university-based knowledge transfer systems to promote innovations and business development locally.

The purpose of this chapter is to shed some light on how local knowledge-intensive high-tech industries have been active in changing innovation culture and renewing local knowledge architecture by enhancing locally produced scientific knowledge, in order to be globally competitive. This chapter further describes these actions in the light of institutional capacity building which creates a 'local buzz', forms shared knowledge arenas and, especially, links the local actors to global pipelines through dynamic innovation networks. Generally the idea is to examine how the development actions taken in these town regions fit the idea of strengthening the institutional capacity to support emerging industries and knowledge entrepreneurship as the central nodes for a globally and locally *networked innovation environment*. Therefore, the following research questions have been included in the study:

- What has been the role of local knowledge-intensive industries when breaking certain types of regional lock-ins and changing the innovation culture, and renewing local knowledge architecture by enhancing locally produced scientific knowledge?
- What actual efforts have been made in these regions to strengthen the 'institutional capacity', and, more precisely, how has the development path followed in these regions affected the manifestation of local knowledge entrepreneurship?

The study is qualitative and based on written materials, statistics, reviews and reports gathered from these town regions, and it also relies heavily on around 70 thematic interviews conducted in two research projects.[2] The technology development processes under study are certain developments of automation and embedded systems technologies in the regions, more particularly the infusion of intelligent products and systems (smart systems), mechatronics and applied software.[3] In the following sections the ideas presented here are described in more detail and examined against the actions taken in the town regions of Pori and Seinäjoki.

Institutional capacity as a building block for supporting innovations in less favoured regions (LFRs)

Recently it has been stated that *institutional capacity* focuses on the webs of relations involved in regional development policies, which in turn interlink public development agencies, firms, and educational and research institutes in collective action (Healey *et al.* 1999; Healey 2004, 2006). In general, researchers working on questions dealing with research and technology policy, regional economic development and competence building have stressed *institutional elements* which are meaningful in the economic development, especially in less favoured regions (Cooke and Morgan 1998; Morgan 1997; Lundvall 1992, 2002; Landabaso *et al.* 1999; Malecki and Oinas 2000; Kautonen and Sotarauta 1999; Kosonen 2001, 2004; Sotarauta and Kosonen 2003, 2004; Sotarauta *et al.* 2003; Virkkala 2003).

Once networks or coalitions are created and formed, actors in networks should further be able to create new spaces and common arenas in which to interact and to manage the resources of institutional capacity (Healey *et al.* 1999). Therefore, institutional capacity is understood in this chapter as a combination of the local needs for knowledge resources and the *partnerships* (coalitions and networks) made by individual actors (e.g. entrepreneurs, development agencies, university units, municipalities, technology centres) in certain institutional settings and certain spaces, in which development processes take place simultaneously. The existence or creation of 'public spaces as shared arenas' is the crucial element of the economic development of LFRs (see Healey *et al.* 1999; also Healey 2004, 2006, Bathelt *et al.* 2004; Sotarauta *et al.* 2003; Amin and Thrift 1995; Henry and Pinch 2001).

This is stressed in the work of Storper and Venables (2002) and Grabher (2002) (and see also Bathelt *et al.* 2004; Sotarauta *et al.* 2003) about the importance of a set of activities called the 'noise' or 'local buzz'.[4] 'Buzz' is used to refer to the information and communication ecology created by face-to-face contacts, co-presence and co-location of people and firms within the same institutional orchestra and place or region (Storper and Venables 2002; see also Bathelt *et al.* 2004). The idea of noise, buzz or perhaps the 'cafeteria effects' lies basically in the simple notion that a certain milieu or agglomeration with closely working actors and individuals can be vibrant and culturally lively with social contacts and interaction in the sense that there are a lot of useful informal and unplanned contacts going on simultaneously and continually, which makes it easier to share information, interpretations, inspiration and motivation among the networks of communication (e.g. knowledge networks) and information linkages internal and external to that milieu (Moulaert and Sekia 2003; Bathelt *et al.* 2004; Maskell *et al.* 2005; Lambooy 2004). In addition, Bathelt *et al.* (2004, p. 42) formulate the advantages of what are called *global pipelines*, which 'are associated with the integration of multiple selection environments that open different potentialities and feed local interpretation and usage of knowledge residing elsewhere'. 'Local buzz' generates necessary opportunities for a variety of spontaneous situations in which firms interact and, more specifically, form interpretative knowledge communities, shared arenas. Summarizing the idea, the key elements of institutional capacity are *institutions* (technological infrastructure), *knowledge resources*, *networks* and, finally, the existence or creation of *'public spaces as shared arenas'*. Therefore, the elements of the institutional capacity are seen in this study as a set of elements intertwined with each other, as presented in Table 5.1.

It can be concluded that in less favoured regions the research communities formed by multiple actors are common knowledge arenas and the forums for local buzz as understood in the work of Storper and Venables (2002) and Bathelt *et al.* (2002, 2004); and see also Maskell and Kebir (2005) and Maskell *et al.* (2005). Local knowledge communities are supposed to be open for transregional influence and interaction, or, as in the LFRs, their main purpose may be to link local actors to global knowledge networks.

Table 5.1 Elements of institutional capacity in the less favoured regions

	Institutions	Resources	Networks	Shared arenas
Elements of institutional capacity	R&D and educational organizations (HEIs)	Knowledge-related resources	Local and non-local innovation networks	Knowledge communities

Challenges for high-tech industries as automation industry and embedded systems providers located in Finnish LFRs

In Finland (which has around 5.3 million inhabitants) there are 20 campus universities in 11 cities, six university filial centres in six other cities and several other university branch units mainly in peripheral areas. In addition, there are 29 polytechnics located in 80 municipalities. The total annual investments in R&D in Finland are around €5 billion, with the share of GDP being around 3.5 per cent (Ministry of Education 2006; OECD 2005, 2006). Generally a large share of the Finnish national R&D investment of GDP is made by the private sector and the ICT sector – to be more specific, the telecom industry (Nokia and its subcontractors). Those companies then are located in main five or six city regions. The volume of publicly funded basic research has remained relatively stable in recent years (Rantanen 2004, p. 36), although the Finnish higher education system has been diversified and widely spread (see, for example, Goddard *et al.* 2003; Ministry of Education 2001, 2006; Ministry of the Interior 2004). Still, there are tendencies towards strong centralization; of all above-mentioned 49 higher educational institutions, 18 are located in the capital city area, the Greater Helsinki Region.[5]

The main university cities (the Helsinki region, the Tampere region, and the Oulu and Turku regions) cover over 80 per cent of the country's research and development activity (Ministry of Education 2006). Growing and innovative city regions are typically large-scale business and industrial centres with strong campus universities and research institutes, and with a population of over 200,000 inhabitants. The Helsinki Metropolitan Area (24 municipalities) alone covers around 1.2 million inhabitants, which amounts to 29 per cent of the entire workforce of Finland and to over 40 per cent of investment in research and product development by Finnish corporations[6] (see Figure 5.1). In the less favoured Finnish regions, the economic recovery process was much slower after the recession, and the ICT cluster did not play a remarkable role in local industrial or business life (see Ministry of Education 2001; Poropudas 2004).

Challenges for knowledge-intensive industries in the Pori and Seinäjoki town regions

The automation industry relies heavily on the Finnish information and communication (ICT) cluster. By the end of the 1990s the economic importance

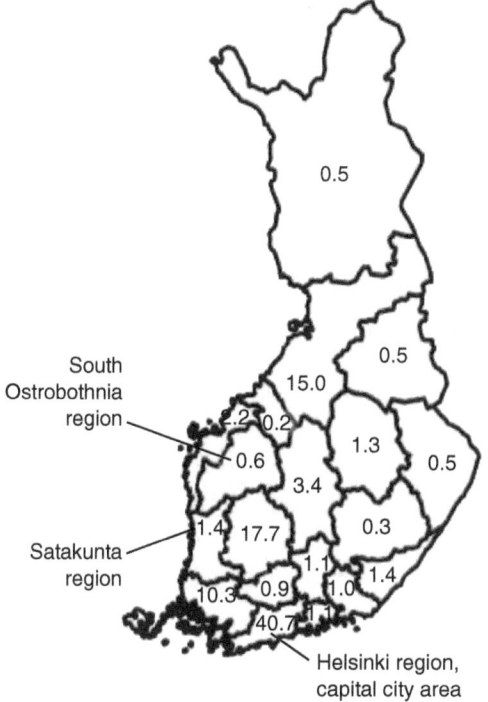

Figure 5.1 The corporate R&D share of Finnish regions in 2004. Sources: Statistics Finland and TEKES 2006

of the ICT cluster grew rapidly, and this multiplied private Finnish R&D investments in the sector. The rapid globalization and the rise to global competitiveness of major Finnish industrial corporations forced their subcontractors and other industrial sectors to invest a much greater share of their profits in R&D than ever before, ranging from less than 1 per cent of their turnover (wood and paper industry) up to around 11 per cent in telecoms in the case of the Nokia Group.[7] Meanwhile, according to both the Finnish government and industry representatives, Finnish basic research seems to be strong mainly in the following fields of study or science: genetics, cancer research, ecology, forestry, neurosciences, theology, and certain fields of technology such as radio technology and material sciences. Also, it is stated by the management[8] of one of the main funding organizations of Finnish basic research, the Academy of Finland (which has an annual budget of around €250 million, which amounts to 14 per cent of public funding in Finland), that research especially in the fields of *construction, machinery* and *automation*, is, internationally speaking, not strong enough. Therefore, the main challenge for the nowadays knowledge-intensive industries located in Finnish LFRs like Pori and Seinäjoki is how to use local, national and international innovation support tools actively to meet the increased requirements for designing, producing and marketing complex and interactive (embedded) systems in global markets.

Most automation and intelligence technology enterprises in the case-study regions were internationalized in the 1990s, or their main market areas or alliances were regarded as being international from the very beginning of the companies' life. The main market areas for the leading companies are the other EU countries, China and other rising Asian economies, Finland and the other Nordic countries, the United States and some of the transition countries in Europe. The firms in the field have specialized expertise in certain areas of mechanical engineering, automation and intelligent engineering solutions,[9] which either operate globally or are located globally, and most often their 'closest' R&D partners are located in other regions in Finland or in other countries. Therefore, the flexibility and the set of innovation capabilities of the local as well as national knowledge-based 'systems' are crucial for these companies classified merely as industrial SMEs.

Pori: automation and embedded systems

The Pori town region is the eighth biggest city region[10] in Finland and one of the four administrative sub-regions (town regions) in the larger region of Satakunta. The town of Ulvila (founded in 1365), where most of the automation technology firms are located, and the town of Pori together have approximately 90,000 inhabitants, thus forming the urban centre for the region.[11] The Satakunta region in turn holds 4.7 per cent of the nation's population, but is responsible for as much as 8 per cent of the total industrial production of Finland. Even so, the unemployment rate in the region is one of the highest for any Finnish region: 16.1 per cent in 2001.

The leading firms in the automation field are classified as whole systems and machinery providers (conductors) and serve as testing and development plants for other industries, mainly for the metal industry and machinery, vehicle manufacturers, oil platforms and shipyards, the electronics industry, the food industry and the pulp and paper industries. The history of the local agglomeration of automation firms dates back to 1853 and can be traced back to the then newly established Rosenlew family business in Pori. The Rosenlew company started producing its agricultural machinery in 1900 and combine harvesters in 1957, but the actual starting point for the automation field was the opening of Rosenlew Tool Factory ('Rosenlew Työkalutehdas') in the 1970s. In the late 1990s and the early 2000s, small automation, software, robotics and electronics companies were booming.[12] Generally, the biggest companies that have their offices or plants in the region are also among the biggest R&D investors of Finnish industry and very often operate globally. They also have private R&D departments with relatively high-scoring R&D outcomes; for example, the Pori town region performs better in terms of patenting (domestic patent applications) than many other Finnish town regions (Oksanen *et al.* 2003).

Seinäjoki: embedded systems and intelligent technology

The Seinäjoki town region (which has approximately 70,000 inhabitants) is a central service centre for a large traditionally agricultural area called South Ostrobothnia (with approximately 200,000 inhabitants). South Ostrobothnia accounts for about 4 per cent of the total population of Finland and its population density is 15 inhabitants per square kilometre. The Seinäjoki town region education level is higher than the Finnish average, but in South Ostrobothnia as a whole, the education level is one of the lowest in Finland. The supporting structures for innovations and innovation culture were weak until the turn of the century, and most of the firms in the region operate on short time horizons. The best firms in the town region, however, operate at a technologically high level and are well internationalized, but their number is estimated to be very low, according to the Technology Strategy of South Ostrobothnia for 2003 and the Regional Development Programme for 1994 and 2002 (see also Sotarauta and Kosonen 2003, 2004).

The leading companies in the field of embedded systems are technology developers (innovators), supporters and service providers (technology transfer and consultancy), or appliers and utilizers. The agglomeration of metal manufacturers and software service providers is the second strongest industrial sector in the South Ostrobothnia region, after the foodstuffs sector (Statistics Finland, 2002[13]). The embedded systems and intelligent solutions (broadly defined) sector is also the largest exporting sector of the industrial branches and enterprises located in the town region. The combined annual turnover of the leading 40 companies in the field of intelligence technology amounts to approximately €600 million.[14]

Strengthening the elements of institutional capacity in Pori and Seinäjoki

In LFRs the capacity to create and share knowledge may be weak for financial and institutional reasons. There are a variety of development-oriented models which are merely based on the local strengths, capabilities and awareness to stimulate the local economic change and strengthen the local innovation environment. The *regional enterprise* comes up especially in the element of building up new shared arenas as spaces for a local buzz in order to get more out of global knowledge 'pipelines', as may be stated following Storper and Venables (2002), Bathelt *et al.* (2004) and Maskell *et al.* (2005). The challenges that SMEs face in the global markets combined with the somewhat weak R&D infrastructure of their location forces them to renew their networks and technology transfer 'pipelines'. If the members of innovation networks are mainly SMEs, as typically in LFR agglomerations might be the case, the linkages may be too difficult for corporations to create and to intensify on their own; other local and national players are then welcomed.

It has been argued in the literature (see, for example, Iammarino 2005) that the dynamism of an economic system is necessarily built on access to knowledge and the absorptive capacity to use it in the valuable and reasonable way. One may argue that this dynamism highlights the role of local knowledge entrepreneurship in knowledge-intensive sectors from a crucial point of view. Iammarino (2005, p. 501), for example, highlights the following main functional dimensions: (1) absorption of new knowledge, technology and adaptation; (2) diffusion of innovations throughout the local agglomeration of industries and businesses and knowledge infrastructure; and (3) generation of new knowledge, technology and innovation with 'global pipelines' (see also Bathelt *et al.* 2004; Maskell *et al.* 2005). As Bathelt *et al.* (2004, p. 39) state,

> shared experience in the same technologies and ongoing attempts to solve the same sort of problems ... support the development of mutual engagement, joint enterprise, shared repertoire and negotiation of meaning. Communities of practise thus lead to the generation of distinct routines, conventions and other institutional arrangements.

Where the local knowledge-intensive innovation environment is not dynamic, the establishment and maintenance of interconnected relations outside the local agglomeration require time and effort from the members of local innovation networks. Maskell and Kebir (2005) state that the main emphasis in strengthening innovation environments and agglomeration instruments (clustering) should lie in widening the communication horizon from a prior local or regional one to transregional and global knowledge sources, whether in the form of permanent knowledge nodes and hubs (Maskell and Kebir 2005) or temporary gatherings we may call 'satellite clusters'[15] (Maskell *et al.* 2005). Therefore, the shared arenas or public spaces are of importance in creating a local innovative milieu where the local buzz is nurtured and evolved together with, and because of, the communication nodes accessed and knowledge gathered from global pipelines.

What actions were taken in Pori and Seinäjoki town regions, then, to create and strengthen the innovation environment in emerging industries? The turning point for these types of development activities was in the late 1990s, when the local leaders and managers (e.g. of companies, polytechnics, university units, regional development agencies and the chamber of commerce) realized the challenging situation and started to strengthen the local innovation environment: 'Something has to be done....' However, the question remains: why did automation and embedded systems firms take action in order to enhance the local innovation environment in the Pori and Seinäjoki town regions?

Altogether, the main strategy was to bring knowledge into the town region by (1) *inducing universities* (and polytechnics) to found new units and *create university filial centres*; and (2) *creating shared arenas* (public spaces). In Seinäjoki the EPANET research network (see below) has been able to transcend disciplinary borders by creating a research community of researchers from different disciplines and universities. In Pori the initiative came from the automation agglomeration; soon afterwards, higher education and university institutions responded rapidly, and municipalities started to assist in financing the institute. From this perspective, change in innovation culture meant change in the 'local way of doing' between 'global' knowledge actors and local organizations from a static institutional view to a process- and partnership-oriented view, in which all parties are seen to influence each other in the local development path.

Inducing universities to found new institutions for university filial centres

From the national point of view, serving the needs of less developed communities (in order to strengthen the country's overall competitiveness) from the early 2000s was the turning point in strengthening the knowledge infrastructure in the LFRs. The third-strand tasks made it possible for the universities to 'review' and to start expanding their institutional structures not only internally, but also *spatially*, namely with other regional and local partners in the surrounding or neighbouring communities.

As an outcome of this, many universities launched university filial consortia in 2001–2004 with less favoured town regions.[16] Together these regions make up approximately 20 per cent of the nation's population, but before the launch of the university filial consortia these regions hosted only a few separate institutes, departments or educational offices. From the beginning of 2004 these umbrella organizations were systemically organized into a local and national network and called 'university filial centres' (Kinnunen *et al.* 2004; Kosonen 2004; Lievonen and Lemola 2004; Poranen 2006).[17] By the end of 2004 the university filial centre network was 'frozen' at the level of these six towns and their filial centres, at least for a while, to see what the impact of these centres on the respective regions, universities and the national higher education system would be. The university filial centre cities and towns are shown in Figure 5.2, showing the linkages from university filial centres to their relative campus universities.

Under the University Filial Centre Network there are approximately 1,000 employees, mainly researchers and project workers (among whom around 75 professors), almost 4,000 degree students and 27,000 students in total (short courses included). Compared to the total number of degree students at all Finnish universities in 2005 (approximately 170,000 students), a relatively small number of university degrees are produced within the network, but when compared to the smallest Finnish campus universities, the number of degrees produced is bigger (Poranen 2006; Ministry of Education 2006).

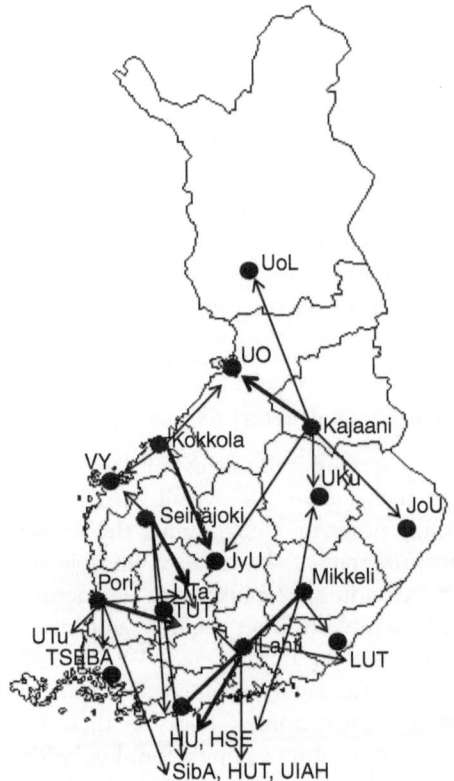

Figure 5.2 Locations of the university filial centres and the administrative linkages to the main universities and the campus cities

Pori town region: strengthening the elements of institutional capacity via the building up of institutions, resources and networks

Strategically, leaders in Pori put emphasis on the wider higher education network; they created and strengthened the Pori University Filial Centre. In their visions, research was seen as a 'logical outcome' of the investments in university units and Pori University Campus. The first university unit was established in the town region as early as 1983. After many years of heavy investments in the higher education infrastructure, and especially in the university centre, the messages from the local industry and business life have stressed the need for increased co-operation between Pori Polytechnic, Pori university units and PrizzTech Ltd (see, for example, Ahmaniemi *et al.* 2001; Poijärvi-Miikkulainen 2004; *Satakunta Vision 2010* 2003). In principle, the Pori University Filial Centre (Pori University Consortium) is the unit for basic research and higher technical education in specific fields, while the polytechnic is a local educational unit with more applied R&D functions.

The Pori University Filial Centre is involved in the field of automation only indirectly. It has some collaborative projects, but its contribution is not a very large one. Adult and extension studies in different fields have been offered since 1987, but in the past five years the university units in Pori (Pori University Filial Centre) have offered an entire degree programme in Pori for high school graduates. The fields of education and expertise (of the centre) do not meet the needs of the local automation industry very clearly, and the university units may have some tendency to regard the agglomeration as being not challenging or big enough to be particularly interesting for academic purposes. Also, as one may judge from the history of the agglomeration, it is more likely to collaborate with Satakunta Polytechnic and the former technical and engineering colleges, which may lead the university units to concentrate on other industries and business areas. However, it is stated in the strategy papers, reports and interviews that the university filial centre has such a generic technical and business administration experience that it could be useful for the polytechnic and the local automation companies.

The Seinäjoki town region: strengthening the elements of institutional capacity via building up institutional base and forums

The Seinäjoki University Filial Centre is among the newest 'university filial centres', as it was officially formed at the end of 2003 and opened as a 'centre' from the beginning of 2004 (Kinnunen *et al.* 2004). The centre was formed from already existing units and university functions operating in the Seinäjoki town region. The first unit was launched in 1981. Leaders in development agencies and the municipalities in the Seinäjoki town region put emphasis on applied research – they created the EPANET research network, as the 'research path' was seen as a 'faster way' to fill major gaps in the region's knowledge infrastructure. The network focuses on applied research and is the main research activity under the university filial centre 'umbrella' and the main research 'community' in South Ostrobothnia.[18]

The EPANET research network works especially on themes found in the local business environment. Therefore, the network is largely accepted and directly invested in by local companies, including in the field of intelligence technology. The centre is coordinated by one of the universities for a period of two to three years at a time. The aim of the centre is to coordinate the classical university tasks mentioned (research, education and the 'third strand' activities). More concretely, the centre's task is to strengthen collaboration between university units and between Seinäjoki units and other universities and research institutions in Finland.

Creating shared arenas

The Automation R&D Consortium in Pori

The starting point for realizing the importance of development of the automation industry perhaps lay in the informal discussion sessions[19] for local automation company leaders that Satakunta Polytechnic organized over a period of some years. Many reports written in the Pori town region (especially in Satakunta Vision for 2010 and Satakunta Region Technology Strategy) realized that the automation industry and related businesses were in need of extended R&D activities in order for the industry to increase its competitiveness. Therefore, the industry, together with Satakunta Polytechnic and the local municipalities, established the Institute of Automation and Information Technologies.

The next step in establishing permanent R&D activities in the automation industry branch was to start the 'Automation Research Project' and to set up the Automation Research Manager Project in Satakunta Polytechnic in 2000. These projects and steps were stated as targets in the Satakunta Vision 2005 report and in its automation industry chapter, discussed and written by local entrepreneurs. In the beginning there were four automation companies involved in the Automation Research Manager Project, which were also responsible for most of the costs of the project. Later this function changed its form; the activities are now covered and funded by a variety of sources: Satakunta Polytechnic itself, regional development and funding institutions with the EU Structural fund, the towns of Pori and Ulvila, the Technology Agency of Finland (Tekes) and the European Commission (through its research programmes and research framework). The companies involved pay their shares out of their individual project budgets.

The third step was to organize these activities in the form of an Automation R&D Centre (within the polytechnic's organizational structure). The activities are still organized merely in the form of projects and R&D programmes. After a couple of years of the automation projects, Satakunta Polytechnic has also started to strengthen the resources of the centre internally. During the process of building up the Institute of Automation and Information Technologies internally in the polytechnic, the informal get-togethers with a larger representation of local automation company (morning coffees) almost stopped for three years. The tradition assumed at the beginning of 2004, when the group invited to the discussions was also enlarged in number and widened in location. Recently there were approximately 20 organizations involved, mainly automation, controlling systems and engineering companies. Especially active organizations at this stage were PrizzTech Ltd (+ Satakunta Centre of Expertise Programme), Porin Seudun Kehittämiskeskus Oy, Posek (the Pori Region Development Centre Ltd, a local business development organization owned by the municipalities), and the Satakunta Chamber of Commerce.

EPANET: a network for applied research in Seinäjoki

The Seinäjoki town region is making a new effort to create a higher educational and research network. The network, called the South Ostrobothnian University Network (EPANET), is a cooperative network of the above-mentioned six Finnish universities in the Seinäjoki town region. The core of the network is a loosely organized group of fixed-term research professors who gather around themselves a group of researchers in turn, but who all have their 'home base'[20] in South Ostrobothnia, mostly in Seinäjoki. By the end of 2005 there were around 16 full-time professors (research chairs), around 40 other researchers and around 50 PhD students and 38 undergraduates in the EPANET network. The EPANET research programmes contribute mostly to applied research in the following fields: information technology applications, economics and business administration, food technology, regions and welfare, and more industry-specific topics.

The EPANET research network has formed a new kind of creative community working especially on themes found in the local business environment. The network is therefore largely accepted and directly invested in by local companies, as the network focuses on applied research. The idea is to obtain a broad understanding of the characteristics and problems of regionally based industry by combining tacit knowledge with theory and by combining the approaches of different disciplines. The idea is not to function as a direct problem-solving and research transfer institution for companies, but merely to search and find new research questions arising from traditional industries and local knowledge-sharing culture in agriculture, foodstuffs, forestry, machinery, furniture and carpets, therefore functioning as a source of 'local buzz'. The EPANET network strengthens the institutional academic infrastructure in South Ostrobothnia by allocating new knowledge and relational resources and forming a new type of research community.

Discussion: entrepreneurial LFRs – creative in development activities?

It is said that the relative competitiveness that Finland has achieved after the serious recession (or slump in reality) at the beginning of the 1990s consists of its well-governed national innovation system and relatively successful intertwining of different major parties of knowledge society: public administration (agency), corporations (business), and universities and higher education institutes (knowledge creation) (see, for example, the OECD 2005). As Sotarauta and Kautonen (2006) note, central government has played the dominant role in science, technological and innovation arenas, especially by implementing new investment tools, such as the Technology Agency of Finland (Tekes), technology programmes, new initiatives by the Finnish Parliament and by the Council of State and several ministries (see also Boschma and Sotarauta 2006).

Sotarauta and Kautonen (2006) state that Finnish local governments and sub-regions then (because of the strong autonomy of single municipalities and cities) had an important role in building local structures and institutions, launching new processes for science, technology and innovation, in order either to strengthen already existing business clusters of the locality or to transplant or create a new local cluster. Usually this was done together with representatives of a business sector in question, regional development agencies, technology centres (which are often owned by the cities or municipalities) and the higher education institutions and sectoral research institutes located in that town region. Generally, Finnish city or town governments often take the place of the private or institutional investors usually found in the capital-area businesses or major European regions and in other OECD countries (OECD 2005).

What has been missing is the mutual understanding of the roles of both regional and local actors on the one hand and national actors and governance on the other hand, but as for a variety of regional and local enterprises and their own incentives, the interplay has become more obvious and visible (see Figure 5.3). This, together with strong investments in the national-level innovation systems and R&D investments generally, over a period of more than ten years, has evolved as an evolutionary interplay between Finnish regional actors (local innovation environments) and national-level governance with top R&D agencies of Finland. In other words, in Finland the medium-sized city regions engage in innovation, technology and science policy in order to fill the possible cognitive, financial and geographical gaps in economic performance compared to growth metropolis.

The (case) industries on their part raised the awareness of the need for new, often scientific, and internationally competitive knowledge. Operating

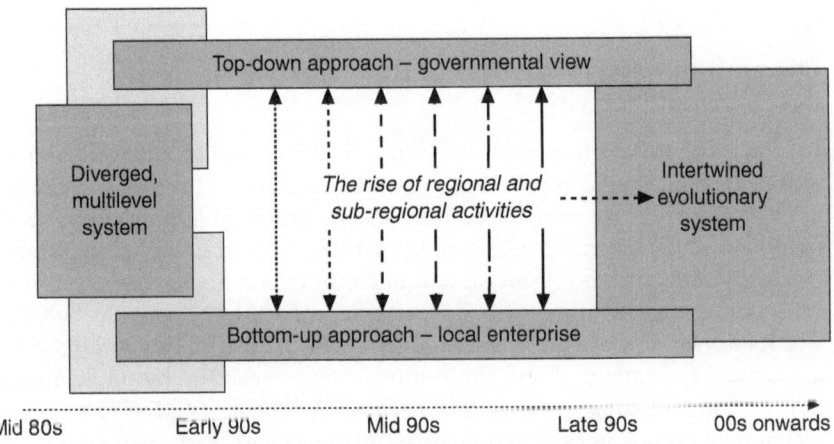

Figure 5.3 The slow shift from separate activities from national and local levels to an evolutionary interplay of both. Source: Own modification

globally in the knowledge economy, industries faced increased challenges to their knowledge capabilities and regarding the speed at which they could integrate new knowledge into their processes and productions; they aimed to be faster and more flexible than their competitors in Asia (China), the United States and Europe in particular. Since R&D in universities, labs and research institutes was not particularly concentrated, the local agglomeration of automation companies (in Pori) and embedded systems (in Seinäjoki) decided to create a new, locally based approach to meet the extended needs of high-quality research.

As the local knowledge infrastructure was unable to offer such knowledge pathways, linkages and practices (innovation culture) to scientific knowledge, the industries became more and more interested in participating in the efforts to strengthen local institutional capacity through development programmes, coalitions, science parks, technology centre activities and, more recently, through local research communities. When the production routines and key personnel were highly embedded in the then current social systems and places, firms put their emphasis on unlocking innovation habits and culture in the regions where they were already located. While a few narrowly focused R&D institutions located there did not offer wide enough transregional or international technology communication enhancing local dynamic development, firms in Pori and Seinäjoki town regions were either in favour of building up and widening local shared arenas or willing to join the exploratory network-building efforts led by a local development network: university and polytechnic unit leaders, members from regional development agencies and policy makers. This implicated a somewhat unusual activeness in participating in renewing processes in education, R&D, programme-based development and, to some extent, politics. In these cases, entrepreneurs and policy makers put their emphasis on locally based research communities (shared arenas), where the new ideas from both the local actors and global partners could flow and be absorbed to innovation processes of the cluster.

Final remarks

This chapter has highlighted how some local knowledge-intensive high-tech industries succeeded in breaking certain types of intellectual and institutional lock-ins, changing innovation culture and renewing local knowledge architecture by enhancing locally produced scientific knowledge, in order to be globally competitive. Furthermore, the study examined what actual efforts were made in these regions to strengthen the institutional capacity, and, more precisely, what efforts were taken to 'bring new knowledge into' the town region. To summarize, the process of building up an innovation environment for emerging industries calls for new organizational modes, new technology and innovation culture as well as actual access to new technology and knowledge, as has been done in the case locations, the town regions of Pori and Seinäjoki.

Table 5.2 presents the combination of the main elements of institutional capacity and the level of social interaction, be it about strengthening the structural level, stabilizing the communication channels or strengthening the R&D culture and all the social aspects included: enabling creativity in the local communities, adapting new information and ideas, unlocking past habits, combining knowledge from various sources and fields, attracting skilled developers, researchers, teachers, consultants, industrialists, public administrators, business angels, etc., and enhancing communication and interaction between all players.[21] As these aspects are actualized in the context of less favoured town regions such as those of Pori and Seinäjoki, the ability to link local players to the global knowledge channels and networks is crucial.

In summary, institutional capacity appears in this study as a combination of the local needs for *knowledge resources* and *global partnerships* (coalitions and networks) formed by individual actors (e.g. entrepreneurs, development agencies, university units, municipalities, technology centres) in certain institutional settings and certain knowledge-oriented spaces, in which development processes take place simultaneously. However, in various lock-in situations, *industries may take up the role of breaking the past paths*. In Pori the automation industry was a very obvious initiator of new organizational modes, and in Seinäjoki the 'smart systems' industry was one of the bravest in investing in the somewhat fuzzy and future-oriented research programmes.

From the perspective of these two less favoured Finnish town regions, it can be concluded that although the thinness of the scientific infrastructure makes development efforts challenging for the local development and innovation networks, it can be an encouraging feature for knowledge entrepreneurship, which is otherwise so weakly manifested in these town regions. In addition, if the national and regional players recognized the so far unrecognized possibilities of these R&D communities – even if their location is not in the major university cities (therefore outside the mental map of

Table 5.2 Elements of the institutional capacity in the less favoured regions such as Pori and Seinäjoki

Elements of institutional capacity	Institutions	Resources	Networks	Shared arenas
Structures	Technological infrastructure	Visible, exchangeable resource base	Local and global innovation networks	Public forums, places to interact
Communication channels	R&D organizations (HEIs)	Knowledge-related resources	Nodes and key individuals	Knowledge communities
Social aspects, 'culture'	Non-org. institutions	Competencies	Interaction	Local buzz

national competitiveness of the leading policy makers as well as business life associations) – the overall quality and level of the research in Finland in these particular fields might rise.

Notes

1 The study strongly relies on the empirical findings from Pori and Seinäjoki, but also to some extent follows the themes highlighted in the 'innovation systems' and 'institutional thickness' approaches in the regional science and economic geographic literature.

2 The Local Innovation Systems Project (LIS, 2002–2005) investigated cases of actual and attempted industrial transformation in about 23 locales in the United States, Finland, Japan, the United Kingdom and Norway. Additional research has been carried out in Ireland, India, Taiwan and Israel. The LIS Project aimed at developing new insights into how regional capabilities can spur innovation and economic growth, and into how to develop new models of innovation-led industrial development. The research partners of the project consisted of the following research groups and institutions: MIT Industrial Performance Center (USA), Sente, Research Unit for Urban and Regional Development Studies, University of Tampere (Finland), Helsinki University of Technology (Finland), the Centre for Business Research, University of Cambridge (UK) and Rogaland Research Institute (Norway). The studies of the Pori and Seinäjoki regions are two of the total of 23 case studies. The second study, The Innovation Capabilities of Innovation Developers in Finland (InnoKom, 2005–2006) focuses mainly on the requirements for innovation capabilities of the innovation developers in regional development agencies and other intermediaries in knowledge society. The aim of the project is (1) to study what type of innovation capabilities are needed among the local authorities, intermediaries and innovation networks that aim to enhance the innovation and knowledge society at the local, regional and national level; and (2) to produce a certain map of the required competencies of regional innovation policy as a learning tool for local and regional developers.

3 Automation, intelligence technology and embedded systems can be defined as 'combinations of hardware and software whose purpose is to control an external process, device or system in order to provide intelligence to a larger system of which they are part' (Tekes Technology Strategy 2002 and *Building ArtEmIs 2004*).

4 This term is used to refer to the information and communication ecology created by face-to-face contacts, co-presence and co-location of people and firms within the same institutional orchestra and place of region. (Storper and Venables 2002; Bathelt 2005; see also Bathelt *et al.* 2004.)

5 The Greater Helsinki Region hosts eight campus universities, several polytechnics, six technology parks and the largest technology campus in the Nordic countries (Ministry of Education 2006; Helsinki Region Marketing Ltd 2006).

6 Source: wwww.helsinkiregion.com (accessed 28 May 2006).

7 In 2005, Nokia's R&D spending totalled 11.8 per cent of its net sales (€3.8 billion), and 36 per cent of its total workforce in 26 countries worked in the field of R&D (Nokia Group, financial statement 2005).

8 Source: a press release of the Academy of Finland dated 29 May 2006.

9 'The development of intelligent products, processes and systems involves the cross-technological application and development of various technologies and application targets, including learning and anticipating systems, systems that adapt to the operating environment, well-functioning and bio-compatible materials, personal and natural interfaces, systems and services based on positioning and

identification as well as solutions based on remote diagnostics, remote operations and virtual reality' (Tekes Technology Strategy 2002, p. 10).

10 Source: Ministry of the Interior, Statistics Finland (2005, 2006), StatFin web service.

11 Sources: the Regional Council of Satakunta, and the Regional Development Centre Programme for Pori Town Region, Statistics Finland, Ministry of the Interior, Suomen Kuntaliitto.

12 Sources: interviews and, in the case of Sampo-Rosenlew, www.samporosenlew. fi/english.htm.

13 The industrial areas are metal product manufacturing, machinery, electronics and optical instruments (manufacturing), vehicles, PC consultancy, software design, programming and consultancy, and research.

14 Source: Seinäjoki Centre of Expertise for Intelligence Technology.

15 A 'satellite cluster' is a form of temporary cluster defined by Maskell *et al.* (2005), very often organized in the form of trade fairs, conventions, professional gatherings, etc.

16 Namely Kajaani, Kokkola, Lahti, Mikkeli, Pori and Seinäjoki.

17 The Finnish way to express the organizational mode is to call it the 'university consortium' (UC). The term 'university filial centre' itself stands for a local structure (a building or some other concentration of academic institutions), where several campus universities have established branch units in the same town or the same town area, in close proximity to science parks, technology centres, polytechnic facilities, for example. The annual budget of all the six filial centres amounts to over €70 million (€72.4 million in 2005) and equals 3.7 per cent of the total funding of Finnish universities (€1,956 million in 2005) (Poranen 2006; Ministry of Education 2006).

18 See Kinnunen *et al.* (2004) and Sotarauta and Kosonen (2003, 2004).

19 The sessions called 'Morning Coffee for the Automation Industry' (*Automaation Aamukahvit*) were organized every one or two months for some years.

20 When the nominations of professorships are confirmed, the home base will be mentioned and entered.

21 Adapted from the results gained in the Local Innovation Systems Project (see, for example, Lester (2005).

References

Ahmaniemi, R., Kautonen, M. and Tulkki, P. (2001) Tietointensiiviset yritysverkostot Porin alueella, Working Papers 63/2001, Work Research Centre, Tampere: University of Tampere.

Amin, A. and Thrift, N. (1995) Globalization, institutional 'thickness' and the local economy, in P. Healey, S. Cameron, S. Davoudi, S. Grahm and A. Madani-Pour (eds) *Managing Cities: The New Urban Context*, New York: John Wiley.

Bathelt, H. (2005) Geographies of production: growth regimes in spatial perspective (II) – knowledge creation and growth in clusters, *Progress in Human Geography*, 29 (2), pp. 204–216.

Bathelt, H., Malmberg, A. and Maskell, P. (2004) Clusters and knowledge: local buzz, global pipelines and the process of knowledge creation, *Progress in Human Geography*, 28 (1), pp. 31–56.

Boschma, R. and Sotarauta, M. (2005) Economic policy from an evolutionary perspective: the case of Finland. To be published in the *International Journal of Entrepreneurship and Innovation Management* (forthcoming).

Building ArtEmIs (June 2004) *A Report by the High Level Group on Embedded Systems.* Advanced Research and Technology for Embedded Intelligence and Systems, European Commission.

Cooke, P. and Leydesdorff, L. (2006) Regional development in the knowledge-based economy: the construction of advantage, Introduction in P. Cooke and L. Leydesdorff (eds) The knowledge-based economy and regional development, special issue of the *Journal of Technology Transfer*, 31 (1).

Cooke, P. and Morgan, K. (1998) *Associational Economy: Firms, Regions and Innovation,* Oxford: Oxford University Press.

Goddard, J. and Chatterton, P. (1999) Regional development agencies and the knowledge economy: harnessing the potential of universities, *Environment and Planning C*, 17, pp. 685–699.

Goddard, J., Asheim, B. T., Cronberg, T. and Virtanen, I. (2003) *Learning Regional Engagement: A Re-evaluation of the Third Role of Eastern Finland Universities,* Publications of the Finnish Higher Education Evaluation Council 11:2003 (Korkeakoulujen arviointineuvoston julkaisuja 11:2003), Helsinki: Edita.

Grabher, G. (2002) Cool projects, boring institutions: temporary collaboration in social context, Introduction to a special issue of *Regional Studies*, 36, pp. 245–262.

Healey, P. (2004) Creativity and urban governance, *Policy Studies*, 25 (2), pp. 87–102.

Healey, P. (2006) Transforming governance: challenges of institutional adaptation and a new politics of space, *European Planning Studies*, 14 (3), pp. 299–320.

Healey, P., de Magalhaes, C. and Madanipour, A. (1999) Institutional capacity-building, urban planning and urban regeneration projects, in M. Sotarauta (ed.) *Urban Futures: A Loss of Shadows in the Flowing Spaces?, Futura – a Quarterly Magazine of Finnish Society for Futures Studies*, 18 (3).

Henry, N. and Pinch, S. (2001) Neo-Marshallian nodes, institutional thickness, and Britain's 'Motor Sport Valley': thick or thin?, *Environment and Planning A*, 33 (7), pp. 1169–1183.

Iammarino, S. (2005) An evolutionary integrated view of regional systems of innovation: concepts, measures and historical perspectives, *European Planning Studies*, 13 (4), pp. 497–519.

Kautonen, K. And Sotarauta, M. (1999) *The Innovation Programme of Seinäjoki Region,* Seinäjoki, Finland: seinangapurien.

Kinnunen, J., Eskelinen, H., Lehto, E. and Karjalainen-Jurvelin, R. (2004) *Etelä-Pohjanmaan korkeakouluverkosto Epanet – enemmän kuin yliopistokeskus,* Arviointiryhmän raportti, Tampere: Tampere University Press.

Kosonen, K.-J. (2001) Institutionaalinen kapasiteetti ja alueellinen innovaatiokyvykkyys, in M. Sotarauta and N. Mustikkamäki (eds) *Alueiden kilpailukyvyn kahdeksan elementtiä*, ACTA no. 137, Helsinki: Finnish Association of Local Authorities.

Kosonen, K.-J. (2004) Institutionaalinen kapasiteetti ja yliopistokeskusseutujen innovaatioympäristöjen vahvistaminen [Strengthening innovation environment and institutional capacity in university filial centre cities], in M Sotarauta and K.-J. Kosonen (eds) *Yksilö, kulttuuri, innovaatioympäristö. Avauksia aluekehityksen näkymättömään dynamiikkaan* [An individual, culture, innovation environment: openings to invisible dynamics of regional development], Tampere: Tampere University Press.

Kosonen, K.-J. (2005) Strengthening the research and educational basis for regional development in less-favoured regions. MIT IPC Local Innovation Systems Working Paper 05-003, http://web.mit.edu/lis/papers.html.

Lambooy, J. (2004) The transmission of knowledge, emerging networks, and the role of the universities: an evolutionary approach, *European Planning Studies*, 12 (5), pp. 643–657.

Landabaso, M., Oughton, C. and Morgan, K. (1999) Learning regions in Europe: theory, policy and practice through the RIS experience, Paper given at the Third International Conference on Technology and Innovation Policy: Assessment, Commercialization and Application of Science and Technology and Management of Knowledge, Austin, TX, 30 August–2 September 1999.

Lester, R. K. (2005) LIS Project – Phase I findings: overview and discussion, *MIT* IPC Local Innovation Systems Working Paper 05-004, http://web.mit.edu/lis/papers.html.

Lievonen, J. and Lemola, T. (2004) *Alueellisen innovaatiopolitiikan haasteita. Tutkimustulosten tulkintaa*, Sisäasiainministeriön Alueiden kehittämisen julkaisuja 16/2004, Helsinki: Sisäasiainministeriö.

Lundvall, B.-Å. (1992) User–producer relationships, national systems of innovation and internationalization, in B.-Å. Lundvall (ed.) *National System of Innovation and Interactive Learning*, New York: Pinter.

Lundvall, B-Å. (2002) The university in the learning economy, DRUID Working Paper no. 02-06, Danish Research Unit for Industrial Dynamics, www. druid.dk (accessed 28 October 2003).

Malecki, E. and Oinas, P. (2000) Technological trajectories in space: from 'national' and 'regional' to 'spatial' innovation systems, CIBER Working Papers 00-35, Center for International Business Education and Research, University of Florida, http://bear.cba.ufl.edu/centers/ciber/papers.html (accessed 31 October 2003).

Maskell, P. and Kebir, L. (2005) What qualifies as a cluster theory?, DRUID Working Paper no. 05-09, Danish Research Unit for Industrial Dynamics, www.druid.dk.

Maskell, P., Bathelt, H. and Malmberg, A. (2005) Building global knowledge pipelines: the role of temporary clusters, DRUID Working Paper no. 05-20, Danish Research Unit for Industrial Dynamics, www.druid.dk.

Ministry of Education (2001) *Korkeakoulujen alueellisen kehittämisen työryhmän muistio* (Opetusministeriön työryhmien muistioita 28:2001), Helsinki: Ministry of Education.

Ministry of Education (2004) *Koulutus, tutkimus ja työllisyys (Valtion tiede-ja teknologianeuvostolle laadittu seurantaraportti)*, Helsinki: Ministry of Education.

Ministry of Education (2006) *Korkeakoulujen rakenteellisen kehittämisen periaatteet* (Keskustelumuistio 8.3.2006), Memorandum, 8 March 2006, Ministry of Education, Finnish Council of State.

Ministry of the Interior (2004) *Alueellisen innovaatiopolitiikan haasteita. Tutkimustulosten tulkintaa (Alueiden kehittäminen 16/2004)*, J. Lievonen and T. Lemola, Helsinki: Ministry of the Interior.

Morgan, K. (1997) The learning region: institutions, innovation and regional renewal, *Regional Studies*, 31 (5), pp. 491–503.

Moulaert, F. and Sekia, F. (2003) Territorial innovation models: a critical survey, *Regional Studies*, 37, pp. 289–302.

OECD (2005) *OECD Territorial Reviews. Finland*, Paris: OECD Publications; and SourceOECD, www.sourceoecd.org.2005.

OECD (2006) *OECD Economic Surveys. Finland*, Paris: OECD Publications; and SourceOECD, www.sourceoecd.org.2006.

Oksanen, T., Lehvo, A. and Nuutinen, A. (2003) Suomen tieteen tila ja taso, Katsaus tutkimustoimintaan ja tutkimuksen vaikutuksiin 2000-luvun alussa, Suomen

Akatemian julkaisuja 9/03, Suomen Akatemia, Painotalo Miktor Oy, Helsinki.

Poijärvi-Miikkulainen, J. (2004) Porin Yliopistokeskuksen alueellinen vaikuttavuus, Turun Kauppakorkeakoulu, Porin yksikkö, Julkaisusarja A5/2004, Kehitys Oy, Pori.

Poranen, A. (2006) *Yliopistokeskusten rahoitusselvitys, Selvitys yliopistokeskusten rahoituksesta sekä rahoituksen kehittämislinjoista*, Oulun yliopisto, Kajaanin yliopistokeskus (University of Oulu, Kajaani University Consortium), Kajaani 2006.

Poropudas, O. (2004) Koulutus, tutkimus ja työllisyys, Valtion tiede- ja teknologianeuvostolle laadittu seurantaraportti *(Koulutus, tutkimus ja työllisyys. Valtion tiede-ja teknologianeuvostolle laadittu seurantaraportti)*, Helsinki: Ministry of Education.

Rantanen, J. (2004) Review of the structure of university and polytechnic research, One-person committee, Reports of the Ministry of Education, 36: 2004 (Yliopistojen ja Ammattikorkeakoulujen tutkimuksen rakenneselvitys, *Opetusministeriön työryhmämuistioita ja selvityksiä* 2004:36), Helsinki: Ministry of Education.

Satakunta Vision 2010 (2003) *A Regional Vision for the Years 2005–2010 for Satakunta Region: A Project Report*, Satakunta Chamber of Commerce, European Union (Structural Funds) and Regional Council for Satakunta, edited by SWOT Consulting.

Sotarauta, M. and Kautonen, M. (2006) Co-evolution of local and national in the Finnish innovation and science arenas: a prerequisite for innovation capacity? To be published in a special issue on Regional Governance and Science Policy of *Regional Studies* (forthcoming).

Sotarauta, M. and Kosonen, K.-J. (2003) Institutional capacity and strategic adaptation in less favored regions: a South Ostrobothnian university network as a case in point, MIT IPC Local Innovation Systems, Working Paper 03-003.

Sotarauta, M. and Kosonen, K. (2004) Strategic adaptation to the knowledge economy in less favoured regions: a South-Ostrobothnian university network as a case in point, in P. Cooke and A. Piccaluga (eds) *Regional Economies as Knowledge Laboratories*, Cheltenham, UK: Edward Elgar.

Sotarauta, M., Linnamaa, R. and Suvinen, N. (2003) *Tulkitseva kehittäminen ja luovat kaupungit, Arvio Tampereen mahdollistavasta mallista ja osaamiskeskusohjelmasta*, Alueellisen kehittämisen tutkimusyksikkö – Sente ja Tekniikan Akateemisten Liitto TEK ry, Sente-julkaisuja 16/2003, Tampere: Tampereen yliopisto.

Storper, M. and Venables, A. (2002) BUZZ: The economic force of the city, Conference presentation at the DRUID Summer Conference 'Industrial Dynamics of the New and Old Economy – Who Is Embracing Whom?', Copenhagen and Elsinore, 6–8 June, www.druid.dk (accessed 7 January 2004).

Tekes Technology Strategy (2002) Helsinki: Tekes (the National Technology Agency of Finland).

Virkkala, S. (2003) *Oppiva alue käsitteen tausta ja sovelluksia alue- ja maaseudun kehittämisessä*, ChyNetti nro 31, Chydenius-Instituutin verkkojulkaisuja, Jyväskylän yliopisto, Chydenius-Instituutti, Kokkola: Jyväskylän yliopisto.

Part II
Cluster evolution, variety and policy

6 Regional innovation clusters: evaluation of the South-East Brabant cluster scheme

Roel Rutten and Frans Boekema

This chapter discusses the evaluation of the South-East Brabant cluster scheme. This scheme, which ran from 1994 through 2005, supported new product development of small and medium-sized enterprises (SMEs) in the above region, also known as the Eindhoven region. In early 2005 a group of Master's students from Tilburg University, under the supervision two staff members from the Department of Organization Studies, carried out an evaluation of the scheme. Of the 102 clusters, 25 were involved in the evaluation. The chapter begins with a sketch of the Eindhoven region and of the cluster scheme, its objectives and origin. It then discusses the framework that was used to evaluate the scheme. Next the results of the evaluation are presented in three subsections: the outcomes of the product development effort in the various clusters, the process of product development within the clusters, and the conditions under which this process took place. Conclusions regarding the effectiveness of the cluster scheme are presented in the final section.

The Eindhoven region

The focus of this evaluation is the southeastern part of the Dutch province of North Brabant. This region is also named after its main city, Eindhoven. South-East Brabant is a small region, even by Dutch standards; barely 750,000 people call the region their home. Yet it harbors a technological powerhouse. Some of the leading R&D-performing companies of the Netherlands are located in this region, for example Philips and ASML. Measured in terms of patents per million population, the Eindhoven region ranks among the European Union's (EU) technological top regions. In addition to several large multinational enterprises (MNEs), the region is home to an army of SMEs, many of which have engineering or even research facilities and maintain international linkages. Networking among companies is traditionally strong in this region, which means that the region possesses strong and well-nurtured social capital (cf. Cooke *et al.* 2005). A final strength of the region is its elaborate knowledge infrastructure, with at its nucleus a technological university and some of the key private research centers of the Netherlands. Following a hike in economic growth in the mid and late 1990s, the gross domestic product (GDP) of the Eindhoven region is now on a par with that of the Netherlands as a whole. This level is well above the EU average but considerably lower than that of Europe's richest regions.

In the early 1990s the economic climate in South-East Brabant was not quite so favorable. Against the background of increased price competition from Eastern Europe and Asia, the manufacturing industry in the region was in sharp decline. Some of the large manufacturing companies, such as Philips and truck manufacturer DAF, laid off thousands of jobs. Down the supply chain, this led to bankruptcies and further lay-offs among the region's SMEs. The economic outlook for the region was sufficiently bad for the EU to give South-East Brabant the status of an Objective 2 region. Depending on the state of economic disarray of a region, the EU has several policy instruments available to provide relief. The most important of these are the Objective 1 and 2 statuses, under which substantial funds for economic development can be transferred to a region. Objective 1 status is for the underdeveloped, or less favored, regions of the European Union. Objective 1 regions have a GDP that is 75 percent or less of the EU average. Regions that have a GDP above the 75 percent benchmark but are facing sharp industrial decline qualify as Objective 2 regions. Objective 2 regions receive less EU funding than Objective 1 regions, and the funds are intended to be spent on different objectives, in particular on strengthening the "industrial tissue" – that is, on strengthening linkages between companies and between companies and knowledge centers. The key aim for Objective 2 regions should be to strengthen existing companies, in particular through the promotion of innovation.

One such policy initiative developed in the Eindhoven region was the cluster scheme (Stimulus 1999), the object of the present evaluation. The objective of the cluster scheme was to support new product development in SMEs. More specifically, the cluster scheme had four objectives:

- developing new products in regional SMEs
- strengthening regional SME competitiveness
- strengthening the innovation networks of regional SMEs
- and, thus, strengthening the regional economy.

A cluster in the sense of the Eindhoven region cluster scheme is a temporary grouping in which several organizations collaborate in order to develop a new product. We will continue to use this definition of clusters throughout this chapter, as it is the definition that the Eindhoven region cluster scheme used. However, we are aware of the fact that "cluster" is a multi-faceted concept that suffers from considerable conceptual ambiguity (cf. Bell 2005; Cook and Morgan 1998; Porter 1998). A cluster counted at least one SME, but usually two to three SMEs were involved. Participation of at least one MNE was encouraged, as was the participation of at least one knowledge center. The reason behind the involvement of MNEs and knowledge centers was twofold. First, it ensured that the cluster had sufficient technological knowledge as well as research facilities at its disposal. Second, it strengthened linkages of SMEs with MNEs and knowledge centers – that is, it strengthened the "industrial tissue" of the region. Often, but not always, a consultant or engineering bureau was involved to coordinate and manage the clusters activities. Again, the aim of the cluster scheme was to support new product development – that is, R&D activities, carried out by SMEs. These are expensive activities, particularly for SMEs, therefore the EU paid a part of the R&D costs for each cluster. The maximum EU contribution was 50 percent of the project costs, but usually it was less. On average it took a cluster approximately two years to complete its activities. In principle, any group of companies from the region that wished to perform R&D was eligible under the cluster scheme. However, the Program Office that coordinated the implementation of the Objective 2 program for South-East Brabant assessed all applications. Promising proposals were usually accepted, provided sufficient funding was available. The Program Office (named Stimulus) was located in Eindhoven and staffed entirely by people from the region. There was no direct involvement from the EU in Stimulus.

Framework for evaluation

The evaluation of the cluster scheme was aimed at assessing the long-term effects of temporary collaboration. After all, the clusters were dissolved on completion of their activities. Typically, *ex post* evaluations of EU regional policy are carried out immediately on completion of the policy program, which does not allow sufficient time for long-term effects to materialize (cf. Baslé 2006). The effect of product development on company performance is one such long-term effect. Evaluating the effectiveness of the cluster scheme in terms of its impact on company performance, therefore, is relevant for two reasons. In the first place, it provides a more complete view on the effectiveness of this particular policy effort. As pressures to spend tax euros effectively are increasing, the outcomes of this evaluation are valuable for the Objective 2 program of South-East Brabant in particular and for regional policy makers in general. Second, longitudinal research on R&D collaboration between companies is rare, not least because it is costly. Given the importance for company performance that the literature attaches to innovation networks, R&D collaboration, etc., the lack of longitudinal studies is very unfortunate (Cooke and Morgan 1998; Grabher 2002; Porter 1998; Landabaso 2000). Though the data that were gathered in this research project can be used for many things, empirical testing of hypotheses among them, this chapter is an evaluation study, and therefore of a more descriptive nature.

The literature on R&D collaboration in temporary networks is very extensive and diverse. Different schools of thought, for example industrial organization theory and resources dependency theory, have developed different, sometimes contradictory, theories on the matter (Grabher 2002; Porter

1998; Powell 1998; Sydow *et al.* 2004). Theoretically, this study chose to keep an open mind to the theoretical diversity that characterizes the field. This is possible, as the same variables appear in different theories. Of course, different theories assume different (causal) relations between the same variables, but an evaluation study is not a causal analysis. Consequently, this study need not concern itself with the causal relations between variables, except on a high level of abstraction. The key assumption underlying this evaluation study is that R&D collaboration is a process, or rather consists of several processes, such as knowledge creation, communication, and management (Burns and Stalker 1961; Nonaka and Takeuchi 1995; Hobday 2000). These processes, which make up the actual collaboration effort, create collaboration outcomes. The processes are the process variables in this study. The outcomes can be broken down in economic effects (for example, the effect of the collaboration on firm competitiveness), technological effects (such as the effect of the collaboration on the technological capabilities of a firm), and learning effects (for example, establishing permanent collaboration networks or the implementation of organizational changes) (cf. Rooks and Oerlemans 2005). The different examples of outcomes are the outcome variables of this study. The assumption is that the more the processes meet certain criteria, the better the outcomes will be. For example, if the process of knowledge creation functions smoothly, a company is more likely to have developed new technological competences. However, on this level, no causal relations are assumed between the individual variables. Nor is this necessary. An evaluation is an assessment on the basis of a criterion. Therefore, it suffices to give definitions and operationalizations of the variables and to specify matching criteria. For example, the variable communication can be operationalized as the frequency of communication between cluster partners and as the intensity of the communication. For both operationalizations the criterion is that it should be "high." The assumption is that more communication contributes to better outcomes in terms of economic and technological effects. Again, the precise causal mechanism responsible for this assumed effect is not object of this, or any, evaluation study.

In addition to "processes" and "outcomes," the framework for evaluation has a third category of variables – that is, "conditions." This pertains to the characteristics of the organizations participating in the clusters and to the characteristics of the clusters themselves. The assumption is that the conditions impact how the processes unfold. For example, in a cluster that is dominated by a large MNE, communication and management of the cluster is likely to function differently than in a cluster where a dominant MNE is lacking (Larson 1992). Note that this evaluation only distinguishes between small and large companies, where a small company is regarded as one employing no more than 50 full-time equivalents.

A detailed theoretical discussion of the framework for evaluation would be too elaborate within the context of this chapter. Instead, the framework is presented in a simple form in Table 6.1. The above-mentioned Master's stu-

Table 6.1 Framework for evaluation

Variables	Indicators	Criterion
Conditions		
Previous experience with collaboration	Previous experience	Favorable
Motives for collaboration	Kinds of motives	Not purely opportunistic
Composition of the cluster	Supplementary of knowledge bases	Yes
Composition of the cluster	Size of the partners	Mixture of large and small
Process		
Use of external knowledge by companies	Access to knowledge base of partners	High
Use of external knowledge by companies	Absorptive capacity of companies	High
Intensity of communication in cluster	Frequency of communication	High
Form of communication in cluster	Participative openness	High
Structure of communication	Formal and informal communication	Mixture
Management of the cluster	Perceived balance of power	Equal
Management of the cluster	Social and formal control mechanisms	Mixture
Trust	Mutual dependency between partners	High
Trust	Opportunistic behavior	Low
Outcomes		
Economic effect	Introduction of new product	Yes
Economic effect	Change of turnover	Increase
Economic effect	Change of competitiveness	Increase
Technological effect	Change of knowledge intensity	Increase
Technological effect	Change of competences	Improved competences
Learning effect	Permanent networks with cluster partners	Yes
Learning effect	More external collaboration with non–cluster partners	Yes
Learning effect	Implementation of organizational changes	Yes

dents and their supervisors developed this framework in the fall semester of 2004, which makes it an original research tool. The method used to develop the framework for evaluation was simple but laborious. After we had decided that the key variables should be "conditions," "process," and "outcomes," a literature survey was conducted in order to determine what different authors based on different literatures had said about them. For example, "trust" was identified as an important variable pertaining to the process. In general, the literature argued that the process unfolds better when more trust is present. However, trust can be approached from several perspectives. The transaction cost approach, for example, an economic perspective, refers to trust as the absence of opportunistic behavior (e.g. Williamson 1993). In the embedded-ness perspective, however, trust refers to the fact that actors may be mutually dependent on one another (e.g. Granovetter 1985). Both perspectives were included in the framework (see Table 6.1). On the basis of more than 120 journal articles, books, and book chapters, the framework for evaluation was developed over the course of several brainstorm sessions with the Master's students and their supervisors. The results of their efforts – that is, the com-plete framework – can be found in Azouz *et al.* (2005).

Because of the focus on, on the one hand, "outcomes" and, on the other hand, on "conditions" and "processes," this evaluation is both a process evaluation and an outcome evaluation. A process evaluation assesses to what extent a particular process, in this case the process of R&D collaboration in clusters, met certain criteria. These criteria were conceived from theory. The process part of this evaluation mainly serves to reflect on theory. As argued, on a higher level of abstraction, namely the level of the three categories of variables, causal relations are assumed between the three categories. There-fore, one expects to find that the more the clusters meet the criteria that pertain to the process variables, the better the performance of the clusters will be. Put differently, comparing the empirical measurements of the variables with the theoretical criteria allows for pattern matching. The outcome evalu-ation is primarily used to determine the effectiveness of the cluster scheme. Did it achieve what it intended to do in terms of economic and technological effects on the level of companies. In other words, what is the value in euros for tax purposes of the cluster scheme?

As argued, Table 6.1 does not show all variables that make up the frame-work for evaluation. For this chapter, a selection of several key variables has been made, in order to keep its length within acceptable margins. From the discussion of the variables in Table 6.1, it follows that all variables were meas-ured on the level of individual companies. However, this chapter presents an evaluation of the cluster scheme. That is, the level of analysis used in this chapter is that of the cluster scheme, not that of the individual companies or clusters that participated in the scheme.

Data collection

Of the 102 clusters, 25 were selected for this evaluation study. The cluster scheme ran from 1994 through 2005 but this period was divided into three, namely, 1994–1996, 1997–1999, and 2000–2005, which corresponds to the three consecutive European regional development programs in the Eindhoven region. Only minor changes were implemented in the cluster scheme during this period. Nevertheless, in order to obtain a representative sample, 25 percent of the clusters that were started in each period were selected (see Table 6.2).

Clusters were further selected on the basis of their financial volume. The distribution of the clusters in the sample accounted for the different financial volumes of the clusters in the population (see Table 6.3). In other words, our sample of 25 clusters is representative of the population of 102 clusters with regard to the three periods of the Eindhoven regional development program and the financial volume of the projects. The vast majority of the companies in the clusters came from the metal, electronics, and information technology sectors, but this was not a selection criterion in our evaluation study.

The data were collected in early 2005, when the group of Master's students interviewed as many companies from the selected clusters as possible. For every cluster, at least two companies were interviewed. The interviews were conducted on the basis of a structured questionnaire, with most questions asked in the form of a five-point Likert scale. Respondents had the opportunity to comment on the position they marked on each scale. General information on the clusters was retrieved from the files at the Program Office (i.e., Stimulus).

Table 6.2 Sampling by period

	1994–1996	*1997–1999*	*2000–2005*	*Total*
Number of clusters	20	32	50	102
Of which completes	19	29	17	65
Cluster in sample	5	8	12	25
Percentage of clusters in sample	25	25	24	25

Table 6.3 Sampling by financial volume

	<250k€	*250–500k€*	*>500k€*	*Total*
Number of completed clusters	42	18	5	65
Of which in sample	18	5	2	25
Completed clusters, 1994–1996	14	5	–	19
Of which in sample	4	1	–	5
Completed clusters, 1997–1999	15	11	3	29
Of which in sample	4	3	1	8
Completed clusters, 2000–2005	13	2	2	17
Of which in sample	10	1	1	12

The conditions

Previous experiences with collaboration

The first condition is the previous experience with collaboration of the companies involved. The assumption is that previous positive experiences are a good preparation of and a good basis for successful collaboration in the present clusters (e.g., Klein Woolthuis *et al.* 2005). In general, the companies involved in the clusters in this study did indeed have favorable previous experiences. For nearly 70 percent of the companies the degree to which previous experiences with collaboration were positive was high to very high (see Figure 6.1).

Composition of the cluster

With regard to the composition of the clusters, companies found that their knowledge bases were largely supplementary. According to 84 percent of the companies, this was the case to a high or very high degree (see Figure 6.1). A supplementary knowledge base means that the companies are not competitors and, more importantly, that they can learn something from their partners, as the partners possess knowledge that they themselves do not have. According to innovation theory (e.g., Freel 2003), this is important. The distribution of the size of the companies in the clusters shows a mixed situation. In 11 out of the 25 clusters in the sample, a mixture of small and large companies were involved. A further 11 clusters had only small companies; the remaining three clusters had only large companies.

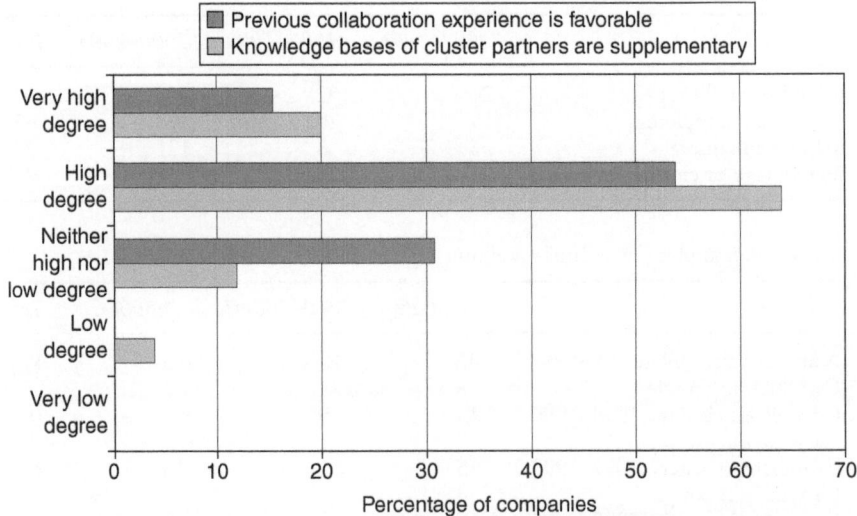

Figure 6.1 Conditions at the start of the cluster collaboration

Motives for collaboration

As to the motives of the companies to engage in the collaboration effort, it turned out that "opportunistic" motives played only a minor role. Fewer than 10 percent of the companies had "sharing risks" as their primary motive for collaborating on product development, and "sharing costs" was the primary motive for fewer than 5 percent of the companies. These two motives may be called "opportunistic" because they indicate that the companies that hold them are interested in benefits for themselves first of all. Many more companies gave "non-opportunistic" motives as their primary reason for collaborating on product development. More than 37 percent pointed at the "opportunity to innovate," nearly a third of the companies gave "access to external knowledge" as their primary motive, and 14 percent mentioned "specialization in knowledge and skills" as a primary motive (see Figure 6.2). These motives are "non-opportunistic" because companies cannot achieve them alone; they signal that companies are thinking in terms of relationships and collaboration. Even allowing for the fact that the split between "opportunistic" and "non-opportunistic" motives is to some extent arbitrary, the data allow the conclusion that the companies that participated in the clusters did not entertain purely opportunistic motives, because the difference between the two categories is very pronounced.

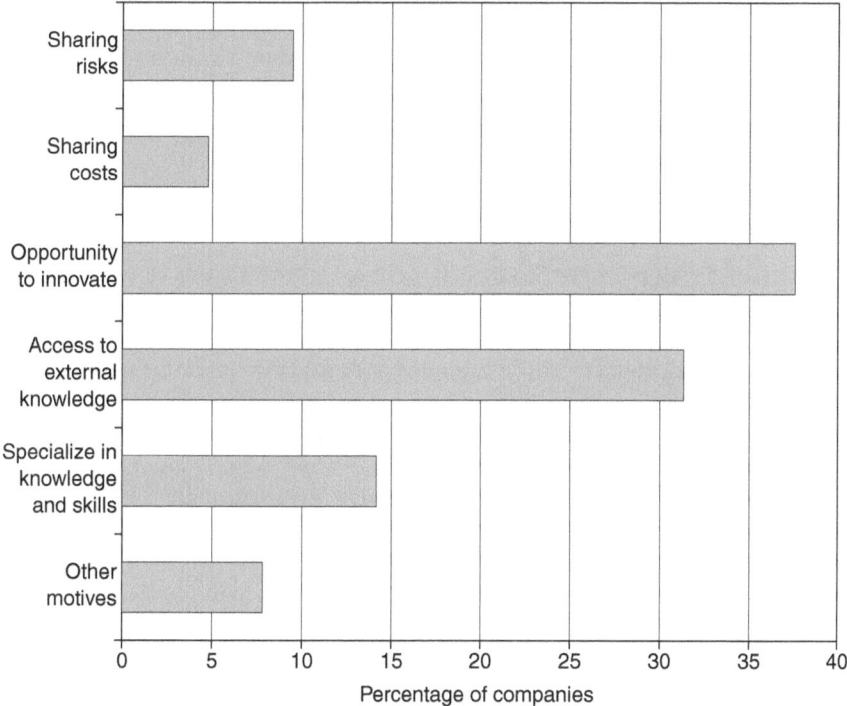

Figure 6.2 Primary motives for collaboration

The process

This section discusses some of the process characteristics of the collaboration in the clusters. Figure 6.3 lists seven such characteristics and shows the degree to which they were present in the various clusters. Access to the knowledge base of the partners, for example, indicates the degree to which the various partners in a cluster were able to tap into each other's knowledge. Absorptive capacity shows the degree to which clusters partners were able to actually use the knowledge of their partners. If innovation is to occur, both should be high. Participation openness shows the degree to which the cluster partners were encouraged to "speak their minds." Communication along formal lines shows the degree to which communication in the clusters was formal and structured. The underlying idea is that although a mixture of formal and informal communication is most conducive to innovation, formal communication is easier to measure. If cluster partners indicate that the level of formal communication in their cluster was low, we assume the level of informal communication to be high, since communication had to take place in one form or another. The balance of power shows a degree to which the clusters partners collaborated as equal partners or if there was a clear leader in a cluster. Mutual dependency shows the degree to which the cluster partners reported that they were dependent on each other to successfully complete their projects. Opportunistic behavior shows the degree into which cluster partners behaved opportunistically rather than in their mutual interest. For both mutual dependency and opportunism, we assume the level of trust among cluster partners to be high if these two process characteristics are low. In the remainder of this section these seven process characteristics are used to shed light on how the cluster partners collaborated with one another in the various clusters.

Use of external knowledge

In order to develop a new product in a cluster, companies have to share the knowledge that each has with their cluster partners. That is, cluster partners need to have access to each other's knowledge bases (Owen-Smith and Powell 2004). In the case of the Stimulus clusters, this condition was met. In 15 out of the 25 clusters in this evaluation, access to the knowledge base of the partners was very high or high (see Figure 6.3). Furthermore, it is important to know to what extent companies have actually used the knowledge of their partners. That is, the absorptive capacity of the clusters must have been high in order for them to actually to develop new products. This condition, too, was met, as 20 of the 25 clusters reported that their absorptive capacity (Cohen and Levinthal 1989) was high or very high (see Figure 6.3).

Form of communication

Important as these two indicators may be, they say little about the quality of the communication within a cluster. One indicator for this is whether the

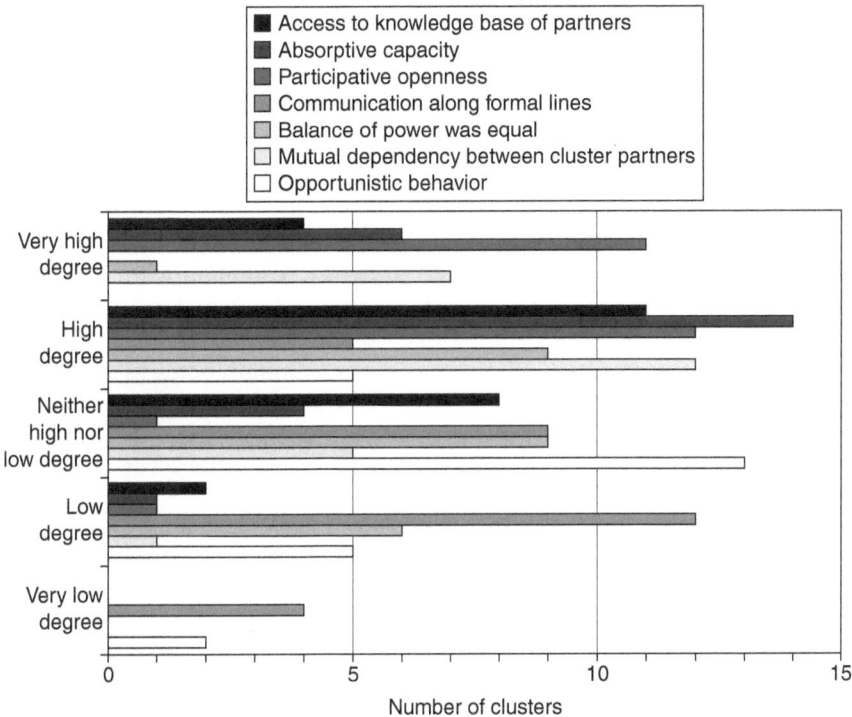

Figure 6.3 Process characteristics of the collaboration in the clusters

communication was open and participative – that is, whether it was customary for the cluster partners to speak freely and to be open and honest toward each other in their communication. The more this is the case, the better the quality of the communication and, consequently, the better the chance of the cluster working productively on product development (Rutten 2003). In the case of the Stimulus clusters this condition, too, was met. In 23 of the 25 clusters, companies reported that communication was open and participative to a high or very high degree (see Figure 6.3).

Structure of communication

Another indicator that says something about the quality of the communication in the clusters is the degree to which a mixture of formal and informal means of communication was used. Although a certain amount of formal communication is necessary, and even helpful, as, for example, progress has to be reported formally to superiors, applications for releasing funds have to be submitted formally, and (dis)agreements between companies have to go through formal channels, too, too much emphasis on formal mechanisms of communication indicates a lack of openness and trust within the cluster (cf. Burns and Stalker 1961; Butler *et al*. 1998). Moreover, formal channels of communication are not ideal for transmitting tacit knowledge, which is a crucial element in product development (cf. Nonaka and Takeuchi 1995; Owen-Smith and Powell 2004). A mixture of formal and informal mechanisms of communication is thus necessary. The Stimulus clusters met this condition, as nine clusters reported that the degree to which communication followed formal lines was neither high nor low. This suggests that these clusters had a balance between formal and informal means of communication. In the remaining 16 clusters the degree to which communication followed formal lines was low or very low, which suggests that informal means of communication were most important in these clusters. Taken together, this means that a mixture of formal and informal means of communication was used, with a tendency towards the latter, in the Stimulus clusters (see Figure 6.3).

Management of the cluster

The next indicator to be discussed here is the balance of power in the clusters. This has to do with the management of the cluster. The assumption is that power should be distributed more or less equally among the cluster partners in order to avoid a hierarchical mode of managing the cluster. According to the literature, this is unfavorable for the process of communication and, therefore, knowledge creation in a cluster (cf. Johannessen *et al*. 1997; Nonaka and Takeuchi 1995; Rutten 2003; Uzzi 1997). A look at the balance of power in the 25 Stimulus clusters in this evaluation reveals a mixed picture. Six of the clusters reported that the degree to which the power balance in their cluster was equal, was only low. Nine clusters reported that the degree to which the power balance in their cluster was equal was neither high nor low. The remaining ten clusters reported that the degree to which the power balance in their cluster was equal was high or very high. So, on the level of the cluster scheme, the degree to which power was balanced equally among cluster partners was only moderate, with a tendency toward high – that is, toward an equal distribution of power. Another indicator that has to do with the management of the clusters is the use of formal and informal control mechanisms. Parallel to the use of formal and informal means of communication, a mixture of formal and informal control mechanisms should be used to manage a cluster. One of the key reasons for this is that professionals, such as the engineers collaborating on product development in the clusters, are managed and motivated best through informal control mechanisms. On the other hand, formal control mechanisms are necessary to give structure to the clusters. This insight is at least as old as the classic work of Burns and Stalker on the management of innovation (1961) and has been reaffirmed by a variety of other authors since (e.g., Butler *et al*. 1998; Johannessen *et al*. 1997; Nonaka and Takeuchi 1995; Florida 2002). The evaluation shows that the use of control mechanisms meets the criterion that a mixture of formal and informal mechanisms is necessary. Only two of the clusters used more formal than informal control mechanisms. In 13 clusters as many formal as informal control mechanisms were used. The remaining ten clusters used more informal than formal control mechanisms. On the level of the cluster scheme this means that a mixture of formal and informal control mechanisms was used, with a tendency toward more use of informal control mechanisms.

Trust

With regard to collaboration in clusters, it is necessary to look at the level of trust between partners. In this chapter, trust is viewed somewhat narrowly in terms of the absence of risk and the absence of opportunistic behavior. Trust, of course, involves much more than that (cf. Granovetter 1985; Field 2003). When mutual dependency between partners is high, this is a strong incentive not to behave opportunistically and to assume an open and honest attitude towards one's partners (Uzzi 1997). Only in that way is it possible to achieve a constructive and fruitful mode of collaboration, which, in turn, is the best guarantee of a result in terms of new product development. Of the 25 clusters in the evaluation, 19 indicated that the mutual dependence between the cluster partners was high or very high. This indicates that on the level of the cluster program, this mutual dependency criterion was met (see Figure 6.3). When we looked at actual opportunistic behavior in the clusters, the evaluation showed that in five clusters the degree of opportunistic behavior was high. In a further 13 clusters, the degree of opportunistic behavior was neither high nor low. In the seven remaining clusters the degree of opportunistic behavior was low or very low. This means that, on the level of the cluster scheme, the degree of opportunistic behavior in the clusters was moderate, whereas, according to the evaluation framework it should have been low (see Figure 6.3).

Intensity of the communication

A final and very important indicator is the actual intensity of the communication between the cluster partners. It should be high, as only through intensive communication can the kind of (tacit) knowledge be exchanged that is necessary for new product development. Furthermore, a variety of modes of communication should be employed; communication should depend not only on electronic forms of communication but also on face-to-face communication. Electronic means of communication provide speed; face-to-face communication provides richness of communication (e.g., Johannessen *et al.* 2001; Nonaka and Takeuchi 1995). Three different phases were distinguished in each cluster project: the start-up phase, the implementation phase, and the finalization phase. The data below concern the implementation phase only, as this is the phase during which the actual new product development took place – that is, where knowledge was exchanged between the cluster partners and new knowledge was created. Of course, to distinguish between what is frequent and what is infrequent communication is purely arbitrarily. The respondents in this evaluation had a choice between six different categories to rank the intensity of their communication: daily, weekly, every other week, monthly, less than monthly, and never. In this chapter, the first three categories are considered to be examples of frequent communication. Looking at the different modes of communication, the following picture emerges. Of the 25 clusters, 16 used email frequently as a mode of communication. The telephone was used frequently in 23 clusters. The fax seems to be nearing its use-before date; only 13 clusters used it frequently. Face-to-face communication seems to remain a popular means of communication, even in the digital age. Considering that people had to travel in order to communicate face to face, the fact that 20 clusters frequently used this mode of communication is testimony of its importance with regard to knowledge creation. Group meetings were used least; in only 12 clusters did people frequently use this mode for their communication. Given that group meetings require prior organization, this outcome is perhaps not very surprising (see Figure 6.4). In sum, the data seem to allow the conclusion that communication within the clusters was, indeed, frequent and that, on the level of the cluster scheme, this condition was met.

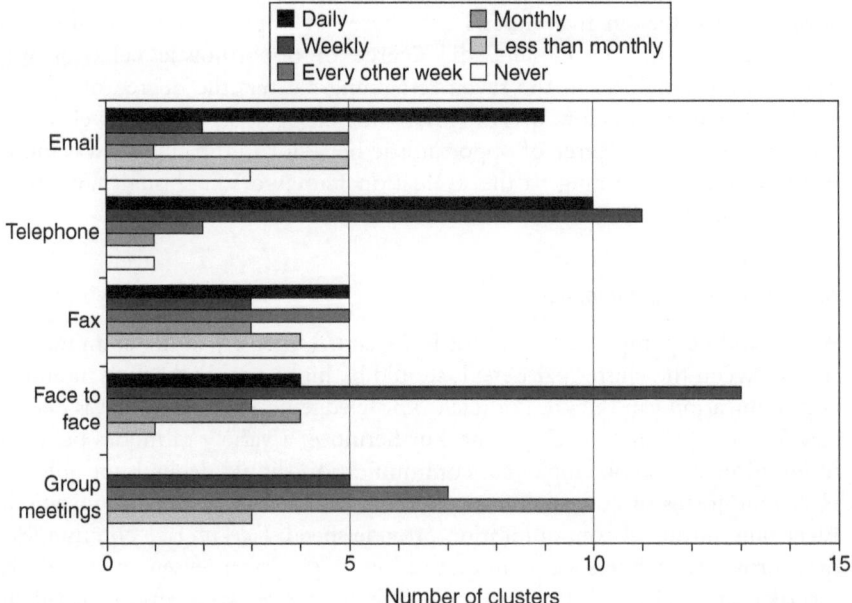

Figure 6.4 Modes and frequency of communication in the implementation phase

Outcomes

The outcomes give an indication as to how successful the cluster scheme was in achieving its goals. In the first place the purpose of the cluster scheme was to have clusters of companies develop new products and sell them on the market. The second goal, which directly followed from the first one, was that companies improve their performance on the basis of the new products. The third goal was to raise companies' awareness that implementing organizational changes in their way of working may enhance their performance.

Economic effects

With regard to the immediate effects of the cluster scheme, 16 clusters reported that they had successfully developed a new product and introduced it on the market. This equals nearly two-thirds of all clusters in this evaluation, which is hardly a poor result. Only two clusters did not successfully conclude their project. In the case of the remaining clusters, conflicting answers from various clusters partners made it possible to draw proper conclusions (see Table 6.4).

Generally, the introduction of the new products on the market has been beneficial to the companies in the clusters. In 18 clusters, companies saw an increase in their turnover one year after the market introduction of their new product. In subsequent years this number dropped; however, this fall was largely caused by the fact that several clusters had only recently concluded their work and were thus unable to give a long-term figure on the change in turnover. The most important outcome with regard to the change in turnover is that turnover increased in the majority of the clusters that did report on turnover change (see Table 6.5).

Another important effect of the cluster scheme was the change in company competitiveness. Again looking at the results per cluster, the effect of the cluster program on this point speaks for itself. Eight clusters reported that the competitiveness of the companies in these clusters had increased to a very high degree, and a further 12 clusters reported an increase to a high degree. The

Table 6.4 Selected outcomes

	New product introduced on market	Permanent collaboration with cluster partners	Permanent collaboration with non-cluster partners	Implementation of organizational changes as result of cluster project
(Unit of observation)	*(Cluster)*	*(Companies)*	*(Companies)*	*(Companies)*
Yes	16	50	42	26
No	2	13	20	37
No answer	6	2	0	0
N valid	24	65	62	63

Table 6.5 Increase of turnover from the new product

Increase of turnover as a percentage[a]	After 1 year	After 2 years	After 3 years	After 4 years
0%	4	2	2	2
0–10%	7	6	5	6
10–20%	5	8	4	1
>20%	6	2	2	4
N valid	22	18	13	13

Note
a Average for all companies in a cluster.

remaining clusters reported that the increase in competitiveness was neither high nor low (see Figure 6.5). With regard to company competences, the aim of the cluster scheme was that these should improve as a result of the collaboration in the cluster. For example, companies should become more skilled in the process of collaborating and in the process of developing new products. The results per cluster show that the cluster scheme was helpful in this respect, though not to the same degree as in the case of competitiveness. Still, the majority of the clusters, 16, reported that the competences of the companies in these clusters improved to a high degree. A further eight clusters reported that the improvement of competences was neither high nor low. Only one cluster reported a low increase in competences (see Figure 6.5).

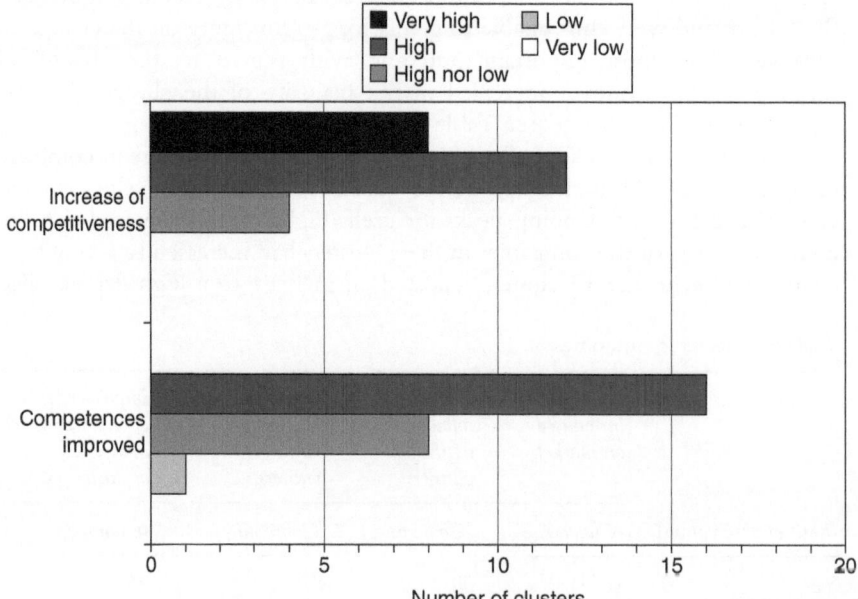

Figure 6.5 Changes in the competitiveness and competences of the clusters as a result of the collaboration

Technological effects

Working on product development in the clusters may trigger awareness on the part of companies that sustained investment in knowledge creation is necessary in order to develop new products in the future as well. An indicator of this awareness is the change in a company's investments in R&D. These investments should increase if companies have become more aware of the need to continuously develop new products. However, if we look at the various clusters in this evaluation, no such general awareness seems to have materialized. One year after completion of the project, ten of the clusters reported that the companies participating in them had increased their R&D spending, while nine reported a decrease. In two clusters no change in R&D spending was observed after one year. In the longer term, after four years the same pattern can be observed, although the number of reporting clusters is lower as not all clusters had ended four or more years ago. After four years, six clusters reported an increase in R&D spending, five clusters reported a decrease, and one cluster reported no change in R&D spending (see Table 6.6). Consequently, on the level of the cluster program – that is, looking at the average for all clusters – a change in R&D spending cannot be observed. On the whole, the cluster scheme did not contribute to an increase of sustained attention for product development.

Learning effects

The final three effects that are discussed here concern learning on the part of the companies that participated in the cluster scheme. One learning effect is the establishment of permanent networks with cluster partners. Increasingly, new product development requires inter-firm collaboration (cf. Grabher 2002; Rutten 2003). If a company establishes more permanent inter-firm relations with its clusters partners, it signals an organizational change from a situation in which temporary collaboration to a situation of a durable inter-firm relationship has been established. Such durable relationships are better "vehicles" for the exchange of tacit knowledge between firms (cf. Uzzi 1997). In all, 65 companies that participated in the clusters reported back on

Table 6.6 Change in R&D investments

Change in R&D investment per cluster[a]	After 1 year	After 2 years	After 3 years	After 4 years
Increase	10	10	8	6
No change	2	2	1	1
Decrease	9	3	4	5
N valid	21	15	13	12

Note
a Average for all companies in a cluster.

this issue. Of those companies, 50 said that they did establish permanent relationships with one or more of their cluster partners, whereas 13 said that they did not establish such links (see Table 6.4). A second learning effect concerns establishing permanent relationships with non-cluster partners. This would indicate that, because of the cluster project, companies have become aware of the need for inter-firm collaboration to such a degree that they started looking for partners in a wider field. Of the 62 companies that reported back on this issue, 42 said that they did establish permanent collaborations with non-cluster partners, whereas 20 reported that they did not do so (see Table 6.4). This means that the cluster scheme did contribute to establishing more durable inter-firm linkages.

A final learning effect concerns the implementation of organizational changes by the companies that participated in the clusters. This would indicate that, as a result of their product development efforts, companies have reflected on their way of working and implemented organizational changes to improve on it. This proved to be the case among a minority of the companies only. Of the 63 companies that reported back on this issue, 26 said that they had implemented organizational changes, whereas 37 said they had not (see Table 6.4). This means that, on the level of the cluster scheme, the conclusion must be that it did not result in organizational changes being implemented in the participating companies.

Limitations

Before we proceed to the conclusion section, it is important to point out some weaknesses of this evaluation study. In the first place, the sample of 25 clusters is small and limited to one region only. Although this probably does not affect the evaluation of the cluster scheme as such, since our sample is representative of the cluster scheme as a whole, it does compromise our ability to draw conclusions beyond this particular cluster scheme. A second limitation is that, for a number of clusters, the time between the end of the cluster and the evaluation was too short for economic effects to materialize. Third, since it was our goal to evaluate the cluster scheme, we have, of necessity, focused on the broad picture. Therefore, many relevant findings on points of detail have been obscured. To explore the wealth of details that our evaluation yielded would require further analyses in the form of surveys and case studies that focus on narrower research questions.

Conclusion

This section presents the actual evaluation of the cluster scheme. That is, it assesses how far the empirical observation for each of the indicators from Table 6.1 meets the criterion that was set for that particular indicator. Both an outcome and a process evaluation will be presented below. A summary of the results of the above descriptive empirical is presented in Table 6.7.

Table 6.7 Results of the evaluation

Indicator	Criterion	Observation
Conditions		
Previous experience	Favorable	Favorable
Kinds of motives	Not purely opportunistic	Not purely opportunistic
Supplementary knowledge base	Yes	Yes
Size of partners	Mixture of large and small	Only some clusters mixed
Process		
Access to knowledge base partners	High	High
Absorptive capacity of companies	High	High
Frequency of communication	High	High
Participative openness	High	High
Formal and informal communication	Mixture	Mixture, tends to informal
Perceived balance of power	Equal	Mixture, tends to equal
Social and formal control mechanisms	Mixture	Mixture, tends to social
Mutual dependency between partner	High	High
Opportunistic behavior	Low	Moderate
Outcomes		
Introduction of new product	Yes	Yes
Change of turnover	Increase	Increase
Change of competitiveness	Increase	Increase
Change of knowledge intensity	Increase	No change
Change of competences	Improved competences	Improved competences
Permanent networks, cluster partners	Yes	Yes
Permanent networks, non-cluster partners	Yes	Yes
Implementation of organizational changes	Yes	Yes

The outcome evaluation looks at the degree to which the results of the cluster scheme correspond to its aims. In Table 6.7 the aims are summarized in the column marked "criterion," whereas the results are in the "observation" column. The bottom row of Table 6.7 concerns the outcome variables. A comparison between the criterions and observations shows that, with the exception of two indicators, the cluster scheme achieved its objectives. This means that the cluster scheme must be considered a success. As the above descriptive empirical part showed, not all clusters and companies in the cluster scheme shared in the scheme's success to an equal degree. However, on nearly every indicator a clear majority of clusters or companies did report favorable results. That, in turn, allows us to conclude that the cluster scheme was a success. The money spent on the cluster scheme did, in fact, contribute to strengthening the performance of the participating companies and, indirectly, to strengthening the economy of the Eindhoven region. Therefore, this cluster scheme merits the attention of other regions trying to strengthen their economy by means of innovation networks. As this has evaluation shown, the Eindhoven region cluster scheme stands out as an example of good practice.

It is interesting to note that the cluster scheme was very successful with regard to the technological outcomes, in that companies did develop new products and acquired new knowledge and skills. However, the economic outcomes, such as market success and increased investments in R&D, were less pronounced, or even absent. As we have argued, this may have been caused by the fact that it takes longer for such effects to materialize than the time between the end of several of the clusters and the evaluation allowed for. But there is another possibility. Prior to receiving funding, each cluster had to be approved by Stimulus. It is conceivable that clusters made up of companies that were innovative already had a higher chance of getting approved because these companies knew how to make a good proposal for a cluster project. If this is the case, innovative companies would be overrepresented in our sample – that is, our sample would be biased. That would explain why the results on some outcome indicators were only moderate, as it is difficult to improve a performance that is good already. Whether our sample was indeed biased and how this may have affected the outcomes of our study is a question for further research. It must be stressed, though, that sample bias need not be a problem. After all, the question is not so much *whether* the cluster scheme facilitates innovation but *how* it does so. That is, the mechanisms are of primary concern. As one is not very likely to observe such mechanisms from non-innovative companies, having many innovative companies in one's sample may actually be an advantage.

The second evaluation, the process evaluation, shows a similar result. On all but a few indicators the cluster scheme met the conditions that were set in the framework for evaluation. This result gives a certain degree of empirical support for the conclusion that the causal relations between "conditions," "process," and "outcomes," as assumed at the beginning of this chapter, were correct. On the basis of the literature, the assumption was made that:

• The outcomes are the result of a process of inter-firm knowledge creation. That is, if the process of inter-firm knowledge creation has certain characteristics, as laid down in the criteria, the outcomes of this process should be more favorable. Moreover,
• the process of inter-firm knowledge creation takes place under certain organizational conditions. The more these conditions meet the criteria as derived from theory, the more productive the process of inter-firm knowledge creation should be.

As it turns out, these theoretical patterns are also observed in the empirical results of the evaluation study. In general, the organizational conditions as observed in the cluster scheme met the criteria. Put differently, the conditions were favorable. Theoretically, therefore, one would expect "process" to meet the criteria as well. As the empirical results showed, this turned out to be the case. As, in theory, a favorable process should lead to favorable outcomes, one would expect to find these favorable outcomes. Indeed, the empirical results showed that the outcomes were favorable. The empirical pattern thus matched the theoretical pattern – that is, the assumed causal relationship between the variables.

Of course, this does not prove that the observed outcomes are actually caused by the processes and conditions. This particular research design does not allow for such conclusions. With regard to causality, all that can be said is that the empirical findings corroborate the theoretical assumptions. Although the significance of this outcome is of limited value with regard to theory development in general, it is very important within the context of this particular study. Because of the fact that the empirical findings corroborate the theoretical assumptions, the outcome evaluation is more robust than if the opposite were true. Moreover, given the fact that the data collection followed a retrospective longitudinal design, the data can be used to make causal inferences and to contribute to theory development.

References

Azouz, A., Van Baarle, S., Blaauw, E., Groneschild, J., and Molnar, M. (2005) In de ban van de clustering: Een evaluatieonderzoek naar de Stimulus Clusterregeling, Master's thesis, Tilburg University, Department of Organization Studies.

Baslé, M. (2006) Strengths and weaknesses of European Union policy evaluation methods: Ex post evaluation of Objective 2, 1994–99, *Regional Studies*, 40 (2), pp. 225–236.

Bell, G. (2005) Clusters, networks and firm innovativeness, *Strategic Management Journal*, 26 (3), pp. 287–295.

Burns, T. and Stalker, G. (1961) *The management of innovation*, Oxford: Oxford University Press.

Butler, R., Price, D., and Coates, P. (1998) Organizing for innovation: Loose or tight control?, *Long Range Planning*, 31 (5), pp. 775–782.

Cohen, W. and Levinthal, D. (1989) Innovation and learning: Two faces of R&D, *Economic Journal*, 99–397, pp. 569–596.

Cooke, P. and Morgan, K. (1998) *The associational economy: Firms, regions, and innovation*, Oxford: Oxford University Press.

Cooke, P., Clifton, N., and Olega, M. (2005) Social capital, firm embeddedness and regional development, *Regional Studies*, 39 (8), pp. 1065–1078.

Field, J. (2003) *Social capital*, London: Routledge.

Florida, R., (2002) *The rise of the creative class*, New York: Basis Books.

Freel, M. (2003) Sectoral patterns of small firm innovation, networking and proximity, *Research Policy*, 32 (5), pp. 751–770.

Grabher, G. (ed.) (2002) Special issue: Productions in projects: Economic geographies of temporary collaboration, *Regional Studies*, 36 (3).

Granovetter, M. (1985) Economic action and social structure: the problem of embeddedness, *American Journal of Sociology*, 91 (3), pp. 481–510.

Hobday, M. (2000) The project-based organisation: An ideal form for managing complex products and systems?, *Research Policy*, 29 (7/8), pp. 871–893.

Johannessen, J., Olaisen, J., and Olson, B. (2001) Mismanagement of tacit knowledge: The importance of tacit knowledge, the danger of information technology, and what to do about it, *International Journal of Information Management*, 21 (1), pp. 3–20.

Johannessen, J., Olsen, B., and Olaisen, J. (1997) Organizing for innovation, *Long Range Planning*, 30 (1), pp. 96–109.

Klein Woolthuis, R., Hillebrand, B., and Nooteboom, B. (2005) Trust, contract and relationship development, *Organization Studies*, 26 (6), pp. 813–840.

Landabaso, M. (2000) Innovation and regional development policy, in Boekema, F., Morgan, K., Bakkers, S., and Rutten, R. (eds) *Knowledge, innovation and economic growth: The theory and practice of learning regions*, Cheltenham, UK: Edward Elgar.

Larson, A. (1992) Network dyads in entrepreneurial settings: A study of the governance of exchange relationships, *Administrative Science Quarterly*, 37, pp. 76–104.

Nonaka, I. and Takeuchi, H. (1995) *The knowledge-creating company: How Japanese companies create the dynamics of innovation*, Oxford: Oxford University Press.

Owen-Smith, J. and Powell, W. (2004) Knowledge networks as channels and conduits: The effects of spillovers in the Boston biotechnology community, *Organization Science*, 15 (1), pp. 5–21.

Porter, M. (1998) Clusters and the new economics of competition, *Harvard Business Review*, 76 (6), pp. 77–90.

Powell, W. (1998) Learning from collaboration: Knowledge and networks in the biotechnology and pharmaceutical industries, *California Management Review*, 40 (3), pp. 228–240.

Rooks, G. and Oerlemans, L. (2005), South Africa: A rising star? Assessing the X-effectiveness of South Africa's national system of innovation, *European Planning Studies*, 13 (8), pp. 1205–1226.

Rutten, R. (2003) *Knowledge and innovation in regional industry: An entrepreneurial coalition*, London: Routledge.

Stimulus (1999) *Cluster regeling* (SCR), Stimulus: Eindhoven.

Sydow, J., Lindkvist, L., and DeFillipi, R. (2004) Project-based organizations, embeddedness and repositories of knowledge: Editorial, *Organization Studies*, 25 (9), pp. 1475–1488.

Uzzi, B. (1997) Social structure and competition in inter-firm networks: The paradox of embeddedness, *Administrative Science Quarterly*, 42 (1), pp. 35–67.

Williamson, O. (1993) Calculativeness, trust, and economic organization, *Journal of Law and Economics*, 86 (1), pp. 453–486.

7 Cluster emergence: a comparative study of two cases in North Jutland, Denmark

Dagmara Stoerring and Bent Dalum

Introduction

Understanding the mechanisms behind the emergence of industrial clusters is of great interest because it may help the emergence of new such clusters. In many countries and regions the political systems and associated actors are seeking best practice in order to copy the success of other, established industrial clusters. This has also been the case in North Jutland, Denmark, where a well-established cluster of wireless telecommunications, NorCOM, was – at least to some extent – the motivation for a cluster initiative in biomedical technology, Biomedico, in the same region. The NorCOM cluster was originally initiated by local companies. It was an industry-driven bottom-up process that can be tracked back to the 1960s, but later it developed further through university–industry interaction. The story of NorCOM illustrates the phenomenon of a high-tech cluster being able to emerge in a peripheral region.[1] At present the North Jutland region, with its approximately 500,000 inhabitants, still has a fairly large share of low-tech industries, such as primary industries, agriculture, food processing, fishery and building materials. The Biomedico initiative was started around 2000 by policy actors, encouraged by and basing their efforts on the world-leading results of the Department of Health Science and Technology of the regional university in North Jutland. It is a top-down initiative in the search for new industries that could supplement the existing NorCOM cluster in order to achieve a more diversified industrial structure. This search gained further momentum when the telecom sector experienced the general downturn of the ICT industries in 2000–2001.

In this chapter we investigate the cluster emergence process based on an analysis of NorCOM and Biomedico. In particular, we use two models on cluster emergence that have recently been presented in the literature: those of Feldman *et al.* (2005) and Carlsson (2006) respectively. These two models describe the process of cluster emergence in stages that are characterised by different mechanisms. In the NorCOM case we use these models to track its development whereas in the Biomedico case the models are used to identify at what stage of cluster development this initiative currently might be.

Furthermore, this theoretical framework makes it possible to compare the two cases, highlighting the differences in their development and finding the connections between them. Finally, we discuss some theoretical contributions about the emergence of clusters, including the role of policy.

We start by presenting the two models in the next section. This theoretical framework is then used to present and analyse the NorCOM cluster in the subsequent section and the Biomedico case in the section after that. There then follows a comparative analysis of the two cases, highlighting the main differences and NorCOM's role in the Biomedico case. Finally, we give our conclusions concerning this study.

Theoretical framework

In this section we present and discuss two models of cluster emergence that attempt to put the development of different cluster into a sequence: the first developed by Feldman *et al.* (2005) and the second, complementary to the first one, presented by Carlsson (2006). These models help us to compare and discuss the two clusters in case studies.

Entrepreneurial event by Feldman

Feldman *et al.*'s study on cluster emergence (2005) may be classified as part of the growing research on inherited organisational capabilities of entrepreneurship (Dahl 2003). It presents a model of cluster formation based on the entrepreneurial event that is illustrated by an interpretative history of emergence of high-technology clusters in biotechnology and information and communications technology (ICT) in the US Capitol region. The paper develops the argument that the location of entrepreneurs with the skills and opportunity to capitalise on an emerging technology determines to a significant extent where high-technology clusters emerge. Feldman *et al.* argue that the literature on regional economic development has ignored the role of the individual change-agent in the development of regional economies. The second argument is the limited understanding of how innovative clusters emerge, hold and transform regional economies. Most studies in the literature are devoted to the conditions of mature industrial clusters, ignoring cluster genesis, whereas it is important from a policy perspective how a cluster is started in a region that previously would not have been characterised as innovative. It is important to study the early attempts of entrepreneurs that lead to cluster development.

The model describes how entrepreneurs are often motivated to take the risk of bringing innovation to the market by a shock to the system of production or discontinuity. Thus, some initial change – whether a crisis or an opportunity – drives latent entrepreneurs to start companies.

Feldman *et al.* (2005) define innovative clusters as complex adaptive systems, where the external resources associated with clusters are developed over time: "Clusters include firms working in related or supporting technologies, and an infrastructure of institutions and social relationships that provide resources and promote the interests of the whole cluster" (Boschma 1999). They assume that cluster emergence is a process that goes through common stages. The model presents the evolution of a high-technology industrial cluster over time in three phases: emergent phase, self-organisation phase and the system maturation phase. The region is characterised by the interdependent relationships between entrepreneurs, government policy and the local environment (including social and commercial institutions, physical and human capital resources). The process of cluster emergence takes place within this framework.

In the first phase the region may have some attributes of an industrial cluster such as human capital or a prominent research university, but there are few start-ups and no venture capital activity in the area. The decisive incentive in the form of an exogenous shock such as employment downsizing, increasing procurement or technology transfer initiatives sparks the entrepreneurial activity, and a classic learning-by-doing process starts. This development is driven by individual activities of entrepreneurs, who start to form initial relationships with region's organisations and institutions.

The second phase is dominated by increased entrepreneurial activity, and self-organisation of the cluster takes place as communities of local entrepreneurs create public and private networks[2] aimed at building or improving relevant infrastructure: attracting physical (venture) and human capital to the area, expanding universities. The entrepreneurs act as engines and stimulators influencing the local environment. This phase is characterised by self-reinforcing feedback among entrepreneurs, institutions and resources, and, most importantly, the cluster characteristics are reinforced, together with the ability to cope with external shocks.

This leads to the third phase, in which the industry matures into a dynamic, self-sustaining and reinforcing system. The successes of the first start-ups generate opportunities for new start-ups; thus, serial entrepreneurs and second-generation start-ups occur, together with secondary industry contractors (related industries, buyer–supplier relationships), and policy makers join the process. The authors present the following events that can be observed at the third maturation stage:

- Incubators and other technology partnerships are created.
- Regional public-sector financing and grant-giving programmes are created.
- Mergers and acquisitions take place.
- Venture capital companies move to the area or open branch offices.
- Universities start offering new programmes for training personnel or entrepreneurship education, or building new branches.
- Policy changes as government joins the efforts to boost high-technology development in the region even more.

Feldman *et al.* describe the region studied (the Capitol area) as being without generally recognised conditions for high-technology development, and say that a set of unrelated events (employment downsizing and introduction of technology transfer policies) created the entrepreneurial opportunity. The nature of innovation makes it difficult to plan clusters; a long-term perspective on policy is needed. According to the model, local government policies tend to be implemented and effective in the later stages of cluster development. In their case study the governmental role was crucial in the legislative area that has fed into infrastructure development, training programmes and tax incentives.

Carlsson's triggering event

The second model of cluster emergence that is used in this chapter is that of Carlsson (2006), where the focus is on the prime mover and the event that sparks the entrepreneurial process. Thus, Carlsson's proposition may be treated as similar or complementary to the approach of Feldman *et al.* Carlsson adds some important additional assumptions, however, and his theory is deduced from a richer case study including examples not only of American clusters but also clusters in other countries such as Ireland, Israel and China. The main assumption in Carlsson's approach is a distinction between an industry cluster and regional agglomeration. A cluster is more limited geographically than regional agglomeration is, and regional agglomeration may contain more than one cluster, or even none.[3] This is illustrated by the examples of Route 128 and Silicon Valley as regions with more than one cluster and Research Triangle Park as a "regional agglomeration of research organisations" with no industry cluster as yet.

Carlsson also agrees that formation of an industry cluster is a very long process that spreads over several decades. He distinguishes only two phases in this process: the early stage, when a regional agglomeration of activity in a certain sector is formed in a rather gradual way; and the second phase, when a cluster emerges in more rapidly. The first phase can be initiated "spontaneously, even randomly or by chance ... or as a result of public policy intervention", and as the outcome, a regional agglomeration is formed that can differ sectorally and geographically. In his approach the second phase is initiated by "some triggering event coupled with an entrepreneurial spark". Like Feldman, he believes that clusters emerge when entrepreneurs react to a technological opportunity. He suggests that the rapid growth of companies can be achieved only when a second generation of start-up occurs, hence knowledge transfer and spillover are important. According to his approach, infrastructure is important in the initial stage. A certain critical mass of activity is supportive, but it is the entrepreneurial culture and spark that, together with competence to transform technological opportunity into commercial success, enables a cluster to emerge. Carlsson provides very different examples of triggering events, for example the diffusion of transistor technology from

Bell Labs to the scientific community via the establishment of Shockley Semiconductors in Silicon Valley (technology driven), and introduction of the single European market, which started a rapid growth in foreign direct investment in Ireland (policy driven).

Discussion of the two models

There is a large variety of triggering events, and one may question the method of identifying that any particular event triggered cluster development. Carlsson's and Feldman's assumption that the entrepreneurial spark should follow the triggering event is important but still leaves some space for interpretation. Another interesting issue is how the notion of triggering event can be used to identify clusters that have just entered the real stage of emergence, or, using another theorist's words, clusters that have just passed a "market test" (Porter 2000). Surely the longer the historical perspective, the easier it is to point to the triggering event. Can we say that a regional agglomeration does not have a cluster yet because we cannot observe a triggering event? In this case Feldman's model is clearer because it gives some characteristics of cluster maturation. Both Feldman and Carlsson emphasise that most clusters emerge when the second-generation start-ups occur. The case studies discussed by Carlsson also show that the triggering event is very often followed by financing, so maybe the rapid growth in venture capital can be used as an indication that the regional agglomeration has entered the second phase of cluster emergence.

The most important is that competence triggers the emergence of clusters. Both authors emphasise that geography matters, in the sense that entrepreneurship is a local activity:

> Many individuals have location inertia due to reasons such as family mobility constraints, location preferences, familiarity of the environment, the relatively higher costs associated with changing residence, or the high cost of establishing a new company in a thickly populated environment where office and housing costs tend to be higher.
>
> (Feldman and Francis 2003, p. 780)

Although the two models of cluster emergence presented above differ at some points, they may be treated as complementary. The question is whether it is possible to have a general theory for clustering. The variation of examples provided by the models might suggest that it is difficult to generalise, and this makes cluster policy difficult. However, a kind of theoretical framework based on the models may be helpful for analysing the NorCOM and Biomedico cases, observing the mechanisms behind emergence of these clusters, and distinguishing differences between them.

From the two models described above we can identify the following stages that a region has to go through in order to "produce" a cluster:

1   Building regional agglomeration when research or/and industrial compe-
     tences are built, so that a region has developed more competences, and
     thus is more specialised in a certain technology than other regions.
2   Starting entrepreneurial activity. The first start-ups occur together with
     further development (improvement) of regional infrastructure; regional
     agglomeration "thickens" (increases), and self-reinforcing processes
     strengthen through technology transfer mechanisms.
3   An actual cluster emerges: a triggering event in the form of an exogenous
     shock occurs and is followed by a rapid growth of start-ups (entrepre-
     neurial spark) or movement of the companies from outside the region.
4   Cluster maturation and further development takes place.

Analysis of the NorCOM cluster in North Jutland

This section provides a short history of the NorCOM cluster, looking at different phases of cluster emergence as they were identified in the previous
section.

First stage: building regional agglomeration of industrial competences, 1948–1980

This cluster may be traced back to 1948 when SP Radio[4] was founded,
which may be treated as a grandparent company of the cluster. SP Radio
managed to switch from producing consumer radios and televisions to producing professional maritime communication equipment for small ships in
the mid-1960s. This transformation may be correlated with incentives from
local demand conditions. North Jutland was an area with a large commercial
fishing fleet and a large fleet of small and medium-sized yachts. This segment
of the market was more or less neglected by large foreign companies, and SP
Radio quickly attained a leading position. In 1973, engineers from SP Radio
founded Dancom, also in maritime communications. Dancom experienced a
turbulent period in the 1970s. In 1977 it became the "seedbed" for the foundation of Shipmate[5] in the field of maritime communications and Dancall in
1983, which began in maritime communications but quickly entered mobile
communication terminals. During the 1980s, Dancall, SP Radio and Shipmate were the three biggest radio communication firms in the region, with
production as well as R&D taking place in North Jutland. Through their
presence the region had some attributes of industry specialising in maritime
communications.

During the mid-1960s the regional infrastructure started to be developed
when several technical knowledge institutions of relevance for the industry
were established in North Jutland – some as affiliates of the Technical University in Copenhagen, resulting in the establishing of Aalborg University
(AAU) in 1974. AAU had the potential to become a fairly powerful education and research institution[6] because there were from the very beginning

around 200 members of academic staff in the Technical Faculty. There was a good match between the few existing radiocommunication firms and the profile of the staff in electronic engineering. This point of departure made AAU visible on the national and international research scene from the early and mid-1980s onwards. During the 1960s and 1970s the direct research spillover effects from AAU (and its predecessors) to the local radiocommunications industry were not of great importance. There was mainly indirect knowledge transfer via newly graduated engineers.

Second stage: starting entrepreneurial activity and thickening of the regional agglomeration, 1980–1990

The 1980s witnessed significant changes: the regional infrastructure strengthened through the university and the very rapid development of the mobile communications industry. The foundation of Dancall and then Dancom coincided with the take-off of the first generation of terminals for the new Nordic Mobile Telephone system (NMT). The emergence of the pan-European Global System for Mobile Communication (GSM) required an innovative effort on a considerably larger scale than was the case with NMT, concerning infrastructure as well as terminals. That led Dancall and Cetelco in early 1988 to establish a joint venture, DC Development, with the aim of developing a second generation of GSM terminals. At the same time (1987–1989), the university, the regional county council and a major local bank had collaborated in establishing a science park, NOVI, at the university campus. Although these two efforts were independent, they became, incidentally, well timed. NOVI needed a high-profile prestige project to become recognised, and Dancall and Cetelco needed "neutral ground" near the university. At its peak in 1992 DC Development employed around 30 engineers.

Third stage: cluster emergence – fundamental change and take-off in the 1990s

During the 1990s the third stage of real cluster emergence started. The region emerged during the mid-late 1990s as a major development "hub" in mobile communications equipment, with eight to ten R&D firms or affiliates and a large manufacturer of mobile terminals, Flextronics.[7] It is difficult to decide what event triggered this process. In 1992 the DC Development joint venture was closed down, although it succeeded in developing the basic technology for a GSM terminal.[8] A knowledge asset, 30 young electronics engineers, had emerged, and provided the entrepreneurial spark. Thus, the closing of DC Development had the same shock, or triggering, effect in North Jutland as employment downsizing in the US Capitol region in the study by Feldman *et al.* (2005).

During the 1980s, Dancall had benefited from "patient" capital from Danish wage earners' pension funds – owing to its high-profile image as a fast-growing high-tech firm. In 1993, Dancall was acquired by the UK company Amstrad. During the period of Amstrad ownership, Dancall's staff increased in number from 200 employees in late 1993 to 650 in 1997, when the company was acquired by the German Robert Bosch Group. As Bosch Telecom Denmark the company experienced a very rapid growth and investment process during 1997–1998. A new manufacturing plant, including major R&D facilities, was built in late 1998 that had 1300 employees. In connection with Bosch's withdrawal from telecommunications, however, Bosch Telecom Denmark was split into two units and sold in 2000. The US electronics assembly specialist Flextronics acquired the production facilities and started with 1300 employees. The R&D and service departments with around 350 employees were taken over by Siemens.

There has been a steady emergence of small firms focusing on the development of mobile communication equipment for international firms, many of them founded by engineers leaving DC Development. Some were acquired during the second half of the 1990s by important international companies such as Texas Instruments and Motorola. During 1999 Ericsson and Nokia established R&D centres. This shows that at this stage foreign ownership had become widespread, and the cluster grew not only as a result of local start-ups but also because of the establishment of MNEs.[9] The establishment of the first private Danish GSM operator, Sonofon, in Aalborg in 1991 (with 750 employees in Aalborg at present) was also very important for the development of this cluster.

Looking at the regional infrastructure, we can observe the growing role of Aalborg University, which was able to deliver engineers and basic research with a sufficiently application-oriented focus. A technical university may thus play a rather direct role in the restructuring process of a region previously dominated by traditional industries. An asset of this region was the awareness of the importance of further developing the technological basis of the cluster. In 1992 the Danish Technical Research Council allocated a large five-year grant to a new Centre for Personal Communication (CPK); in 1997 this was extended for another five years. CPK involved around 40 researchers and included basic research in radio communications technology and speech recognition.

NOVI started to be a place hosting new bridging institutions, which may be seen as an example of self-organisation by a cluster. In 1997, NorCOM was established as a club for wireless communications firms. Since 2000, NorCOM has been transformed into a business association, with university researchers as invited members. Financial support schemes, such as various Danish national and EU programmes, have also been beneficial. These diversities of actions have gradually evolved into a rather coherent policy framework. In conclusion, in this stage of cluster emergence initiated by the triggering event of closing DC Development we can observe building and "thickening" of the regional institutions, as well as the creation of public and private networks by local entrepreneurs.

Fourth stage: cluster maturation since 2001 – echoing the turbulence of the global telecom sector

Since 2001 the telecommunications industry has experienced fairly hard conditions globally, and so, since 2001–2002, have the NorCOM firms. Several companies have been closed, e.g. Shima, Ericsson Aalborg and Telital. The most dramatic case has been the closing of Flextronics in 2004. Employment in this factory peaked at 1700 in 2002. The Flextronics production facilities have partially been taken over by the Danish company Orion, producing flat-screen TVs. Another group of companies have, however, experienced remarkable growth in employment, such as Texas Instruments, Danish Wireless Design, ETI, GateHouse, SpaceCom and TTPCom.

Cordless phone technology has also played a significant role in the cluster, with one company, RTX Research, having experienced rapid growth since 1993. It now employs 230 people. More recently, RTX has entered other fields of wireless communication. RTX Healthcare focuses entirely on developing world-class wireless connectivity products and solutions for medical and healthcare devices and may be treated as a player in the Biomedico initiative The "founding father" of the entire cluster, SP Radio, was taken over by a foreign firm in the 1990s, and in 2004 bought by the Danish firm Thrane & Thrane. This new company went through a very rapid growth process in the 1990s, having around 600 employees in 1998, and consolidated its position as one of the world leaders in its field.

These events show that the NorCOM cluster may have entered the fourth stage of cluster development, when it has to overcome external shocks. This may result in strengthening of the cluster's position, or alternatively in its disappearance. NorCOM's net employment has actually increased by at least 250 people in the past couple of years, leaving aside the closing of Flextronics. This period has seen dramatic changes, but the dynamism of the NorCOM cluster – considered as an international development hub – has been maintained and strengthened. Also, Aalborg University's research capacity has been strengthened through the establishment of CTIF (Center for TeleInfrastructur), supported by national and international companies and research funds. CTIF focuses on the convergence between wireless and wired technologies as a main research field for the future.

The Biomedico cluster initiative in North Jutland

In this section we analyse the Biomedico cluster initiative in order to identify which stage of cluster emergence the region might be in at present. We start by looking at the main competences in biomedical technology in the region and then we investigate whether there are already technology transfer processes in the region.

Cluster competences

During the past three years, several projects have been initiated to promote regional development in biomedical technology in North Jutland. These projects resulted in a political initiative to promote and develop in north Denmark a cluster within the life sciences that is also presented as a Biomedico cluster. The initiative was started in 2000 by the Aalborg Commercial Council[10] together with the Industrial Liaisons Office at Aalborg University, after which other actors – policy makers such as North Jutland County and Aalborg Community, and finally industry representatives – joined. This initiative was formalised in 2003, with the BioMed Community: Science and Innovation for the Living representing the main actors in the region.[11] The genesis of this initiative may be found in the fact that at the regional level, actors interested in local economic development were looking for new industries that could supplement the existing NorCOM cluster.

The main actors involved in the Biomedico initiative are as follows:

- Aalborg University (AAU) – research within the medico-technical area at the Centre for Sensory Motor Interaction (SMI) has developed new methods for stimulating and treating electrical signals from muscles, measuring and activating the motor system and locating pain. Moreover, the university has started a centre for research into stem cell technology to determine how stem cells may be used to develop human "spare parts". Another potential research field at AAU is biotechnology, and the cluster initiative actors also include nanotechnology as part of the cluster's competence.
- Aalborg Hospital, Århus University Hospital[12] – obtained university hospital status in 2003 as an affiliate of Århus University on the basis of its own research and its tradition of cooperation with Århus University and Aalborg University.[13] The cooperation with Aalborg University is formalised in the HEALTHnTECH Research Centre, established in 2003, which should also offer support and evaluation of product ideas and applications developed by the industry.
- Companies and industries in the region – the initiative actors identified in the Competences Catalogue number about 35 companies. The catalogue states that the agglomeration has not achieved the critical mass as far as the number of companies is concerned. The profiles of these companies, however, can be described as being in the field of biomedical technology only to a limited degree. On this list there is a company producing cosmetics (Beauté Pacifique), a very few companies developing medical devices and some IT companies; a considerable proportion are distributors and wholesalers of health care equipment. In employment terms the biggest companies are subsidiaries and production facilities for large firms from other parts of Denmark: Oticon, Novo Nordisk, Coloplast and Bang and Olufsen Medicom. If we exclude these companies from the list, the rest are

mostly very small development companies employing up to five people. Some of them are spin-offs from university research and should therefore rather be called development projects. Among these firms there are only three companies specialising in medical technology and employing more than five people (Judex, Neurodan, RTX Healthcare).

The publishing of promotional material, marketing, attracting new firms to the region and the promotion of new and established companies have been the main activities of Biomedico during its first three years. The so-called "Firms club" was established for companies from the whole of north Denmark and should enhance the cooperation between the region's companies, Aalborg University and Aalborg Hospital. The cluster initiative actors have mobilised some financial resources for consulting activities and the "Research House" initiative – a kind of incubator based next to the hospital.

Investigating technology transfer mechanisms in the region

The above description of the cluster initiative shows that there are attempts in the region to build regional infrastructure that according to the model is more typical of later stages of cluster development. The main element of a cluster, however, is missing, namely companies. There are many similar initiatives to "establish" clusters in different regions of the world. What may distinguish the Biomedico cluster initiative in North Jutland from many similar policy initiatives are the competences at Aalborg University's Department of Health Science and Technology. Recognition by politicians and university researchers of research carried out at this department was one of the reasons for starting this initiative. The few companies within the biomedical technology have their roots in the university. Thus, in order to investigate whether there are already technology transfer processes in the region that are, according to the model, typical of the second phase of cluster emergence, we have to take a closer look at university research. This section is based on interviews carried out at Aalborg University. In the interviews the main processes of technology transfer that, according to the literature, might lead to the emergence of an industrial cluster were investigated.

University research history and financing

The research at the oldest part of the department, the Centre for Sensory Motor Interaction (SMI), started in 1978, when a professor of medical technology was employed. This professor was the supervisor of the present leaders of SMI. The department, with SMI at its core but also other growing research groups, especially in medical informatics and decision support modelling, has been building its reputation since then on the basis of the key researchers, approximately four people. SMI started as the first group in its

area in Denmark and in many European countries, and has grown from
having eight to having around 70 researchers at the moment (Figure 7.1). In
1997 the group started an International Research School for the education of
international PhD students. At present it has about 45 PhD students.

In 1993 SMI was awarded a yearly grant of €1,500,000 from the Danish
Basic Research Fund, but this expires in 2006. This funding is considered as a
decisive factor in aiding SMI's development and the ability to attract other
finance.

University start-ups

The first university spin-off, Judex, was established 20 years ago by one of the
leading university researchers. The next spin-offs from the university research
related to medical technology started in the mid-1990s: Neurodan, Neuro-
con (bought by Neurodan) and Neurotrain, JNI Biomedical, and Index.
Neurodan is the most successful of these start-ups. It has recently been
bought by a big German concern, Otto Bock, specialising in orthopaedic
devices. According to the head of SMI, this acquisition should lead to:

> the movement of an important part of OB's R&D to Aalborg. OB's
> choice of Aalborg for their R&D is mainly due to the company's wish to
> get close to the research and the candidates from SMI and Department of
> Health Science and Technology. Furthermore, this will also strengthen
> the biomedical innovation in the area and in Denmark. OB would also
> like to strengthen the cooperation with SMI, Aalborg University, in the
> form of research projects, etc.

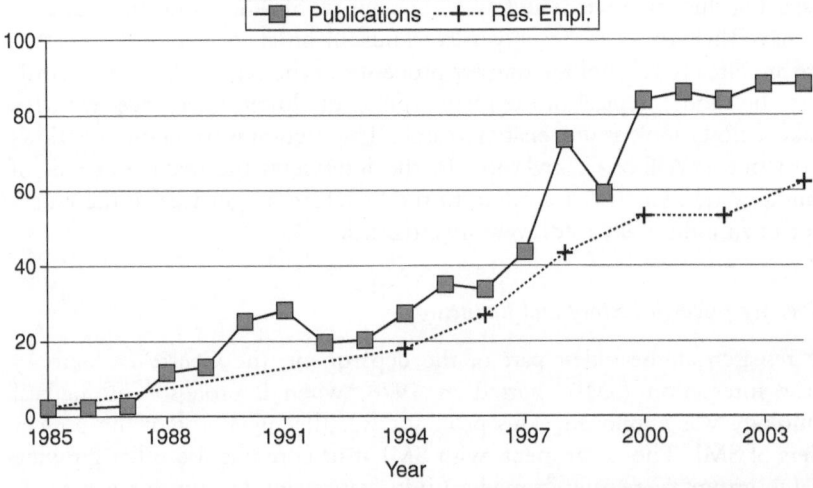

Figure 7.1 Number of researchers at Denmark's Department of Health Science and
Technology and number of publications during the period 1985–2004

We have also found a further five other university spin-offs, which are rather small start-ups established to hold rights to their patents. Most of these start-ups stem from SMI.

Cooperation partners

The research groups at health technology have many different kinds of cooperation types ranging from academic cooperation with other research institutions (universities involved in European projects), clinical contacts with institutions in the public sector (hospitals), and industry collaborations. This includes both national and international partners, the latter enhancing the international visibility and prestige of their research. As for the industry partners, there are names of the world's most important industry players: Siemens/Maquet, GE Healthcare, Novo Nordisk, Radiometer, etc. as well as a few small local companies: Judex, RTX Healthcare, Neurodan. For the SMI it is not easy to find Danish industrial collaboration partners that are interested in its technology, as it is a relatively new area. The Danish companies it works with are mostly small companies that are involved in financing PhDs, but an increasing number of Danish companies have expressed their interest in the SMI's research. Some of the NorCOM cluster companies are also interested in this area.

Patenting clinical research

In the period 1993–2004 the university filed 13 patents within the field of medical technology (Figure 7.2). Two patents were sold to a big international company, GE Healthcare. Both were the results of students' projects.

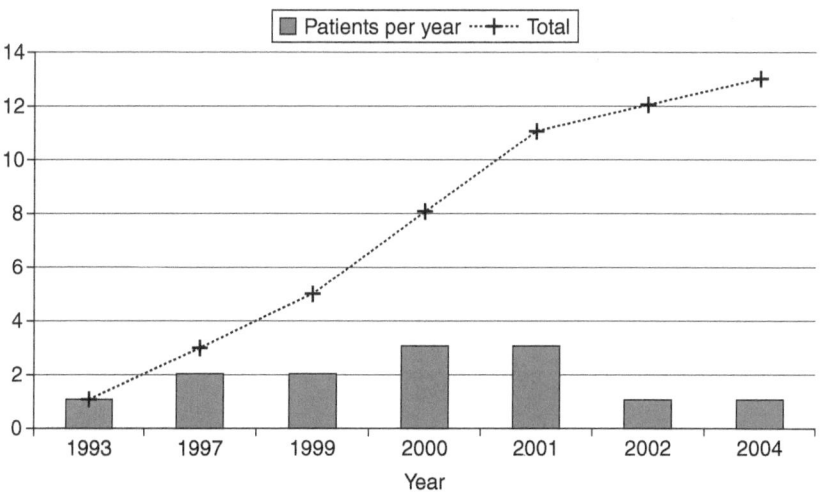

Figure 7.2 Patents at Denmark's Department of Health Science and Technology in the period 1993–2004

An important factor for a long-term development of the innovative part of the university's research is clinical support in the form of closer cooperation with hospitals and doctors. According to the university professors, they need clinical people who are more enthusiastic about technological possibilities. Most of the people interviewed, including the university authorities, believe that the situation might improve if Aalborg University had the right to educate medical doctors.[14]

Human capital: students/graduates and PhD students

The education of engineers in health technology (as a separate degree) started in 2000, with the first graduates qualifying in summer 2005. Before that, there were specialisations within Electrical Engineering: Master of Biomedical Engineering, Master of Medical Informatics, with five to 25 graduates per year. In the past two years, 40 to 50 students started a five-year education in Medical Technology (for a Master's degree in Biomedical Engineering and Informatics). According to the head of SMI, the engineers can be easily absorbed by the health sector in Denmark.

The majority of the more than 40 PhDs come from the International Research School at SMI and are of foreign origin. The university is not able to keep all its PhD graduates, so they very often move to other research institutions in Denmark and abroad.

Another important aspect is the possibility of finding a job in North Jutland. It is almost impossible to find employment in medical industry in the region, as there are very few companies and they are very small. Most of the graduates move to other regions of Denmark, such as Århus or Copenhagen. Those who stay in North Jutland mostly find employment in the NorCOM companies.

Entrepreneurial culture and spirit among researchers and students

There is an entrepreneurial spirit among researchers, as the examples of persons interviewed show (altogether they are behind approximately eight start-ups). Asked about the entrepreneurial spirit among the students, the professors are convinced that the students (especially the youngest generation) are very interested in technology applications but that it will take some time for them to become involved in start-ups. A semester spent at the hospital, which is a compulsory part of the new education in medical technology, has already resulted in many promising projects. The students are closer to the patients and the problems at the hospital.

Discussion of the Biomedico story

The interviews show that some of the technology transfer processes are already present in the region, which has reached a certain level of attractiveness as far as the university research and human capital are concerned. Comparing our findings to the model, we conclude that for the moment one might talk about a regional agglomeration of research competences within medical technology, thus implying that stage 1 or perhaps 2 of the model has been reached. However, this does not necessarily result in a cluster. A triggering event that will start the entrepreneurial spark is needed. Another possibility is the entry of companies from other regions that may be attracted by the university. The acquisition of Neurodan by a German company may be a sign of this process and might also turn out to be a triggering event. At the moment the region offers human capital in the form of university graduates. These students face problems with finding local employment, and starting their own companies may be one of the possibilities through which they can stay in the region, which they might want to do as according to the theory, geography matters in the sense that entrepreneurship is a local activity.[15] However, it is very uncertain how likely it is that a cluster will develop through the university graduates becoming successful entrepreneurs.[16]

The regional infrastructure in the form of the cluster organisation, public–private networks, is already present, though this might be perceived as a problem of this cluster, as the cluster initiative represents a typical top-down model where it is attempted to create a cluster as a matter of policy. This approach is characterised by the attempts to copy the success of the earlier cluster, as attracting big companies from outside is the aim of the branding activities. The belief in the strategy of "importing companies" from outside might be related to the important role of multinationals in the development of the NorCOM cluster. On the other hand, the politicians also appreciate the role of start-ups, as they have appointed a consultant to support small start-up companies. Finally, networking seems to be the main instrument, and a tradition of cooperation has been gained with the emergence of NorCOM. The policy makers coordinate all these activities, and it is difficult to see the industry part being an engine. It is rather that the companies are gathered by politicians.

Comparing NorCOM and Biomedico

The NorCOM story illustrates how an underdeveloped peripheral region has been transformed into a regional system specialising in wireless technologies. The story highlights the region's ability to adapt to shocks and accommodate to new demands. Furthermore, the case stresses the role of entrepreneurs who actively interacted with their local environment and adapted to new situations and crises, and were able to capitalise on new technological opportunities. The NorCOM story shows that coping with external shocks is the way to strengthen a cluster. The story of NorCOM emergence is also the story of building regional institutions. First of all, Aalborg University was created and grew to become an important technical university. The university has expanded its research competences to new areas that may be a source through which new clusters can develop. Our investigation shows that there may be another cluster emerging in the region within the medical technology area. However, potential clusters has very different conditions in which to emerge, as the region has changed in parallel with the process of NorCOM's development.

Theoretical model

We have tried to analyse development of the two clusters according to the same theoretical model of cluster emergence. We have seen that it is not easy, as the clusters are at different stages of development. NorCOM is an older cluster that has entered the last stage of cluster maturation, and this cluster has clearly defined competences: expertise on wireless data transmission. With Biomedico we might be in one of the preliminary stages of cluster emergence. The first problem with Biomedico is defining what is the core competence. Here we can learn from NorCOM that it takes time to define the core competence of a cluster and that new technologies are mostly created at the interfaces between different disciplines.

Cluster genesis

The two clusters have a different starting point. In the NorCOM case the entrepreneurs were the leading mechanism – it was a spontaneous process initiated by the companies. In Biomedico we have politicians trying to create a cluster. On the other hand, looking at the history of biomedical research in the region, we can say that Biomedico has a starting point at the university. In the case of NorCOM the university was established after initial cluster development. For Biomedico the university and its competences matter in the sense that entrepreneurship is a local activity. The region, with its strong research competences, might attract firms from other regions wishing to take advantage of the university researchers, who are not so easy to persuade to move to other regions. Moreover, it is not only the presence of the university but the existence of institutions at the university that should facilitate technology transfer (the patent office, the liaison office). This makes the circumstances of emergence of the Biomedico cluster easier than in the case of NorCOM. Finally, looking at the genesis of the two clusters, we have to note that today the North Jutland region has a different national and international reputation as compared with 30 years ago – that is, it is known for being capable of developing a high-tech cluster. This means that it might be easier to brand North Jutland as a high-tech region for the Biomedico initiative.

Technology implications

The difference in technology is another reason why it is not easy to compare the two cases. First, biomedical technology is an example of a science-push story whereas the NorCOM case was more industry driven. Although the use–producer interaction plays an important role in mobile communications technology, in the NorCOM story there was no end user–producer interaction in the region; rather, Nordic consumers in the 1980s were decisive. Only in the very early phase did NorCOM have lead users in the region: the fishing fleet and sailors for SP Radio. On the regional level a vertical user–producer interaction within a cluster played a more important role. Cluster development is characterised by the building of vertical relations and user–producer interaction within a cluster. In Biomedico this kind of development might be starting with the specific product of Neurodan when they get a regional subcontractor (e.g. Judex). But the medical technology is still more a science push in relation to the end user. The public sector is important, as the products must be developed in close interaction with clinical trials. The Research House initiative, which should be a kind of incubator near the hospital, might bring the companies closer to the patients and improve user–producer interaction.

Second, the two fields have different product development periods. In the case of biomedical technology, product development takes place over the very long term (up to ten years, as the example of Neurodan shows), whereas the products in mobile telecommunications change very quickly.

Third, patenting plays a very important role in biomedical technology. By contrast, patenting did not play a big role in NorCOM until recently.[17]

Labour supply

Looking at the differences between the two cases, we have to emphasise the different supply of labour. In the NorCOM story it was the shortage of the engineers that drove the university development whereas in the Biomedico case we have an "excess supply" of university graduates vis-à-vis the regional labour market, and the problem is the absence of the companies that could take advantage of it.

NorCOM's role for Biomedico

We can see that these clusters cannot be treated separately. The NorCOM cluster was very important in transforming the region into one with a more research-driven profile. The regional agglomeration of research competences in medical technology was created together with the emergence of the NorCOM cluster. Even from the policy point of view, NorCOM was an inspiration for starting the political initiative to support emergence of another cluster. The psychological spillover of NorCOM's success was one of the factors leading to the starting of the cluster initiative in biomedical technology. One may also define it as social capital. It is the presence of social capital in the form of the tradition of cooperation between different regional actors that has been gained during the decades together with NorCOM's emergence that has an important implication for Biomedico. The regional actors have noticed the regional agglomeration of research competences quite early and have started initiatives that may help the industrial cluster to emerge.

Conclusions and further research

In this chapter we have discussed the cluster emergence process. Using two recent models on cluster emergence by Feldman *et al.* and Carlsson respectively, we have studied and compared two cases of different clusters in the same region. From the theory we identified four common stages of cluster emergence that a region goes through. First, a region specialises in research or a certain type of industry, building regional agglomeration of research or industry competences. In the second stage we observe the start of entrepreneurial activity with further development of regional infrastructure, self-reinforcing processes that strengthen through technology transfer mechanisms. The third stage, when a real cluster emerges, is initiated by a triggering event that determines a sudden increase in the number of companies by a rapid growth of start-ups or by movement of the companies from other regions. Finally, in the fourth stage we have cluster maturation and further development.

We have analysed and compared an established cluster of wireless telecommunications, NorCOM, with a Biomedico cluster, or rather a cluster initiative in biomedical technology in the same region. The older NorCOM cluster matches fairly closely the model of cluster development, as in the first stage a regional agglomeration of industry competences had been developed and then a triggering event – the closing of a research joint venture – led to a rapid increase in the number of companies in the 1990s. At present the NorCOM cluster is in the phase of cluster maturation.

In the case of the much younger Biomedico cluster initiative there is considerable political motivation to build a cluster. This is inspired by the success of the NorCOM cluster; thus, this is different from what the model suggests will bring about cluster development. The political action in the region has already developed a regional infrastructure of cluster organisation involving public and private networks, which is usually more characteristic of later stages of cluster development. However, the very small number of companies in the area indicates that cluster development is in its preliminary stages. We have analysed in detail the region's main competences in biomedical technology, which are to be found at the university. We investigated the mechanisms of technology transfer that occur in the second stage of the model such as university start-ups, cooperation, financing, patenting, human capital, an entrepreneurial spirit, etc. This analysis leads us to the conclusion that the Biomedico cluster initiative is rather a regional agglomeration of research competences and it may already have started the second stage of the model, but the number of companies is still very small, and a triggering event that will start the entrepreneurial spark is needed. According to the model, another possibility is the entry by companies from other regions that may be attracted by the university. The acquisition of Neurodan by a German company may be a sign of this process and might also turn out to be a triggering event.

It is difficult to compare a cluster that has developed over a period of 40 years with a cluster initiative that is in a very preliminary stage. Additionally, we have to bear in mind different technologies, as in case of the NorCOM we have an industry-driven cluster whereas medical technology is a typical science-push story. However, comparison of the two clusters leads to some conclusions.

First, we cannot treat these clusters separately. The NorCOM cluster was very important in giving the region a more research-driven profile. The regional agglomeration of research competences in medical technology (or even in IT and medico-related activities) was created together with the emergence of the NorCOM cluster. We have observed that some of the NorCOM companies are interested in medical technology research at Aalborg University. There are potential synergies between the two fields and these should be investigated more closely in future research.

Second, the study shows that the regional phenomenon of North Jutland is very closely related to Aalborg University, which was established during the first stage of NorCOM's emergence and became a crucial element of the cluster, and is a starting point of the Biomedico cluster initiative. This may be a lesson for other regions that try to accelerate regional development: one way of strengthening economic development in the periphery region is through a growing university gaining an international reputation. The new engineering and high-tech profile of the region is to a large extent related to the presence of Aalborg University.

On a more general level our study shows the problems of analysing the cluster emergence process, and in particular those of analysing a potential cluster that is still in its infancy. This is probably also the reason that there are rather few studies in the literature on the very beginning of clusters. We mostly find studies of fully functioning systems that represent further stages of cluster development, which policy makers then attempt to transfer to other regions. Our study shows also the difficult role of policy in this process. The nature of innovation makes it difficult to plan clusters. According to the models discussed, local government policies tend to be implemented and effective in later stages of cluster development. In the case studies presented by Feldman *et al.* and Carlsson the governmental role was crucial in the legislative area, which has fed into infrastructure development, training programmes and tax incentives. The Biomedico cluster initiative illustrates the problem of misunderstanding of the cluster concept by politicians. The presentation of the cluster initiative as a broad life science clusters in north Denmark might have important consequences: the current financial support activities may be too broadly distributed, thus taking away necessary financial support from the most competent area. On the other hand, policy is very unpredictable, as policy making is an interactive process where variables change all the time.

Finally, we would like to emphasise that as our examples show, the cluster emergence is a very long-term process, and policy makers might have problems in sticking to such long-term goals.

Notes

1 For details, see Dalum (1995), Pedersen (2005), Dahl *et al.* (2003) and Dalum *et al.* (2005).
2 They give the example of the emergence of the membership organisations (which supported social capital and were formed to promote networking) such as Mid-Atlantic Venture Capital Association.
3 In line with his statement that a regional agglomeration is a necessary but not sufficient condition for an industry cluster to emerge.
4 Today Thrane & Thrane Aalborg.
5 Today Simrad Stovring.
6 At present, two main institutions in Denmark produce MSc graduates in the engineering disciplines, the Danish Technical University in Copenhagen (DTU) and AAU. During the 1990s, AAU's share of MSc graduates was around 50 per cent in electronic engineering.
7 Dancall was acquired by Amstrad in 1993, then by Bosch in 1997. In 2000 it was divided between production by Flextronics and development by Siemens Mobile Phones; cf. below. Finally, Flextronics in North Jutland was shut down at the end of 2004.
8 Dancall and Cetelco became members of a small but exclusive club of approximately five firms, aiming to be the first of which Motorola, Ericsson and Nokia were the most prominent, in the world to develop GSM terminals.
9 Multinational enterprises.
10 Aalborg Commercial Council provides services to more than 5000 companies, including counselling of business start-ups, financing, export, import, staff and management development, marketing, subsidies, and so on (the Competence Catalogue).
11 BioMed Community is a kind of cluster organisation with the objective of developing and promoting a cluster within the life sciences in north Denmark (www.biomedcom.dk).
12 Denmark has a public health system and hospitals are administered by county authorities (in this case North Jutland County jurisdiction).
13 Cooperation with the hospital gave Aalborg University access to perform clinical tests and provide documentation, thus it also plays a very important role in the university's research.
14 Until sometime in the future this will not be the case, and Aalborg Hospital will only have the right to educate doctors in the latter part of their studies as an affiliate of Århus University.
15 "Many individuals have location inertia due to reasons such as family mobility constraints, location preferences, familiarity of the environment, the relatively higher costs associated with changing residence, or the high cost of establishing a new company in a thickly populated environment where office and housing costs tend to be higher" (Feldman and Francis 2003, p. 780).
16 As we have seen in the models, the emergence of a cluster is often associated with the emergence of the second-generation start-ups.
17 However, there has been a growing awareness and increasing interest in patenting in the companies and at the university during the last few years.

References

Boschma, R. A. (1999) The rise of clusters of innovative industries in Belgium during the industrial epoch, *Research Policy* 28 (8): 853–871.

Carlsson, B. (2006) The role of public policy in emerging clusters., in P. Braunerhjelm and M. P. Feldman (eds) *Cluster Genesis: Technology-Based Industrial Development*, Oxford: Blackwell.

Cooke, P. (2002) *Knowledge Economies: Clusters, learning and cooperative advantage.* London and New York, Routledge.

Dahl, M. S. (2003) Knowledge diffusion and regional clusters: lessons from the Danish ICT industry Department of Business Studies, Aalborg University.

Dahl, M., C. Ø. R. Pedersen and B. Dalum (2003) entry by spinoff in a high-tech cluster, Druid Working Paper no. 11-2003.

Dalum, B. (1995) Local and global linkages: the radiocommunications cluster in northern Denmark, *Journal of Industry Studies* 2 (2): 89–109.

Dalum, B., C. Ø. R. Pedersen and G. Villumsen (2005). Technological life cycles: lessons from a cluster facing disruption, *European Urban and Regional Studies* 12 (3): 229–246.

Feldman, M. and J. Francis (2003) Fortune favours the prepared region: the case of entrepreneurship and the Capitol region biotechnology cluster, *European Planning Studies* 11 (7): 765–788.

Feldman, M. P., J. Francis and J. Bercovitz (2005) Creating a cluster while building a firm: entrepreneurs and the formation of industrial clusters, *Regional Studies* 39 (1): 129–141.

Porter, M. E. (2000) Location, competition, and economic development: local clusters in a global economy, *Economic Development Quarterly* 14 (1): 15–34.

8 The knowledge–space dynamic in the British biotechnology industry: function, relation, and association

Kean Birch

Overspecialise and you breed in weakness.

Major Kusanagi, *Ghost in the Shell* (1995)

Introduction

Within economic geography and regional studies there are a plethora of theoretical approaches and perspectives that seek to explain the relationship between knowledge and space, and its importance to economic development and growth. Just a few examples include the popular 'cluster' theories of Michael Porter (1990, 2000), the theories of the California School concerning transaction costs and networks of small producers (Storper and Scott 1995), and the work of researchers on national or regional systems of innovation (Lundvall 1992; Cooke 1998). There are at least two characteristics that these, and other, theories have in common. First, they have all been criticised in both theoretical and empirical terms (e.g. Lovering 1999; Lagendijk and Cornford 2000; Malmberg and Power 2005). Second, they have a tendency to conflate theoretical description with policy prescription – that is, there are 'clusters' of successful industries, therefore in order to be successful, industries need to cluster. Anthony Giddens termed such tendencies in social science the 'double hermeneutic' in which concepts become self-fulfilling because they are considered an accurate description of the world and are therefore acted upon in policy and theory (see Ferraro *et al.* 2005; Ghoshal 2005). Consequently, such concepts can become self-reinforcing, leading to the promotion and maintenance of specific theories that constitute the world in particular ways.

In relation to regional theories on the biotechnology industry – defined as firms that carry out biological research and development – there has been a very clear emphasis on the role of clustering and concentrations in the successful development of the sector, within both policy and academic discourse. In the latter literature there is a wide, and growing, range of research on the industry, covering, among others, the dynamic role of knowledge in the sectoral innovation process (see Nesta and Dibaggio 2003; Coenen *et al.* 2004) as well as importance of specific national institutional frameworks (e.g.

Kettler and Casper 2000). There is further interest in the importance of 'regional knowledge capabilities' (Cooke 2004) and location generally (Leibovitz 2004), as well as in the expansion of the 'triple helix' or 'competition state' (Etzkowitz and Leydesdorff 2000; Loeppky 2005). However, there is particular academic interest in the concept of 'biotechnology clusters' covering countries, such as Canada, the United Kingdom and Germany, as well as smaller regions such as Scotland (Leibovitz 2004), Maryland (Feldman and Francis 2003), Cambridgeshire, UK (Casper and Karamanos 2003), and Lombardy (Breschi *et al.* 2001).

The cluster concept itself is derived from the work of Michael Porter who identified *competitive advantage* as a localised process dependent upon 'home base' characteristics embedded in 'a network of activities, connected by linkages' both within the firm itself and with activities performed outside the firm (1990: 41). In later work, Porter (2000) concentrated more on spatial clusters (i.e. localised connections), as opposed to functional ones (i.e. sectoral connections), and the increasing importance of such clusters concurrent with the rise of knowledge and innovation intensity. According to Malmberg (2003), the greater clarity from this focus means that Porter's cluster perspective concentrates on studying the 'qualitative' difference between local (i.e. clustered) and global (i.e. non-clustered) interactions. Throughout, whether the spatial focus is national, regional, or local, the academic literature characterises such biotech clusters as constituted by localised informal and tacit knowledge exchange between cognate firms, supply and service organisations, and public-sector institutions (see Ryan and Phillips 2004 for a typology).

Alongside this cluster perspective on the biotech industry, as with other industries, however, is the need to recognise the importance of extra-local connections, especially in relation to market demand and multi-scale, interorganisational interaction (see Simmie 2003, 2004; Cooke and Piccaluga 2004; Malmberg and Power 2005). This has led to concepts such as spatial 'nodes of excellence' or 'megacentres' that are keyed into a global network of biotechnological R&D capabilities (Coenen *et al.* 2004; Cooke 2004; see also Bathelt *et al.* 2004). In both the cluster and more recent nodal approaches there is an emphasis on the importance of knowledge exchange and transfer for the development of innovative capabilities and capacities; consequently, it is important to consider the range of spatial theories that incorporate these facets in an explanatory framework of regional innovation.

I have sought to incorporate a number of the so-called territorial innovation models (see Moulaert and Sekia 2003) into an overall theoretical perspective in order to produce a heuristic device – the *knowledge-space dynamic* – that can accommodate a number of different, although equally important, aspects of spatially embedded innovation processes. These consist of functional, relational, *and* associational features that I have subsequently represented in the knowledge–space heuristic as constituting elements of the innovation process. I have done this in order to explore the scale at which these processes operate, not to argue that certain features of a location are the determining element in the innovation process. As such, the heuristic device reveals a number of perhaps 'obvious' relationships within the innovation system (i.e. between external relations and complementary organisations), although the purpose here is to see at what scale these relationships are strongest. Before examining these scalar relationships, I will briefly outline the extent of the biotechnology industry in the United Kingdom in 2003 in order to illustrate how the 'cluster' perspective appears to be, at first sight, intuitively plausible. However, the later analysis of knowledge relationships across different scales illustrates that the focus on localised relationships is inaccurate at best, although the quantitative basis of the data I consider may not reveal the intricacies that a qualitatively deeper and more extensive research programme could draw out.

Theories of territorial innovation

If a region's factor conditions, its inherent qualities, cannot be characterised as a priori either advantageous or disadvantageous (see Malmberg 2003), then current arguments about regional advantage and 'territorial innovation', as based on a location's endogenous 'assets', are debatable (see Hudson 1999; Lovering 1999; Lagendijk and Cornford 2000 for critiques). In one circumstance such assets may be advantageous, but in another, not too dissimilar context they could lead to a number of problems. The primary focus on the importance of innovation, largely construed in terms of technological progress, may even lead to a problematic link being made between technical change (i.e. innovation) and economic development (i.e. regional prosperity). Both Myrdal and Hirschman, for example, posited that economic growth also entails negative effects in the form of 'backwash' and 'polarisation' respectively (see Cooke 2002: 27–8). As for the relationship between knowledge and space, John Parr (2002) reiterates the same point in his discussion of the different types of agglomeration economies – that is, proximity or concentration may not necessarily be an unqualified good. Consequently, it is useful to reconsider current theories on regional advantage in order to develop a new approach combining elements from a number of earlier theories in order to represent the relationship between knowledge and space in non-prescriptive terms.

Functional, relational, and associational theories

Current conceptualisations of regional innovative advantage can be rather crudely split between two existing models: one functional and one relational. There are several useful reviews of these theories in the broad regional studies and economic geography field (e.g. Malmberg 1997; Yeung 2000; Malmberg and Maskell 2002; MacKinnon *et al.* 2002; Moulaert and Sekia 2003; Simmie 2005). By introducing a third model – associational – I want to complement the previous two and seek to represent the role of location in neutral terms, so that the relationship between proximity and knowledge does not imply advantage. The third model is largely derived from work in science studies, particularly that of actor-network theory (ANT), and focuses on regional associative capacities – that is, on the importance of institutional isomorphism in relation to proximity, interaction, and imitation of spatially embedded organisations (see DiMaggio and Powell 2004). By combining all three models I can produce a heuristic device that incorporates a number of importance facets of the spatially embedded innovation process, rather than contending that any particular facet is more important than another in this dynamic. However, before I introduce the associational model I will outline the other two.

The *functional* perspective can be broadly summarised as a concern with 'material linkages and transaction costs', where these usually concern the positive externalities for innovation engendered by a shared labour pool, shared inputs, and knowledge spillovers (MacKinnon *et al.* 2002: 294). Alfred Marshall was one of the earliest theorists in the functional perspective with his work on external economies, although the later work of Hoover and Jane Jacobs added complexity by distinguishing between internal economies of scale, as well as 'localisation' and 'urbanisation' economies (see Harrison *et al.* 1996; Simmie 2005). The latter two theories concern the relation between particular locations and the benefits they provide firms. The former suggests that similar firms profit from similar inputs, such as knowledge, that can be pooled, while the latter contends that diverse firms benefit from locating next to each other as they then have access to diverse inputs, including varied knowledge resources. Other functional theories include those of Porter (1990) in relation to 'clusters' and knowledge linkages along value chains; Simmie (2004) in relation to the importance of urban centres and external linkages, particularly knowledge from extra-local customers; and the original work in new industrial spaces that focused on transaction costs and shared knowledge within a local production network (Scott 1989; Storper and Scott 1995), before a subsequent shift to a more relational perspective (see Moulaert and Sekia 2003).

In contrast to the functional perspective, the *relational* approach concerns the social and institutional environment of a particular geographic location. As such, it focuses on the importance of knowledge networks, engendered through collaboration, specialisation, and trust, to the innovation process. Some of the earliest relational theories were developed out of the concept of flexible specialisation, particularly in relation to new industrial districts (Piore and Sabel 1984), while later theories emphasised the network-based system of firms that had its origins in the sociological theory of embeddedness (Granovetter 1985). The work of GREMI,[1] for example, stressed the role played by collective learning in promoting an innovative milieu and the benefit this has in lowering information costs (Camagni 1995). Furthermore, the new industrial spaces approach (or the California School), once it shifted to a relational perspective, focused instead on the externalisation of production and the reduction of transaction costs via proximity and 'untraded interdependences' (Storper 1995; Storper and Scott 1995). More recent research has proceeded from concepts derived from a 'systems of innovation' (Freeman 1982) approach that focuses on the role of knowledge – especially tacit variants – and learning, which has been particularly influential in theories of learning regions (Morgan 1997) and regional innovation systems (Cooke 1998).

While the functional and relational territorial models of innovation can be fairly clearly, if crudely, mapped on to existing research and theories, the final *associational* model cannot. It bears a similar name to the work of Cooke and Morgan (1998) in their conception of the 'associational economy', although the name is largely the limit of any similarity. In contrast, the associational model builds upon work in actor–network theory by Bruno Latour (2002), although in this case departing from treating inanimate objects (i.e. roads) as active agents. Because of ANT's roots in ethnomethodology, it focuses on how actors *achieve* the social rather than on the social *link* between actors, as classical definitions of society tend to (Strum and Latour 1999). Consequently, the model presupposes that innovation arises as a direct result of actors' performances and the constitution of institutions through this performance.

The main theoretical point to draw from ANT is that actors choose to locate in proximity to one another and change their behaviour to do so; thus, they perform society through associational processes that embed institutional expectations. This theory can be applied to regional studies in the following way:

- Firms (i.e. actors) choose to locate near other firms so that they can engage in economic activity.
- In order to be economically active, firms are willing to change their actions (i.e. behaviour) so that they accommodate other firms.
- Firms accommodate other firms through social performance, in terms of developing common or uniform institutions.

In ANT the achievement of isomorphism through performance is premised upon the idea that actors define the 'social' (i.e. group, society) through their knowledge of others; although how they accrue information is not considered, despite the need for access to information and, crucially, other actors. Thus, an ANT approach actually requires that there be an existing structure already in process before performance; it is a structure that is then strengthened or weakened as actors reinforce or contradict the original choice to co-locate by adapting to proximity. For example, actors imitate each other, anticipate each other's actions, and act in accordance with those expectations (see Strathern 2002).

A central aspect of combining these three perspectives is an attempt to construct a theoretical approach that does not seek to offer prescriptions based upon analytical descriptions; in short, it is an attempt to avoid the normative assumptions that are implicit in previous territorial innovation theories. In such theories the importance of knowledge tends to be presented in terms that both describe a location and prescribe policy for that location, thereby leading to the problematic 'double hermeneutic', or self-fulfilling prophecy, inherent in social science research (Ferraro *et al.* 2005; Ghoshal 2005). This means that conceptualising locations in terms of development or growth not only has to be cast in neutral terms that do not present such change as either 'progressive' or 'optimal' (Grabher and Stark 1997), but also has to acknowledge the multi-layered and multi-scalar facets of territorial innovation processes. Instead, change is largely driven by existing actors within existing institutional fields, which leads to path dependency, as W. Brian Arthur (1999: 106) argues, because 'technologies show increasing returns to adoption' resulting from 'lock-in' as external knowledge economies (i.e. economies of scope) lead to complementarities across organisations.

The knowledge–space dynamic

The preceding theoretical discussion informs the construction of the analytical heuristic device I outline here and call the 'knowledge–space dynamic'. The central characteristic of the following analytical framework is that knowledge cannot be represented by a single factor at a single scale. The aim of the framework is therefore to illustrate how different aspects of knowledge relate within a spatial dynamic that can then be used to explore the varying scalar dimensions of knowledge in the biotechnology innovation process. The relationships that are explored later are somewhat obvious, but the point here is to show how these functional (i.e. firm-level), relational (i.e. sector-level), and associational (i.e. institution-level) relationships are embedded in and operate across different spatial scales. In contrast to Coenen *et al.* (2004: 1010), who argue that 'physical proximity as a causal variable is meaningless' in that 'space as such cannot have any explanatory value as it lacks substance or process' (ibid.: 1005–6), this knowledge–space dynamic seeks to represent the effects of space in terms of functional, relational and associational processes.

The framework incorporates all three theoretical strands highlighted in the 'territorial innovation' section above. A number of important elements from these theories are outlined in Table 8.1, which covers both the knowledge and the spatial aspects of regional change.

The first of these is represented in the interpretation of knowledge as internally, externally, or iteratively produced (see Ernst and Kim 2002). Internal production occurs within the firm and as such is conditioned by the material circumstances of that firm. External production operates outwith the firm and as such its importance to said firm is determined by the relational and absorptive capacities of the firm (Malecki 1997). Finally, iterative

Table 8.1 Territorial innovation theories

	Functional	Relational	Associational
Growth conditions	Material	Social and institutional	Interaction
Advantage	Comparative	Competitive	Complementary
Agglomeration	Scale	Scope	Complexity
Institutional	Hierarchy	Market	Network
Knowledge	Internal	External	Iterative
Proximity	Spatial	Organisational	Social

production occurs in the operation of relationships, rather than the contents of the relationship (i.e. the process of interaction). As such, it is dependent upon the interaction and imitation of two or more firms with one another and cannot be produced or acquired in isolation or from an external partner.

The second aspect of regional change can be characterised in terms of the organisation of transaction costs and the importance of different types of proximity (see Boschma 2005). The former refers in particular to the work of Ronald Coase and later Oliver Williamson, but also that of George Richardson on networks, although without the assumption that each firm will adopt the most efficient organisational form. Consequently, firms can be considered as organising their activities in terms of hierarchies, markets, *and* networks, not one or the other, across a variety of spatial scales. For example, a firm may find it necessary to produce knowledge internally, yet also acquire knowledge on the market, and through this collaboration with other firms or organisations create yet more knowledge. The first could be located wherever the firm has facilities, while the second will be dependent upon both information about external knowledge and ease of access. With the final source it may be necessary for a firm to collaborate with an organisation that is physically distant.

The framework itself consists of six elements that interact in a circular process; initially this is conceived in uni-directional terms, but it could just as easily be considered in bi-directional terms as well. The framework is meant as a heuristic device to illustrate how the innovation process operates across spatial scales. Each element represents an aspect of knowledge production in spatial terms. The overall dynamic is outlined in Figure 8.1, and each aspect is explained below.

Spillovers illustrate the importance of university research in a region, characterised in terms of organisational units (i.e. departments) rather than whole organisations. Spillovers affect regional knowledge production in terms of providing freely accessible and appropriable knowledge resources, as well as human capital and training. Such spillovers consist of all transfers of knowledge between organisations on the continuum from explicit knowledge embedded in artefacts, through embodied knowledge in skills, to the tacit knowledge of 'know-who' and 'know-how' (see Johnson and Lundvall 2001). Spillovers are also strongly influenced by the current knowledge stock

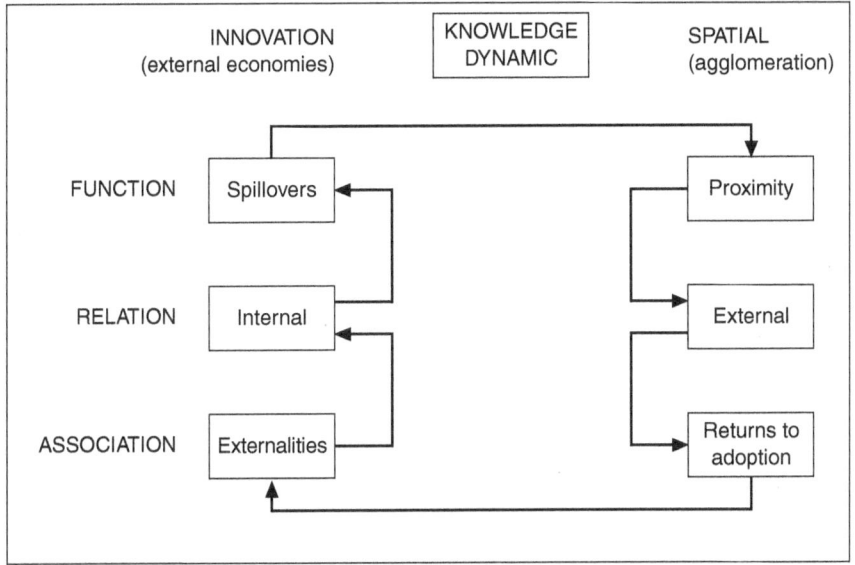

Figure 8.1 The knowledge–space dynamic

within a location; that is, previously produced knowledge held across all organisations.

Proximity represents the population of firms in a region and the likelihood that new firms will locate themselves in a territory because it already contains other firms. This affects regional knowledge production by encouraging the concentration of firms with existing firms so that through imitation, replication, and uniformity they can develop organisational knowledge and social bonds. Proximity relies on a concentration of existing firms that provides new firms with a motivation to choose a particular location over another; that is, uneven development is necessary for the promotion of such concentrations.

External relations reveal the extent to which knowledge production is affected by extra-organisational interaction that connects locations across different spatial dimensions – that is, the 'local buzz' and 'global pipelines' of Bathelt *et al.* (2004). The dimensions used are split between national and international to illustrate the importance of both scales. Such external relationships are determined by proximity in that firms, because of their decision to locate next to each other for information purposes, also need to connect with other organisations that may not be located in the same place but still provide a significant input into knowledge production.

Returns to adoption indicate the extent to which knowledge production is influenced by the expansion of a particular technological trajectory as multiple type organisations become 'locked in' to a specific technological paradigm (Boschma and Lambooy 1999). This affects regional knowledge production through focusing inter-organisational efforts on a specific field, thereby encouraging the adoption of similar organisational, cognitive, and social institutions in order to facilitate interaction between different organisations. The promotion of such institutional frameworks depends on the initial disparity between said organisations, because the difference leads to the development of uniformity.

Externalities of scale represents the tendency of knowledge production to be either internalised in a small number of large firms (i.e. representing internal economies of scale) or externalised in a large number of small firms (i.e. representing dispersed externalities of scale). The two types of production affect regional knowledge production in different ways; for example, a large number of small firms will produce more interactions between potentially different organisational forms (as with urbanisation economies), whereas a few large firms will produce fewer inter-organisational interactions, but more organisational focus.

Internal relations signify the internal production of knowledge (and thus the knowledge stock) of firms in a particular location, split between three varieties: codified (US patents), codified (EU patents), and appropriable (i.e. articles). The level of internal knowledge stock is derived from the level of the externalities of scale in a location in that both small and large firms produce knowledge, but for small firms to do so, greater interaction between organisations is reached, while for large firms, greater interaction within organisations is required.

Methodological note

The research discussed in this chapter refers to secondary data on the UK biotech industry collected from several primary sources. It is collated in a database of geographically aggregated information on knowledge indicators related to (1) biotech firms, (2) public research organisations (PROs), (3) universities, and (4) service providers (i.e. lawyers, consultants). The information on these organisations is used to map concentrations of the biotech industry in the United Kingdom, while the knowledge indicators are chosen in accordance with the different elements of the knowledge–space dynamic outlined in the previous section. All the collected data were input into SPSS on a territorial basis covering three scales derived from Eurostat regional designations:

* NUTS1 (equivalent to regional development agencies' scope)
* NUTS2 (equivalent to two or three counties)
* NUTS3 (equivalent to one county or city).

The choice of Eurostat scales is so that the data are easily scalable, in that one scale leads to another, and therefore the data can be easily compared across the scales.

Each biotech organisation is identified according to different criteria. Firms are identified in terms of their 'knowledge-based' activity – that is, they had to perform biotechnology research. Consequently, pharmaceutical companies, major corporations, and other specialist supplier and service firms are excluded. Public research is identified in relation to both the number of 'biotech' university departments and the number of PROs in a region. Initial data on these organisations are derived from regional development and biotechnology associations, e.g. ERBI, the Oxfordshire Biosciences Network, and Scottish Enterprise. After this, biotech-specific databases are used to refine and filter the organisations concerned; these include both free-access (i.e. Biospace, the Biotechnology Industry Association) and subscription services (i.e. BioWorld, BioCommerce). Finally, organisations are cross-checked to filter out any unsuitable organisations.

Space and knowledge in the UK biotechnology industry

An outline of the UK biotech industry

The DTI *Biotech Clusters Report* (1999a) identified four locations with high concentrations of biotechnology capacity: Cambridgeshire, Oxfordshire, London, and central Scotland. In the data collected on the biotech industry here, there is a similar concentration of biotechnology firms, service providers, and public research entities. The data are presented using the European Union NUTS2 regions to provide a common reference system that is not based on regional administrative boundaries (i.e. local authorities). There are 37 NUTS2 regions in Britain, most of which cover approximately three counties or one city.

The first and main descriptive indicator mapped is the number of biotech firms in a region. As was explained in the last section, these are firms that engage in biotechnology R&D. There were 436 identifiable biotech firms in 2003 with a mean of 11.78 firms and a median of 6 firms per region. Only four regions had twice the mean number, and these represent over half of all British biotech firms:

- Berkshire, Buckinghamshire, and Oxfordshire (68 firms)
- East Anglia (65)
- Inner London (62)
- eastern Scotland (39).

The second indicator is the number of complementary organisations (i.e. service providers), including lawyers, financial services, investors, networking organisations, and so on. In 2001 there were a total of 470 service providers, with a mean of 12.7 and a median of 5 per region. There are only three regions with twice the mean, and of those Inner London dominates, with nearly a third of all service providers (31.7 per cent):

- Inner London (149 service providers)
- Berkshire, Buckinghamshire, and Oxfordshire (50)
- East Anglia (49).

The third indicator is the number of university departments that relate to biotechnology research. In 2001 there were a total of 255 'biotech' departments, with a mean of 6.89 and a median of 6 per region. There were four regions with more than twice the mean, representing over a third of all departments (34.1 per cent):

- Inner London (46 departments)
- Berkshire, Buckinghamshire, and Oxfordshire (14)
- Eastern Scotland (14)
- East Anglia (13).

The final indicator is the number of public research organisations (PROs) across UK regions. There were a total of 106 PROs in 2003, with a mean of 2.86 and a median of 1 per region. There were a number of regions with twice the mean number, but only four regions with more than 10 PROs in them, representing around 60 per cent of the UK total:

- East Anglia (15 PROs)
- eastern Scotland (13)
- Inner London (12)
- Berkshire, Buckinghamshire, and Oxfordshire (12).

It is evident from the descriptive data above and that from the knowledge–space dynamic below that the UK biotech industry is heavily concentrated in four specific regions. These correspond to the clusters identified by the DTI (1999a), although each concentration also tends to differ from the others across the range of indicators, especially when the indicators are considered in a more evolutionary model that incorporates changes over time. These differences are outlined in Table 8.2.

Using these indicators, each regional concentration can be described in terms of its base and driver of economic growth. The base of each region refers to the source of knowledge and capacity, while the driver refers to the main source of company expansion. Thus, the four regional concentrations can be classified in the following way:

- East Anglia: old, SME and university based; SME driven
- Inner London: new, university based and driven
- Berkshire, Buckinghamshire, and Oxfordshire: global, large firm based; firm and university driven
- eastern Scotland: old, university based; emergent.

Table 8.2 Characteristics of the main British biotechnology concentrations

Indicator	East Anglia	Inner London	Berks, Bucks, and Oxfordshire	Eastern Scotland
Applied public research	New	Old	New	Old
Research council funding	High	High	–	–
Five-star departments	High	High	–	–
HEI departments	Old	Old, New	Old	Old
Service providers	New	Old, New	New	–
Firms	Old, New	New	New	New
Firm sizes	Small, Mid	Micro, Mid	Micro, Mid, Large	Micro
Spin–outs	–	High	High	–
Firms' alliances	Total, UK	UK	Total, UK, International	–
Firms' patents	EU	EU, US	EU, US	–
Firms' articles	High	–	High	–

Note
Where a cell has been left blank, this indicates that the indicator was considered less relevant for identifying the specific features of that particular concentration.

When considered in isolation, each concentration represents a different type of agglomeration that has not necessarily been replicated across other regions. The descriptive data also illustrate the small absolute and relative size of the biotech industry, at least in terms of knowledge-driven firms and organisations. So, having described the UK biotech industry, I want now to explore how the knowledge–space dynamic operates at different scales by analysing the correlative relationships between a series of knowledge indicators.

The knowledge–space dynamic

The knowledge indicators I use here are split across three scales using the Eurostat NUTS designation. The largest scale is NUTS1, representing the equivalent of a British government office region (GOR), such as the South-East, or Scotland. The middle-range scale is NUTS2, representing around three counties or a large urban area. The final scale is the smallest and covers a single county or urban area. The relationship between knowledge indicators is represented in terms of correlations along the knowledge–space dynamic framework in the previous section (see Figure 8.1). In each case the knowledge indicator is correlated with the proceeding (and therefore preceding) indicator to determine the strength of the relationship and at which scale the relationship holds the greatest affinity. There are a total of six relationships. One important point needs to be made initially: a statistical artefact of the design may mean that the small number of NUTS1 regions (there are only 11 in the United Kingdom) leads to more, or less, significant relationships between indicators at this scale than for the others.

The first relationship under consideration is that between knowledge spillovers, in the form of public research, and proximity, in terms of firms (see Table 8.3). There is a strong relationship that is localised up to a medium-level scale (NUTS2, NUTS3) without a significant relationship on a large scale (i.e. NUTS1). These results would suggest that knowledge spillovers are limited to a specific regional scale, as has been argued by previous research, although the scale is not as localised as previously thought. Consequently, it is useful to think of the importance of knowledge in terms of physical proximity, although the strength of the relationship is not 'very strong', implying that the importance of such spillovers to the biotech industry is not as crucial as may be thought, especially as there is little difference between the small and the medium scale.

Table 8.3 Relationship between knowledge spillovers and proximity

	Spillovers – proximity
NUTS1	0.557
NUTS2	0.676★★
NUTS3	0.652★★

Note
★★ Correlation significant to 0.01 level.

The relationship between proximity and external knowledge, identified as the number of firm alliances in a region, is extremely strong, verging on a direct correlation (see Table 8.4). The relationship is also strongest further up the spatial scale, with the NUTS1 scale showing the strongest relationship. The weakest relationship is at the NUTS3 scale for international external knowledge. The extremely strong relationship suggests that proximity (i.e. concentrations of firms) is an important factor in encouraging external relations, whether national or international, since there is little difference in the strength of relationship between these two indicators. External relations cannot exist prior to the firms and therefore it is unlikely that they encourage proximity or concentration. However, this relationship is weaker as concentrations become more localised. Thus, larger spatial concentrations ensure greater engagement with external organisations, perhaps indicating a wider search and imitation environment in which firms participate, although the overall similarity between scales suggests that scale plays only a relatively small role in promoting external orientation.

There is a very strong relationship between external knowledge relations and returns to adoption, characterised in terms of numbers of service providers (SP) and public research organisations (PRO) (see Table 8.5). The relationship is almost equally strong across all spatial scales for SP organisations and for NUTS1 and NUTS2 scales for PROs. The weakest relationship is at the NUTS3 scale for PRO, although it is still a strong relationship. These data indicate that external knowledge relations are an important factor in the promotion of inter-organisational networks that create and develop benefits for those involved in the same innovation processes. There is relatively little difference between national and international external relations, although

Table 8.4 Relationship between proximity and external knowledge

	Proximity – total allies	Proximity – UK allies	Proximity – international allies
NUTS1	0.947	0.957	0.937
NUTS2	0.926	0.937	0.910
NUTS3	0.848	0.886	0.817

Note
All correlations significant to 0.01 level.

Table 8.5 Relationship between external knowledge and returns to adoption

	External – private adoption	External – public adoption
NUTS1	0.767	0.766
NUTS2	0.710	0.776
NUTS3	0.727	0.693

Note
All correlations significant to 0.01 level.

international relations are weaker than national ones across all scales, particularly SP for NUTS2 and PRO for NUTS3. Again the equal strength relation across the scales (slightly weaker at smaller scales) suggests that firms' external orientation operates within a relatively wide environment and across multiple scales.

There are a range of strong and very strong relationships between returns to adoption and either (1) externalities or (2) economies of scale (see Table 8.6). The former consist of micro and small enterprises (MSE) in that they represent specialised knowledge nodes that have not been incorporated into larger organisations. The latter consist of medium and large enterprises (MLE), which represent strongly internalised knowledge centres. Returns to adoption can also be split between public (PRO) and private (SP) returns. At the largest spatial scale there are less significant relationships between externalities and private adoption, and economies and both public and private adoption. The strongest relationships for private adoption are at the smallest scale (NUTS3), although these are fairly similar across scales (if significance is ignored). In terms of public adoption and externalities there appears to be a stronger relationship at wider scales, while for economies it is at a medium scale. Overall, then, it would appear that returns to public adoption are consistent with externalities (i.e. multiple knowledge nodes) and, to a lesser extent, with economies (i.e. few knowledge nodes) of scale. Private returns to adoption limit this relationship to a medium and especially a small scale. Thus, multiple nodes are promoted through smaller-scale private returns to adoption, but medium or larger public returns to adoption – that is, smaller firms are reliant on localised service providers, but all firms are reliant on widely located PROs, although smaller firms are most reliant.

Externalities and economies of scale have very strong relationships with quantities of internal knowledge forms (i.e. US and EU patents, and journal articles) in a location at a medium scale where there are many small firms and at a large scale where there are large firms (see Table 8.7). Localised relationships are limited across both externalities and economies of scale, particularly the latter. The largest scale (NUTS1) appears to produce the strongest relationship for both externalities and economies of scale and non-excludable internal knowledge (i.e. articles), suggesting that both small and large firms

Table 8.6 Relationship between returns to adoption and scale economies

	Private returns to adoption		Public returns to adoption	
	MSE	*MLE*	*MSE*	*MLE*
NUTS1	0.615*	0.678	0.955**	0.684*
NUTS2	0.665**	0.666**	0.919**	0.769**
NUTS3	0.672**	0.672**	0.852**	0.703**

Note
* Correlation significant to 0.05 level; ** correlation significant to 0.01 level; MSE: micro and small enterprises; MLE: medium-sized and large enterprises.

produce internal knowledge at a non-localised or concentrated scale. However, codified internal knowledge (i.e. patents) has the strongest relationship with multiple nodes (i.e. externalities) at a medium scale, suggesting that such nodes produce such knowledge at a more localised and concentrated scale. In contrast, all internal knowledge appears to be produced by large firms at a large scale (i.e. one that is not reliant on localised or concentrated spatial proximity), suggesting that large firms have little reliance on localised or concentrated proximity.

The relationship between internal knowledge and knowledge spillovers (i.e. number of university departments) is strongest at the mid scale for codified knowledge (see Table 8.8). It is weaker at both wider – where there is limited significance – and smaller scales for codified knowledge. The relationship is very weak and lacks significance across all scales for non-excludable knowledge, suggesting that there is no spatial dimension between such internal knowledge and (public) knowledge spillovers. The clearest spatial dimension is in the relationship between internal knowledge and spillovers at the mid scale, which implies that there is some form of connection between a firm's internal knowledge (codified) production and public knowledge spillovers at that scale; this is strongest in relation to US patents.

Discussion and conclusion

As the late 1990s DTI reports claim (DTI 1999a, b), the UK biotech industry is mainly concentrated in four locations: East Anglia; Inner London; Berkshire,

Table 8.7 Relationship between scale economies and internal knowledge

	Externalities of scale			Economies of scale		
	US patent	EU patent	Article	US patent	EU patent	Article
NUTS1	0.668	0.795**	0.844**	0.838**	0.931**	0.954**
NUTS2	0.722**	0.836**	0.662**	0.827**	0.896**	0.842**
NUTS3	0.495**	0.769**	0.485**	0.637**	0.782**	0.681**

Note
** Correlation significant to 0.01 level.

Table 8.8 Relationship between internal knowledge and spillovers

	Internal codified (US) – spillovers	Internal codified (EU) – spillovers	Internal public – spillovers
NUTS1	0.620*	0.455	0.237
NUTS2	0.716**	0.622**	0.304
NUTS3	0.427**	0.439**	0.236

Note
* Correlation significant to 0.05 level; ** correlation significant to 0.01 level.

Buckinghamshire, and Oxfordshire; and eastern Scotland. Using just organisational indicators to identify these concentrations, it is also possible to identify two more regions with above-average organisational concentrations: Greater Manchester and south-western Scotland. However, apart from these six regions the UK biotech industry is almost non-existent, since most regions have fewer than a dozen firms that carry out biotechnology R&D, while the regions with higher concentrations number far fewer firms than is suggested in the DTI reports (i.e. less than 70 firms). Therefore, it is plausible to assume that these regional concentrations do not (and possibly could not) rely on an inward-looking approach, by either firms or regional government, during the innovation process.

My aim here has been to show how these territorial innovation processes need to be conceptualised in terms of the functional, relational, *and* associational features of particular places, rather than suggesting that any one of these elements is more important than any other. It is also meant to show how these three features play out across a number of spatial scales that are not limited to the localised level, but, in fact, necessitate an approach that can accommodate extra-localised linkages at the national and international scales as well (see Malmberg and Power 2005). The analysis of the *knowledge–space dynamic* heuristic device illustrates the degree to which the UK biotechnology industry does rely upon knowledge inter-linkages across different scales. These findings can be represented in the knowledge–space dynamic diagram as shown in Figure 8.2, illustrating at which scale each relationship between knowledge factors is strongest and therefore most important.

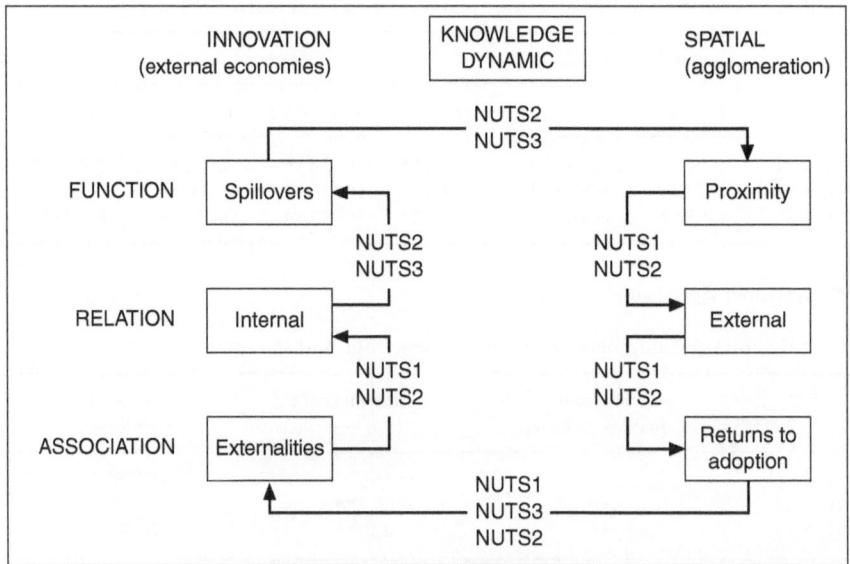

Figure 8.2 Scale relations in the knowledge–space dynamic

First, there are no strong relationships only at the smallest scale (NUTS3), suggesting that such localised concentrations are not dynamic by themselves. Second, there are two strong relationships at a small and medium scale (NUTS3 and NUTS2). These include relationships between (1) spillovers from public research and proximity of firms, suggesting that firms do concentrate fairly near public-sector research organisations; and (2) internal knowledge and spillovers, suggesting that mid-scale locations where firms produce knowledge affect the production of public science.

Third, there are three strong relationships at a medium and large scale (NUTS2 and NUTS1). These include relationships between (1) proximity of firms and external knowledge orientation, suggesting that a loose concentration of firms also promotes connections with external organisations; (2) external knowledge and returns to adoption, suggesting that external orientation loosely promotes the development of complementary organisations; and (3) externalities of scale and internal knowledge, suggesting that (i) many small firms promote internal knowledge production (at a medium scale), and (ii) many large firms promote knowledge production (at a large scale). Finally, there is a strong relationship across all scales between complementary organisations and externalities of scale, which differs between private and public organisational forms.

An analysis of the disaggregated parts of the three elements of the territorial innovation process shows that the *functional* aspects (i.e. spillovers and proximity) of regions are localised, whereas *relational* aspects (i.e. internal and external knowledge) are not localised and have a relationship that is strongest at the widest scale (NUTS1); the relational aspects also have stronger relationships than functional ones. Finally, the *associational* aspects (i.e. externalities and returns to adoption) are mixed both in strength and in relevance of localised and non-localised relations.

Overall, then, it would appear that to explain the whole knowledge–space dynamic that occurs within the UK biotechnology industry there is a need to appreciate several different scales across several different theoretical approaches that incorporate material, relational, and associational factors into the innovation process. This in no way should be taken to mean that these different aspects of the knowledge dynamic can be reproduced in policy prescriptions (or that they should be), but rather that it is important to interpret economic growth and development in 'neutral' terms that do not imply that progress or optimality is either the main driver or the main result of innovation. All this would imply that policy prescriptions focusing on the endogenous features of regions, such as the British government's 2000 Innovative Clusters Fund and 2001 Regional Innovation Fund, designed to help regional development agencies in the financing of incubation and cluster infrastructure projects (DTI 2003), would not necessarily benefit the innovation process. Furthermore, the broader promotion of 'clusters' across government departments (i.e. DETR 2000; DTI 2001) does not address the crucial role of extra-regional linkages, especially in relation to a biotechnology industry dependent upon national- and international-level connections, and may result in policy makers 'breeding in weakness' through an emphasis on specialisation at the regional level.

Acknowledgement

I would like to thank the anonymous referees for their comments.

Note

1 GREMI stands for Groupe de Recherche Européen sur les Milieux Innovateurs.

References

Arthur, W. B. (1999) Competing technologies and economic prediction, in D. MacKenzie and J. Wajcman (eds) *The Social Shaping of Technology*, Buckingham, UK: Open University Press.

Bathelt, H., Malmberg, A., and Maskell, P. (2004) Clusters and knowledge: local buzz, global pipelines and the process of knowledge creation, *Progress in Human Geography* 28 (1), pp. 31–56.

Boschma, R. (2005) Proximity and innovation: a critical assessment, *Regional Studies* 39 (1), pp. 61–74.

Boschma, R. and Lambooy, J. (1999) Evolutionary economics and economic geography, *Journal of Evolutionary Economics* 9, pp. 411–29.

Breschi, S., Lissoni, F. and Orsenigo, L. (2001) Success and failure in the development of biotechnology clusters: the case of Lombardy, in G. Fuchs (ed.) *Comparing the Development of Biotechnology Clusters*, London: Harwood Academic Publishers.

Camagni, R. (1995) The concept of *innovative milieu* and its relevance for public policies in European lagging regions', *Papers in Regional Science* 74 (4), pp. 317–40.

Casper, S. and Karamanos, A. (2003) Commercializing science in Europe: the Cambridge biotechnology cluster, *European Planning Studies* 11 (7), pp. 805–22.

Coenen, L., Moodysson, J., and Asheim, B. (2004) Nodes, networks and proximities: on the knowledge dynamics of the Medicon Valley biotech cluster, *European Planning Studies* 12 (7), pp. 1003–18.

Cooke, P. (1998) Introduction: origins of the concept, in H.-J. Braczyk, P. Cooke, and M. Heidenreich (eds) *Regional Innovation Systems*, London: UCL Press.

Cooke, P. (2002) *Knowledge Economies*, London: Routledge.

Cooke, P. (2004) The molecular biology revolution and the rise of bioscience megacentres in North America and Europe, *Environment and Planning C* 22, pp. 161–77.

Cooke, P. and Morgan, K. (1998) *The Associational Economy: Firms, Regions, and Innovation*, Oxford: Oxford University Press.

Cooke, P. and Piccaluga, A. (eds) (2004) *Regional Economies as Knowledge Laboratories*, Cheltenham, UK: Edward Elgar.

DETR (2000) *Planning for Clusters: A Research Report*, London: Department of the Environment, Transport and the Regions.

DiMaggio, P. and Powell, W. (2004) The iron cage revisited: institutional isomorphism and collective rationality in organizational fields, in F. Dobbin (ed.) *The New Economic Sociology: A Reader*, Oxford: Princeton University Press.

DTI (1999a) *Biotechnology Clusters Report*, London: Department of Trade and Industry.

DTI (1999b) *Genome Valley: The Economic Potential and Strategic Importance of Biotechnology in the UK*, London: Department of Trade and Industry.

DTI (2001) *Business Clusters in the UK – A First Assessment*, vol. 1: *Main Report*, London: Trends Business Consortium.

DTI (2003) *Innovation Report – Competing in the global economy: The innovation challenge*, London: Department of Trade and Industry.

Ernst, D. and Kim, L. (2002) Global production networks, knowledge diffusion, and local capability formation, *Research Policy* 31, pp. 1417–29.

Etzkowitz, H. and Leydesdorff, L. (2000) The dynamics of innovation: from National Systems and 'Mode 2' to a triple helix of university–industry–government relations, *Research Policy* 29, pp. 109–23.

Feldman, M. and Francis, J. (2003) Fortune favours the prepared region: the case of entrepreneurship and the Capitol region biotechnology cluster, *European Planning Studies* 11 (7), pp. 765–88.

Ferraro, F., Pfeffer, J., and Sutton, R. (2005) Economics language and assumptions: how theories can become self-fulfilling, *Academy of Management Review* 30 (1), pp. 8–24.

Freeman, C. (1982) *The Economics of Industrial Innovation*, London: Pinter.

Ghoshal, S. (2005) Bad management theories are destroying good management practices, *Academy of Management Learning and Education* 4 (1), pp. 75–91.

Grabher, G. and Stark, D. (1997) Organizing diversity: evolutionary theory, network analysis and postcolonialism, *Regional Studies* 31 (5), pp. 533–44.

Granovetter, M. (1985) Economic action and social structure: the problem of embeddedness, *American Journal of Sociology* 91 (3), pp. 481–510.

Harrison, B., Kelley, M., and Gant, J. (1996) Innovative firm behaviour and local milieu: exploring the intersection of agglomeration, firm effects, and technological change, *Economic Geography* 72 (3), pp, 233–58.

Hudson, R. (1999) The learning economy, the learning firm and the learning region: a sympathetic critique of the limits to learning, *European Urban and Regional Studies* 6 (1), pp. 59–72.

Johnson, B. and Lundvall, B. A. (2001) Why all this fuss about codified and tacit knowledge?, Paper presented at the DRUID Winter Conference, 18–21 January, University of Aalborg.

Kettler, H. and Casper, S. (2000) *The Road to Sustainability in the UK and German Biotechnology Industries*, London: Office of Health Economics.

Lagendijk, A. and Cornford, J. (2000) Regional institutions and knowledge: tracking new forms of regional development policy, *Geoforum* 31, pp. 209–18.

Latour, B. (2002) Information and organisation, *The Clarendon Lectures in Management Studies 2002*, Said Business School, University of Oxford.

Leibovitz, J. (2004) Embryonic knowledge-based clusters and cities: the case of biotechnology in Scotland, *Urban Studies* 41 (5/6), pp. 1133–55.

Loeppky, R. (2005) History, technology, and the capitalist state: the comparative political economy of biotechnology and genomics, *Review of International Political Economy* 12 (2), pp. 264–86.

Lovering, J. (1999) Theory led by policy: the inadequacies of the new regionalism, *International Journal of Urban and Regional Research* 23 (2), pp. 379–95.

Lundvall, B.-A. (ed.) 1992 *National Systems of Innovation*, London: Pinter.

MacKinnon, D., Cumbers, A., and Chapman, K. (2002) Learning, innovation and regional development: a critical appraisal of recent debates, *Progress in Human Geography* 26 (3), pp. 293–311.

Malecki, E. (1997) *Technology and Economic Development*, Harlow, UK: Longman.

Malmberg, A. (1997) Industrial geography: location and learning, *Progress in Human Geography* 21 (4), pp. 573–82.

Malmberg, A. (2003) Beyond the cluster: local milieus and global connections, in J. Peck and H. Yeung (eds) *Remaking the Global Economy*, London: Sage.

Malmberg, A. and Maskell, P. (2002) The elusive concept of localization economies: towards a knowledge-based theory of spatial clustering, *Environment and Planning A* 34, pp. 429–49.

Malmberg, A. and Power, D. (2005) (How) do (firms in) clusters create knowledge?, *Industry and Innovation* 12 (4), pp. 409–31.

Morgan, K. (1997) The learning region: institutions, innovation and regional renewal, *Regional Studies* 31 (5), pp. 491–503.

Moulaert, F. and Sekia, F. (2003) Territorial innovation models: a critical survey, *Regional Studies* 37 (3), pp. 289–302.

Nesta, L. and Dibiaggio, L. (2003) Technology strategy and knowledge dynamics: the case of biotech, *Industry and Innovation* 10 (3), pp. 329–47.

Parr, J. (2002) Agglomeration economies: ambiguities and confusions, *Environment and Planning A*, 34 (4), pp, 717–36.

Piore, M. and Sabel, C. (1984) *The Second Industrial Divide: Possibilities for Prosperity*, New York: Basic Books.

Porter, M. (1990) *The Competitive Advantage of Nations*, London: Macmillan.

Porter, M. (2000) Location, competition, and economic development: local clusters in a global economy, *Economic Development Quarterly* 14 (1), pp. 15–34.

Ryan, C. and Phillips, P. (2004) Knowledge management in advanced technology industries: an examination of international agricultural biotechnology clusters, *Environment and Planning C* 22, pp. 217–32.

Scott, A. (1989) High technology industry and territorial development: the rise of the Orange County complex, 1955–1984, *Urban Geography* 7, pp. 3–45.

Simmie, J. (2003) Innovation and urban regions as national and international nodes for the transfer and sharing of knowledge, *Regional Studies* 37 (6–7), pp. 607–20.

Simmie, J. (2004) Innovation and clustering in the globalised international economy, *Urban Studies* 41 (5/6), pp. 1095–112.

Simmie, J. (2005) Innovation and space: a critical review of the literature, *Regional Studies* 39 (6), pp. 789–804.

Storper, M. (1995) The resurgence of regional economies, ten years later: the region as a nexus of untraded interdependencies, *European Urban and Regional Studies* 2 (3), pp. 191–221.

Storper, M. and Scott, A. (1995) The wealth of regions: market forces and policy imperatives in local and global context, *Futures* 27 (5), pp. 505–26.

Strathern, M. (2002) Externalities in comparative guise, *Economy and Society* 31 (2), pp. 250–67.

Strum, S. and Latour, B. (1999) Redefining the social link: from baboons to humans, in D. MacKenzie and J. Wajcman (eds) *The Social Shaping of Technology*, Buckingham, UK: Open University Press.

Yeung, H. (2000) Organizing 'the firm' in industrial geography I: Networks, institutions and regional development, *Progress in Human Geography* 24 (2), pp. 301–15.

9 Culture, creativity and local economic development: evidence from creative industries in Florence

Luciana Lazzeretti

Introduction

Creativity is progressively becoming a source of competitive advantage, not only for firms but especially for local systems, and is also becoming an important objective for development policies paying particular attention to the territorial dimension of innovative processes. An appropriate policy centred on the development of a creative economy, that stimulates and rejuvenates the various forms of knowledge, can offer new opportunities from an economic, social and cultural point of view. Such opportunities can trigger an increase of competitiveness, attractivity and social welfare at a local level. This is the more so if we consider how places can attract human capital and talent as a result not only of market forces but also of non-market forces (Glaeser, 1999) – not only through new economic opportunities but also through 'lifestyle amenities' and a high quality of life (Gottlieb, 1995; Kotlin, 2000; Lloyd and Clark, 2001).

In this perspective, a 'city' can be seen as a territorial unit where economic, social, cultural and political networks develop because of favourable conditions that stimulate different forms of creativity (Florida, 2002). The urban concentration of firms improves productivity for two reasons: on the one hand, it constitutes a source of competitiveness as far as it stimulates the spread of specialised products. On the other hand, it strengthens creativity and innovation thanks to the flows of new ideas and know-how (Scott *et al.*, 2001). The opinion of several economists is that the city of tomorrow will be the real engine of national economies. The implication is that we will no longer talk of national economies, but of a global economy built upon a galaxy of regional economies whose main force is urban planning, centred, to a greater extent, on productivity (Pierce, 2002). Specifically, a city that can host significant concentrations of creative industries produces enormous economic benefits in terms of both high-paying and entry-level jobs (Wu, 2005), as well as in terms of 'a creative and impulse-giving socio-economic environment, and the availability of a broad and diversified supply of highly qualified personnel', the so-called 'soft and quasi-soft factors' (Dziembowska-Kowalska and Funck, 2000, p. 2).[1]

The importance of the creative economy has been acknowledged to a greater extent, and creative industries have 'moved from the fringes to the mainstream economics' (DCMS, 2001, p. 3). In the light of this new trend, a topical subject is the link between culture and creativity and the role they play as factors of local economic development. These issues were brought up also by cultural economists, who have long been studying the relationship that creativity entertains with factors such as innovation, human capital, talents and copyrights (Towse, 2004).

The specific line of research followed in this chapter is the analysis of the economic enhancement of art and culture, approached within the branch of studies on cultural districts (Scott, 2000; Lazzeretti 2003; Santagata 2005), and specifically applied to cities of art, with a focus on the analysis of the existing cultural clusters and of the firms involved in the economic enhancement of artistic and cultural resources.

This work is meant to give initial careful consideration to the relationship between culture, creativity and innovation, widening the subject of our previous research by regarding the cultural, artistic, human and environmental resources of places of high culture (HC) not only as production factors that can produce wealth and growth, but also as 'sources' of innovation and creativity that can trigger new activities, sectors or *filières*, or revitalise the existing ones. This direction is very close to the stances maintained in a recent OECD report on culture and economic development (OECD, 2005).

Thus, the attempt is to integrate our line of study with suggestions coming from the 'creative economy' approach (Simonton 1999; Cooke 2005; Pierce 2002), and particularly from Florida's contribution on creative cities (Florida 2002). The creative industries here investigated are considered, at a first approximation, as proxies for local economic development of a culture-driven type, and separated between traditional (visual arts, performing arts, publishing, music, etc.) and non-traditional (software and computer services, graphic design, advertising, etc.). An exploratory analysis of the case of Florence was carried out, since this city had already been defined in previous studies (Lazzeretti, 2004, 2005) as an HC place, being characterised by a considerable endowment of cultural, artistic, human and environmental resources and also by cultural clusters and firms (art restoration, museums and music firms) that advance its economy. The research questions to be answered are the following: can Florence be considered a creative city and, if so, is it characterised by a traditional or a non-traditional kind of creativity?

The chapter is divided into five sections. After this introduction, the second section briefly recalls the general theoretical background for the relationships between culture, creativity and local economic development. The focus is on the cultural district approach, which has been referred to in our previous studies, and on the creative economy model suggested by Richard Florida. The link between these two approaches is illustrated and a first notion of what creativity means for HC places is put forward. The third section, after a short review of the different definitions of creative industry used in empirical analyses, describes the taxonomy, data collection and methodology followed in the present research. The fourth section presents the main results obtained for the creative industries in the city and province of Florence. Finally, the fifth section offers some comments on results, conclusions and suggestions for future lines of research.

The cultural/creative district approach and creative economy

General theoretical background

The issues of creativity and economic development are dealt with from many different viewpoints in economic analysis. In this chapter we place the accent on cultural economics (McCain, 2004), because the relationship between culture, creativity and local economic development constitutes one of the main frontiers for new economies in the third millennium, a viewpoint debated by cultural, urban, regional, industrial and managerial economists.

Concurrently with the advance of research aimed at exploring specific segments of the cultural industry, with either an economic or a managerial outlook (Evrard and Colbert, 2000; Throsby, 2001; Chong, 2002; Towse, 2002; Benhamou, 2004), the debate runs to the inference that culture is a possible flywheel of economic development for cities and places rich in cultural resources (Mossetto, 1992; Wynne, 1992; Zukin, 1995; Greffe, 2003), as well as to the networks between cities of knowledge (Trullèn and Boix, 2001). The positive impact of cultural industries over the 'regional creative climate' (Kunzmann, 1994; Dziembowska-Kowalska and Funck, 2000), and the way in which culture tends to concentrate in places that have a high level of knowledge 'diversity' (Spencer, 2006), were also explored, while the cognitive economic approach looked into the relation between creativity and economy of art (McCain, 2004). On one side the attempt is to focus on the organisation of creative industries with the ambition of linking together with an ideal thread different industries whose common feature is that of conveying new knowledge (such as those founded on culture, design or science; see Belussi and Sedita, 2005). On the other side the attempt is to identify key elements in order to understand and manage creativity in cultural industries (Jeffcutt and Pratt, 2002), also asserting that their distinctive elements must basically be found in their dependence on creativity itself (Caves, 2000).

The recent OECD report (2005) underlines the strategic role of culture as a factor of sustainable economic development (see also Pilotti and Rinaldin, 2004), and emphasises its ability to activate new *filières* of innovation (Sacco and Pedrini, 2003) and to contribute to the rejuvenation of traditional sectors. The report stresses culture's potential for the revitalisation of European urban historical centres and regions, in the context of the issues of urban regeneration and governance (Mommaas, 2004); and also the significant implications of the clustering in systems of firms or institutions, such as districts, clusters, or cultural or creative neighbourhoods (Cooke, 2005).

Creative economy is tied in with culture as they both stress, on the one hand, the human capital (Florida, 2002), the creativeness of specific professions (Barron and Harrington, 1981) and the impact of context-related variables (Amabile, 1988; Woodman *et al.*, 1993), and, on the other hand, the belonging to social networks (Krackhart and Porter, 1985).

Other important contributions are those that give special attention to the risk that cultural activities be considered a bulwark for the renovation (or creation *ex-novo*) of a city's image on the short-term perspective, without creating the grounds for future sustainability (Leslie, 2005), and generating situations of 'hard-branding' (Evans, 2003). A possible answer to this drawback is suggested by the cultural cluster/district approach, as it underlines how the roots of economic development must be found in the historical, economic, social and cultural evolution of local community. Consequently, the actions aimed at local development and centred on the enhancement of culture can be the more successful as they take place in highly path-dependent local systems, which cannot be created from scratch (OECD, 2005; Wu, 2005). In order to avoid culture becoming a mere appeal for territorial marketing, through which cities 'brand themselves as "creative cities"' (Leslie, 2005, p. 403), it becomes interesting to understand whether these local systems are capable of stimulating and supporting creativity.

The above review of contributions constitutes the ideal thread that will guide the following reasoning over culture, creativity and economic development in their reciprocal relationships. The next section summons up the main characteristics of our model of analysis (the Cultural Districtualisation model) and of Florida's model, marking some possible points of integration and cross-fertilisation between them; finally, it introduces a first notion of creativity for HC places, which will direct this analysis of the processes of culture-driven economic enhancement in cities of art from cultural to 'creative' firms.

The cultural districtualisation (CD) model

The model of economic enhancement that was developed and applied in previous studies of cultural clusters and districts (Lazzeretti, 2001) is the cultural districtualisation (CD) model. This approach combines the categories of district theory (Marshall, 1920; Becattini, 2004) with those of the cluster theory (Porter, 1998). The CD is a model for sustainable economic development based on the trinomial culture–economy–society and on the resource–actors–community axis that tends to enhance the cultural, artistic, human and environmental differences of localities. Its fulcrum is the representation of culture 'as a resource' for the economic development of European cities and regions, and the perspective employed is a replication and implementation of the Becattini–Marshall approach to industrial districts.

The concepts the CD model employs are essentially two: the cultural, artistic, human and environmental heritage – that is, the set of resources necessary to define an HC place[2] – and the HC cluster – that is, the set of actors involved in the economic enhancement of resources. Both these concepts will flow into the synthetic definition represented by the 'high culture local system', which is characterised by the presence, in its territory, of a large endowment of a set of artistic, natural and cultural resources that identify it as an HC place, and of a network of economic, non-economic and institutional actors who carry out activities concerning the conservation, enhancement and economic management of these resources and which represent in their totality the HC city cluster (Lazzeretti, 2003, p. 638).

Unlike in other work (Frost Kumpf, 1998; Brooks and Kushner, 2001; Santagata, 2005), in this model there is no attempt at formulating an exact definition of the cultural district, since it specifically focuses on the 'cultural *districtualisation* processes', following a dynamic approach to processes (of a physiological type) that, while viewing the district, allows the possibility of encountering different degrees of *districtualisation* (as suggested by Becattini for industrial district). According to this perspective, the cultural district is considered as a more complex and articulated form of reference, but not the only one, inasmuch as it is possible to encounter more simplified forms that do not show the presence of all the typical district processes to which industrial district literature makes reference.[3]

Creative economy and the Florida model

According to Florida and Tinagli, creativity is the true engine that drives economic development and human change:

> Every single human being is creative and houses creative potential: Every single human being is creative in some way. Creative geniuses play their role, but creativity is a broad social process and requires teamwork. It's stimulated by human exchange and networks; it takes place in real communities and places. We can no longer prosper and grow by tapping and rewarding the creative talents of a minority. If we are to truly prosper, everyone must be brought fully into the system by employing them to do more value-adding creative work. Doing so will raise people's wages and strengthen our national economy, while also helping to bring our regional economies – and our lives – into better balance. Global competition in the creative economy is a wide-open game.
>
> (2004, p. 11)

Following Florida's approach (2002), at the basis of local, and particularly urban, growth, there are three factors, called the 3Ts of economic development: Talent, Technology and Tolerance. These are the key elements of the Florida model (FM), and help us understand creative economy and its possible effects on regional development in terms of attracting of creative people. Also, their presence should necessarily be equally promoted if we want to stimulate innovation and development, because these elements are tightly interdependent.[4]

The factor of talent is determined by people who can influence economic growth, both directly and by the promotion of technological innovation, which in turn improves the city's performance. Talent is measured on the following indicators: the percentage of citizens with a university certificate or degree, the incidence of researchers in the total workforce and the weight of the creative class. The latter class includes entrepreneurs, public and private executives, managers, researchers, lawyers, business and commercial consultants, doctors, engineers, architects, and highly specialised technical and artistic professionals (Florida and Tinagli, 2005).

Technology is appraised in terms of the figure shown for high-tech industries,[5] innovation capacity[6] and the eventual presence of communication and information technology (ADSL and UMTS).

The tolerance component allows the development of a society open to diversity, and therefore also open to the new ideas that can generate innovation in the different economic sectors.[7]

The assumption at the basis of this theory is the existence of a relationship between a metropolitan area in which the so-called creative class developed, with its people of outstanding intellect and ability, and the presence of creative industries. The global capacity of a city to develop creativity is measured by the Creativity Index, which is a synthesis of the 3Ts indexes (Florida, 2002) and is a good indicator with which to compensate for the shortcomings of other indicators that measure competitiveness levels.

The Florida model and cultural districtualisation model

It is possible to find several links between the FM and CD approaches, as they are in fact prone to develop forms of interrelation and cross-fertilisation. The main points of convergence can be summarised as follows:

- The suggestion that *economic development is mainly associated with the socio-cultural component*. The relevance attributed to the role of society can be found, on the one hand, in one of Florida 3Ts factors, namely 'tolerance'; and on the other, in the importance given to social capital and to the territorial rooting of economic and institutional actors that is typical of Marshallian district models.
- *The significance of the place as a living and working space and the strategic role assigned to localities*. Florida, starting from the observation of how competition is increasingly played out at a global level, maintains that localities can create an economic and social microclimate capable of activating and mobilising the resources that are crucial to growth (Florida and Tinagli, 2005). He also underlines how the choice one makes of 'where to live' is as important as that of 'who to work for': in other words, 'the city is a creative habitat' (Gertler *et al.*, 2002, pp. 1–2). At the

same time, the CD model considers local development for HC places as a process in which a set of economic, non-economic and institutional actors decide to utilise some of the shared idiosyncratic resources (artistic/cultural/human/environmental) in order to develop a common project that is simultaneously an economic project and a life project (Lazzeretti, 2005).

- *The role assigned to diversity.* Florida points out the creative potential of diversity, with particular reference to human capital and the ability to generate innovation (Florida and Lee, 2001); correspondingly, the CD model, with its culture-driven model of development, draws attention to the opportunity of gaining creative competitive advantages from differentiations based on the improvement of the idiosyncratic capital of HC places.

- *The focus on the city as a unit of analysis.* The preferred object of both approaches is the city *à la* Jacobs (1984), seen as the flywheel of economic development, the agent of innovation and the advancer of new mixes of resources. Therefore, Florida focuses his attention on the American technological conurbations with their varied and versatile population, while the CD model looks at European cities, where the prevailing element is art and cultural heritage. Apart from the different vocations of these two types of cities, there is a discrepancy due to their size, and thus to the dimension of the units of analysis; as a matter of fact, when he came to replicate his analysis in Europe, Florida's geographical object was not the town, but the province (Florida and Tinagli, 2005).

- *The role of human capital.* This constitutes the pre-eminent strategic resource in the Florida model of creativity. In fact, particular emphasis is laid on the creative class and the talented for their ability to single out and solve problems, and to generate and transmit knowledge. An indicator of their importance is also devised, the so-called 'Bohemian index'.[8] Human capital also constitutes one of the four idiosyncratic resources of HC places, although it does not necessarily prevail over the others. However, the European cities studied in this approach, which are usually rich in artistic and cultural resources, might rather be considered as meeting points of the past and future cultures. A new renaissance might take place in which the role of human capital will be not only the architect of tomorrow's culture, but also the economic advocate of yesterday's culture, with its remains deposited in the city's idiosyncratic heritages.

- *The role of technology.* Compared to the previous points of convergence, technology appears to constitute a significant element of divergence. In fact, it constitutes one of the strategic elements of FM, in so far as this model describes a third-millennium creative economy closely associated with the knowledge economy and the use of the latest technologies. The CD model does not specifically deal with this issue, although new technologies are considered one of the driving forces in the passage from cultural districts to creative districts, as also mentioned in the OECD report (2005).

Creativity and creative industries for High Culture local places

Creativity for HC places

As we have already recalled, one of the main results of previous studies on the economic enhancement of art and culture (Lazzeretti, 2005) was to show its strength and worth, and to register the diffuse presence of the clustering phenomena, at least for the cities and regions specifically investigated. The present study is meant to broaden this perspective by taking into account the relationship between culture, creativity and local development.

The creativity issue is not new to industrial district literature, which is the theoretical background on which we formulated the CD model. The concept of industrial district 'atmosphere' refers to an environment that advances the circulation of knowledge and mutual learning so as to stimulate industrial creativity (Becattini, 1989; Bellandi, 1992). This process reinforces itself to the point of setting up real creative economies and continuous innovation (Dei Ottati, 2005) that does not refer to human capital alone, but to the multiplicative effect that the district atmosphere can generate with its manifold relations between individuals and economic and social networks. A creativity of this kind is chiefly associated to the know-how in the traditional productive sectors of 'made in Italy', which can be revitalised by a fertilisation based on cultural activities and new technologies (Belussi and Sedita, 2005). An emblematic case is that of fashion and multimedia, which converged in the fashion design sector (Zanni and Bianchi, 2004).

In the case of cities, district creativity is associated not only with resourcefulness and know-how but also with a place's ability to attend those belonging to the creative class. This is just what the lesson of Florida teaches us: the city is a 'creative habitat'.

> [The] 'quality of place' must be understood in broader terms than we have traditionally been accustomed to: while the attractiveness and condition of the natural environment and built form are certainly important, so too is the presence of a rich cultural scene and a high concentration of people working in cultural and creative occupations ('bohemians' or the 'creative class').
>
> (Gertler *et al.*, 2002, pp. 1–2)

This special kind of creativity involves not only industrial or institutional creativity but also professional creativity, and creativity of places and environments that is fuelled by contextual factors (Woodman *et al.*, 1993; Amabile, 1988) and relational factors (Krackhart and Porter, 1985), and that is embodied in the human capital of cities.

A good example is given by the Florentine art restoration worker who has been the object of previous work (Lazzeretti and Cinti, 2001; Lazzeretti, 2003). The artisan firm considered as both an actor and a resource for the economic enhancement of culture and art. In fact, at first the artisan worker was associated with the traditional cultural firm, therefore with an *actor* in the cluster of art restoration, whose activity contributes to the enhancement of a territorial resource, art. Still, this artisan firm can also be regarded as a cultural *resource* for the urban neighbourhood it occupies, because it contributes to revitalising that special place. Another fact worthy of note is that most Florence restoration workers employ sophisticated technologies and have a high professional qualification acquired either in the workshop or by attending high-level training courses in local centres (such as the Opificio delle Pietre Dure[9]). Therefore, they can be considered as belonging to the creative industries. This is a case in which art, culture and technology intermingle and give new impulse to an old profession, and also to the neighbourhoods of the Florence historical centre in which the workshops are located, as they enliven their creative atmosphere.

The implication of what has been observed above is that there is a deep connection between culture, creativity and local development, with regard not only to human capital but also to the other idiosyncratic resources of the city – that is, the artistic, cultural and environmental components – as far as they point to culture as a source of innovation and revitalisation both for the economy and for the city as a whole.

Our aim in this context is to recognise these interactions. The first analysis we need to carry out is the identification and mapping of creative industries in the art city of Florence, separating the traditional and the non-traditional creative industries. Therefore, we will widen the notion of cultural industry we have employed till now, and consider as proxies for economic enhancement the creative firms – that is, those firms which consider the cultural resource as a production factor as well as a source of innovation.

The creative industries: a definition

What approach should we follow for the definition of creative industries, and what typology of creative firms should be taken into account?

As to the first question, studies on creative industries follow at least three kinds of approach. According to the *cultural* approach, creative industries are those which supply goods and services associated with cultural, artistic or entertainment values (Caves, 2000). The *copyright* approach describes creative industries as being those carrying out activities protected by the laws on copyright (Towse, 2002; Howkins, 2001), comprising core activities, partially copyrighted activities that are related to totally copyrighted ones, and distribution and retail activities (Allen Consulting Group, 2001).[10] Finally, according to the *creative* approach, creative industries cross-cut different sectors, in particular those 'which have their origin in individual creativity, skill and talent, and which have a potential for wealth and job creation through the generation and exploitation of intellectual property' (DCMS, 2001, p. 5).

In due course, however, the approaches above described have converged, so that 'industries protected by copyrights' have become virtually synonymous with the cultural (creative) industries (Towse, 2002, p. 171). In fact, when measuring its growth rate, 'creative economy [is] usually defined as the cultural industries plus the creative and performing arts – "low" and "high" culture' (Ibid., pp. 171–172).

Moreover, the term 'creative industry' has been entering the agenda of political planning in many countries. The British government was the first to make use of this term to broaden the cultural sector and include, for instance, the multi-media sector. Therefore, the creative industry marks the natural evolution of the cultural industry, matching the structural change due to the establishment of new technologies and new products in the sphere of the entertainment industry (DCMS, 2001; Demel *et al.*, 2004). According to the definition given in a report by the European Commission (2001, p. 22), the creative industry can be considered a 'digital culture', a sectoral area whose boundaries are difficult to draw, because of the manifold synergies and interactions of the traditional cultural sector with information technologies.

Wyszomirsky (2004) proposes a taxonomy of different approaches and defines four sets of criteria for the definition of creative industries. Each approach focuses on a single distinctive factor: (1) the product/service supplied, (2) the organisation of production, (3) the main production process, and (4) the occupational/workforce group. Wyszomirsky claims that most of the initiatives for the development of cultural industry employ the second approach. In fact, the first step for such initiatives is to draw a list of the types of organisations that should be included according to their field of activity and to the industry they belong to. The second step is to gather information on their key dimensions: size, distribution, revenue, export activity, employment and output. In this contribution we will follow this approach for building a working definition of creative industry and for mapping its firms.

In particular, the present research refers to *The Creative Industries Mapping Document* (DCMS, 1998, 2001; from now on, *CIMD*) presented in 1998 and updated in 2001 by the United Kingdom's Department for Culture, Media and Sport, which has already been the main reference point for many studies that involved the definition of creative industries (Arts Council England, 2003; García *et al.*, 2003; UNESCO, 2003).[11] The first step was to make a distinction between activities that are 'purely creative' (core activities) and activities 'connected with them'. The focus is only on the first category: the activities selected are those in which creativity is the distinctive feature of the production and creation processes. A table of correspondence was built (see Table 9.1) between the 'core' activities selected in the *CIMD* and the 'ATECO 2002' classification of Italian economic activities supplied by the Italian Central Statistics Institute (ISTAT, 2002). The creative activities identified are the following: advertising; architecture; graphic design; fashion design; software design; music and performing arts; visual arts (antiques and artisan), art, antique markets and crafts; software and computer services; film and video; television and radio; and publishing.

Another interesting contribution to the definition of creative industries

Table 9.1 Correspondence between core creative activities (BCMS, 201) and ATECD 2002 codification

Creative industries (CIMD)	Economic activities (ATECO)
Advertising	74.4 ADVERTISING *74.40.1 Advertising studios* *74.40.2 Agencies for the concession of advertising;* *intermediation in advertising services*
Architecture	73.1 R&D EXPERIMENTATION IN THE FIELD OF NATURAL SCIENCES AND ENGINEERING *74.20.1 Architecture and engineering studies* *74.20.55 Other technological activities* *74.30.1 Testing and technical analysis of products*
Graphic design	*74.20.52 Technical drafting* *74.87.5 Design and styling related to textiles, clothing,* *footwear, jewellery, furniture and other personal goods and* *household products*
Fashion design	*74.87.5 Design and styling related to textiles, clothing,* *footwear, jewellery, furniture and other personal goods and* *household products*
Software design	*72.60.03 Graphic creation on the web*
Music and performing arts	22.14 Editing of sound registrations 22.31 Reproductions from original sound registrations 73.2 R&D EXPERIMENTING IN THE FIELD OF SOCIAL AND HUMANISTIC SCIENCES 92.31.01 Artist and literary creations and interpretations *92.31.02 Performance organization* 92.32 Theatre, concert halls, and other performance halls, excepting cinema; box offices *92.34.12 Dance schools* *92.34.3 Other activities of entertainment and shows*
Visual arts (antiques and artisan) art, antique markets and crafts	*22.15 Other editions* *22.22 Other graphic arts prints* 26.12 Glass workmanship 26.15.2 Handmade glass workmanship, including glass blowing and decoration 26.21 Manufacture of ceramic products for home use, including decoration and glaze 26.70.2 Artistic workmanship of stones and marbles *36.22.1 Manufacture of jewellery and smithing of precious* *metals or of pieces plated with precious metals* 36.3 MANUFACTURE AND REPAIR OF MUSICAL INSTRUMENTS 36.61 Manufacture of costume jewellery. 52.48.61 Retail of artworks, excluding the antique *52.48.62 Galleries for the exposition with sale of artworks,* *including business agencies for sale* *52.48.63 Retail of handcrafted goods and decorations*

continued

Table 9.1 continued

Creative industries (CIMD)	Economic activities (ATECO)
	52.50.2 Retail of used furniture, antique furniture and objects, including antique books 52.62.65 Itinerant retail of artworks and antique objects 74.81 Photographic activities *74.81.1 Photography studies* *74.81.2 Photography labs for development and prints* *74.81.3 Aerography* *74.87.61 Organization of festivals, shows, expositions, conventions, and similar events* *92.31.03 Restoration and conservation of artworks*
Software and computer services	36.50.1 Manufacture of games, including video games 72.21 Production of non-personalized software 72.22 Production of personalized software
Film and video	22.32 Reproductions from original video registrations 92.1 PRODUCTION AND DISTRIBUTION OF VIDEOS AND FILMS; FILM PROJECTION 92.11 Production of films and videos, including related services, sound registration studios 92.12 Distribution of films and videos, including related services 92.13 Film projection
Television and radio	64.20.3 Management of television and radio network transmissions, excluding the transmitting of programs *64.20.65 Other activities related to telecommunications* 92.2 RADIO AND TELEVISION ACTIVITIES, EXCLUDING THE MANAGEMENT OF THE RADIO AND TELEVISION TRANSMISSION NETWORKS
Publishing	22.11 Publishing of books, pamphlets, flyers, music books, roadmaps and similar 22.12 Newspaper editing 22.13 Publishing and magazines and periodicals 22.21 Newspaper printing 22.25 Auxiliary work related to printing *92.31.01 Artistic and literary creations*

comes from Jeffcutt and Pratt (2002, p. 227), who argue that three kinds of 'convergences' are crucial in the field of cultural industries: *intersectoral*, 'between the media/information industries and the cultural/arts sector'; *inter-professional*, 'between diverse domains ... of creative endeavour (i.e. visual art, craft, print, video, music, etc.) that are brought together by new opportunities for the use of digital media technologies'; and *transgovernmental*, as they 'bring together a complex network of stakeholders – departments of culture and departments of industry, trade, professional and educational bodies – in an attempt to carry out effective "joined up" governance'.

The above considerations are particularly useful for integrating the notion of culture with its qualifications as a production factor *and* as a source of innovation. After the identification of the economic activities falling within the definition of creative industries, the second step of this investigation has been to separate creative industries into two macro groups: the traditional and the non-traditional ones. This procedure helps us to grasp the specificities and the possible trends of this particular mode of interaction and convergence between the different sectors and the various forms of creativity. The distinction between traditional and non-traditional creative industries will be presented separately for Florence, in an attempt to understand the peculiar typology of creativity of this city (see the subsection 'The traditional and non-traditional creative industries ... in Florence', below).

Data collection and methodology

Starting from the 'ATECO 2002' classification of Italian economic activities, the relevant figures for each code identified within the eleven groups of creative industries were extrapolated from the Florence Chamber of Commerce databank (CCIAA 2004). These figures are the number of firms in force and the number of workers for each firm. Data were collected for the city of Florence and for every single municipality in its province.[12] The data are updated to the third trimester of 2004.

The databank used is *Stock View*, which is built on a system made by the Infocamere Society, and provides three-monthly information on the structural features of all the firms registered at the Chamber of Commerce. The data are available both for the province as a whole and for each municipality in the province. As far as institutional sources are concerned, it must be noted that the phenomenon under study presents high levels of hidden, or black, economic activity that it was not possible to record. A refinement of these data might come from their combination with those regarding Florence artisan firms, figures about Florence artists, and any other source that might help to embrace the complexity of the creative class. For the moment, the creative workers taken into account are those that are detectable in the creative firms registered in the above-mentioned databank.

On the basis of these data, the city of Florence was analysed and compared with the other towns in the province, following a process already experimented with in previous studies.[13] Absolute and percentage values were recorded, the location quotient was calculated for each municipality and the territory subsequently mapped; finally, the municipalities with higher concentrations of creative industries and workers were detected.

The location quotients were calculated using the equation

$$LQ_{Ce,m} = \frac{E_{Ce,m}}{E_m} \bigg/ \frac{E_{Ce}}{E}$$

where $E_{Ce,m}$ is the number of enterprises (workers) in creative industries in the municipality m within the province of Florence; E_m is the total number of enterprises (workers) in the municipality m within the province of Florence; E_{Ce} is the number of enterprises (workers) in creative industries in the whole province of Florence; and E is the total number of enterprises (workers) of the whole province of Florence.

The location quotient can be used to compare the distribution of creative industries and workers across different urban areas in the provincial territory. Values greater than 1 represent a concentration of the variable under examination (creative firms and workers) that exceeds the average distribution of the province (Mood *et al.*, 1988).

As has already been mentioned, another empirical analysis was carried out on the typologies of creative industries present in Florence; the aim was to answer the question of whether these are strictly associated with culture or rather with a wider notion of creativity. To this purpose, a comparison was made between the definition of 'cultural industries' assumed in a report on cultural economy in Italy by Bodo and Spada (2004) and the definition of 'creative industries' given in the *CIMD*. In this way, the creative industries present in the *CIMD* were separated into two more detailed and practical groups: one designates the 'traditional cultural sectors' (that are basically those present in the Italian report) and the other includes the non-traditional creative activities. In particular, the *traditional creative activities* include visual arts (antiques and artisan); art, antique markets and crafts; film and video; music and performing arts; publishing; and television and radio. The *non-traditional creative activities* include advertising, architecture, graphic design, fashion design, software design, interactive leisure software, and software and computer services.

Florence as a creative city: some evidence from creative industries

Previous studies and researches on Florence as a creative city

Recently, different studies have been made of Florence, and other Italian towns, as creative cities. A few pieces of information supplied by these analyses employing different methodologies are particularly useful for the outline of a general picture of the phenomenon of creativity in this city.

Florida and Tinagli (2005) have analysed creativity in 103 Italian provinces and calculated a general Index of Italian Creativity for each of them. Florence ranks fifth in Italy in their analysis. In detail, it ranks as the seventh province on the Creative Class Index (22.8 per cent of creative talents), sixth on the Technology Index and third on the Tolerance Index.

To compare the situation of Florence with those of the other 19 regional capitals, a good reference is the work by Amadasi and Salvemini (2005, p. 31), who calculated a High-Symbolic Index (HSI)[14] to assess the presence of firms belonging to sectors with a high-symbolic intensity. The HS indicator was also worked out for seven subgroups relating to different economic sectors.[15] On the overall HSI, Florence was third in 1991, but moved down to sixth in 2001.

Finally, Capone (2006) examined a wider unit of analysis than the one chosen here. In fact, he considered the Local Labour Market Areas (LLMAs) identified in Italy by the 2001 census. Capone analyses the employees of creative industries according to a taxonomy that is similar to the one used in previous studies (e.g. Lazzeretti and Nencioni, 2005) but less detailed, as it stops within the two digits in the NACE 1.1 classification. His study brings to light 50 creative local systems, among which Florence ranks 11th in terms of location quotient, but holds one of the leading positions in terms of its contribution to national employment.[16]

These studies show that Florence is very well positioned in terms of creativity compared to Italy as a whole, and also in relation to each unit of analysis employed (town, province and local labour system). The common data source of these research is the ISTAT 2001 census, whose data can be compared with data from the 1991 census. The situation of ten years previously gives evidence of a positive trend for creative activities in general, which have definitely experienced growth. In the period 1991–2001 the creative class in Italy has in fact grown by 128 per cent: creative people numbered 1,900,000 in 1991, and 4,300,000 in 2001. The proportion of the creative class within the total workforce has grown from 9 to 21 per cent in the same period (Florida and Tinagli, 2005, p. 12). Also, employment in creative firms experienced a 68 per cent growth in the past decade (Capone, 2006).

In conclusion, these three contributions use the same source (ISTAT 2001 census) but refer to different territorial units, and consequently stick to dissimilar concepts of creativity. By comparison, the present contribution should

Table 9.2 Ranking of Italian creative cities, first eleven positions

Rank	Index of creativity in Italy (ICI)[a]	High symbolic index[b]	Location quotient of creative local systems[c]
1	Rome	Rome	Ivrea
2	Milan	Milan	Rome
3	Bologna	Catanzaro	Milan
4	Trieste	Palermo	Turin
5	Florence	Naples, Bologna	Città di Castello
6	Genova	Florence	Pisa
7	Turin	Bari	Verona
8	Parma	Genova	Padua
9	Rimini	Potenza	Tolentino
10	Perugia	Cagliari	Bologna
11	Modena	Campobasso	Florence

Source: a Florida and Tinagli (2005); b Amadasi and Salvemini (2005); c Capone (2006).

be seen as a first, explorative attempt to meet the following criteria: (1) it has recourse to a more updated source (data from the Florence Chamber of Commerce updated to 2004); (2) it proposes a benchmarking between the towns of the Florentine province; and (3) it goes deeper into the investigation of Florence's creative industries by making reference to the *CIMD*.

Benchmarking between creative industries in the province of Florence

What are the creative industries in Florence and its province? How many of them are there? How does the city rank compared to other towns? Which is the prevailing type, traditional or non-traditional creative industries? This section attempts to answer these questions on the basis of the first survey results.

If we look at the percentage of creative firms within industry as a whole in the province, we record a rate of 4.5 (corresponding to 4,024 firms), which seems a rather meagre figure. If we look at the proportion of workers employed in these firms, we find a slightly higher proportion at 5 per cent (corresponding to 9,113 units) (see Tables 9.3 and 9.4).

Although these percentages are unremarkable,[17] if we match them with the percentages given by firms classified according to economic activities (ATECO 2002–NACE 1.1), we find that creative firms have a relevant share. There are sectors of great importance in a city like Florence, where tourism is deeply rooted, that register a slightly lower percentage – particularly, 'Hotels and restaurants' (4.3%), and likewise 'Transport and communication' (4.0%) and 'Financial intermediation' (2.4%). 'Creative industries' are overshadowed by 'Agriculture' (8.1%), 'Business activities' (13.6%), 'Construction' (14.0%), 'Manufacturing' (19.5%) and 'Trade' (28.9%) (see Table 9.5).

Table 9.3 Creative firms, province and municipality of Florence

	Creative firms (A)	Total firms (B)	Percentage (A/B)	Location quotient
Province of Florence	4,024	89,360	4.5	1
Municipality of Florence	2,319	35,853	6.47	1.43

Source: Elaboration on CCIAA (2004).

Table 9.4 Employees in creative firms, province and municipality of Florence

	Creative employees (A)	Total employees (B)	Percentage (A/B)	Location quotient
Province of Florence	9,113	181,008	5.03	1
Municipality of Florence	3,768	70,006	5.38	1.07

Source: Elaboration on CCIAA (2004).

Table 9.5 Percentage of firms per economic activities, province of Florence, 2002.

Economic activities	Percentage of firms
Agriculture	8.14
Manufacturing	19.45
Construction	14.03
Wholesale and retail trade	28.85
Hotels and restaurants	4.29
Transport and communication	4.03
Financial intermediation	2.36
Real estate, renting and business activities	13.61
Public services	4.17
Creative industries	4.58
Other sectors	1.03

Source: Elaboration on CCIAA (2004).

The results of the analysis will now be presented, first as to the location quotient of creative firms for the municipalities in the province of Florence, and later as to the traditional and non-traditional creative industries for the municipality of Florence alone.

The calculation of the location quotient for the creative firms in the 44 urban areas of the province reveals that six municipalities (14% of the total) have a location quotient greater than 1; 24 municipalities (54%) have an index between 0.5 and 1; and 14 municipalities (32%) have an index lower than 0.5 (Figure 9.1).

The mapping of the indexes places Impruneta and Montelupo Fiorentino in first and second place, with location quotients ranging from 1.5 to 2 (this can be basically explained on the basis of their ceramics and terracotta sectors respectively). Florence takes third position, followed by Fiesole, Calenzano, and Capraia and Limite (see Figures 9.1 and 9.2).

The municipality of Florence has a percentage of creative firms of about 6.5% (2,319 firms), which is pretty high compared to the average for the province of 5% (Table 9.3). Its concentration index is high-ranking (1.43), being the third for the province; in other words, in the city of Florence are to be found more than 50 per cent of the total creative firms present in the whole Florentine province. Therefore, on the one hand, Florence lies third at the provincial level, but on the other hand, the Tuscan capital is the main centre of creative development since it houses 57 per cent of the province's creative firms.

If we examine the number of employees in order to find a size variable for firms, some discrepant elements come into view (Table 9.3). For example, even though it retains one of the highest concentrations of firms, the city of Florence does not show a similar high concentration of employees. In general, the two distributions do not show the same pattern, even if the higher indexes can be found more or less in the same urban areas, except

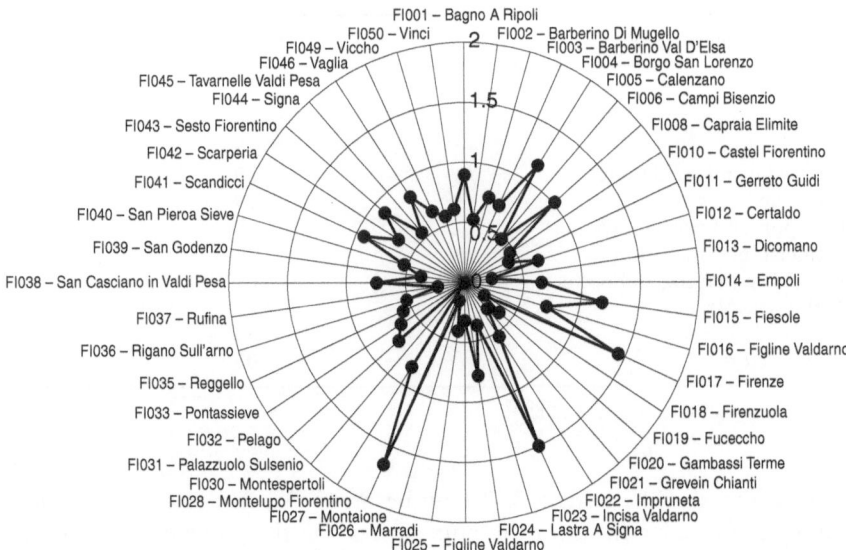

Figure 9.1 Location quotient of creative firms in the 44 municipalities of the province of Florence

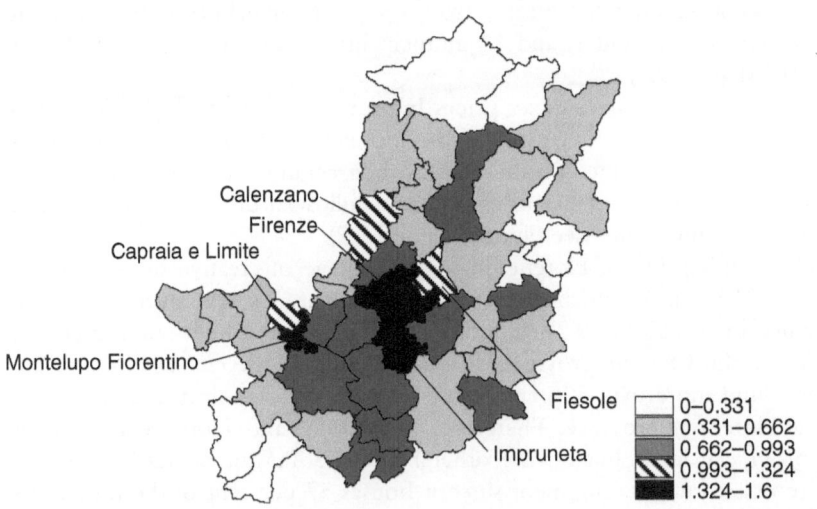

Figure 9.2 Map of creative firms in the province of Florence: inequality coefficient per municipality, 2004. Source: elaboration on CCIAA (2004)

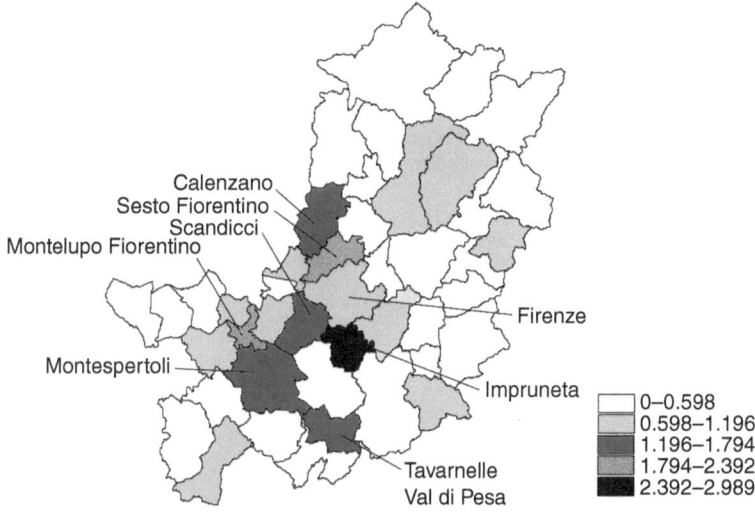

Calenzano
Sesto Fiorentino
Scandicci
Montelupo Fiorentino

Firenze

Montespertoli

Impruneta

	0–0.598
	0.598–1.196
	1.196–1.794
	1.794–2.392
	2.392–2.989

Tavarnelle
Val di Pesa

Figure 9.3 Map of employees in creative firms per municipality, province of Florence, inequality coefficient per municipality, 2004. Source: elaboration on CCIAA (2004)

for a few 'new entries' (Scandicci, Montespertoli, Tavernelle and Sesto Fiorentino).

The average number of employees in Florence's creative firms is lower than the provincial average, probably because of the medium-sized and large specialised firms that work in the ceramics, porcelain and terracotta industries.[18] Nevertheless, Florence appears to be the central pole for the province in terms of employees, since the 3,768 units working in the town represent more than 40 per cent of the total workers in the province (9,113).

Traditional and non-traditional creative industries in the municipality of Florence

The typologies of creative firms that characterise the city of Florence will now be investigated. With regard to the taxonomy of traditional and non-traditional creative industries, which takes into consideration the 11 ATECO code activities (Figure 9.4), Florence – as might be expected – is characterised by a cultural-traditional vocation.

As is shown in Figure 9.4, the percentage of cultural-traditional activities over the overall creative industry is about 72 (corresponding to 1,660 out of 2,319 firms), while that of creative, non-traditional sectors is about 28 per cent (659 over 2,319 firms). In particular, the grouping with a most relevant share is that of 'Visual arts, art, antique markets and crafts' (54.1%, with 1,252 firms), which includes handicraft and antique trade. This result is not

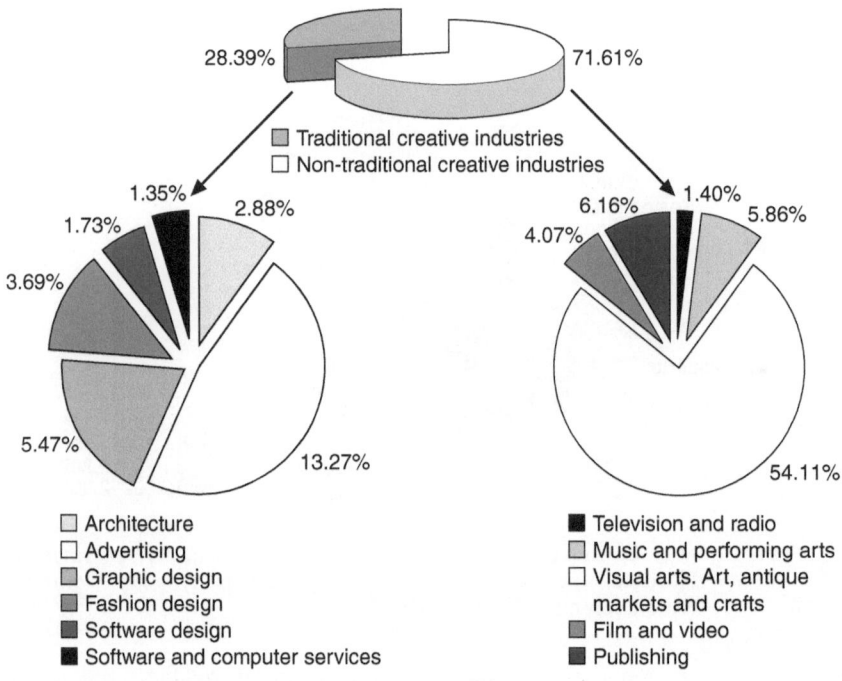

Figure 9.4 Traditional and non-traditional cultural creative industries. Source: elaboration on CCIAA (2004)

surprising for a city like Florence, in which these sectors poll first in terms of local artistic production. The most advanced sector for the non-traditional area is 'Advertising', with a share of 13.3 per cent of creative industry as a whole (308 firms).

These results indicate that Florence essentially emerges as a city of art strongly attached to its artistic and cultural heritage, since the most developed sector of creative industry is the cultural one. Some clearing up on this point is therefore needed. First of all, development may easily be held back by the growth of traditional industries alone, if the city sticks to an exploitation of art and culture resting simply on cultural tourism, and fails to look at them as possible flywheels of more extensive growth. If the city fails to fuel 'new creativity' and merely makes use of the revenues gained from the historical cultural heritage, the risk will be that it will fail to exploit its potential for creativity in full. At the present time this seems to be case for Florence, or at least for its industry, which is deficient in terms of high-tech firms or firms belonging to sectors characterised by technical expertise.

However, some recent studies (Capone, 2006) have shown that there is a stronger trend towards the development of non-traditional creative industries than of the traditional ones. In Florence, in the past decade the number of employees of traditional creative industries has grown by about 30 per cent, while those of non-traditional creative industries increased at the high rate of 100 per cent. The future seems to promise that the non-traditional will surpass the traditional creative industries. Such a trend might signal that Florence is gradually moving towards those sectors in which culture can play a primary role as a source of innovation, and not merely as a production factor. Also, as already mentioned, it is possible that some of the sectors here included among the traditional creative activities can be revitalised by interaction with high-tech sectors. This is certainly the case for art restoration in Florence, which constitutes an important urban cultural cluster (Lazzeretti, 2003), and it is even the object of a project – which is part of the Strategic Plan for the metropolitan area of Florence – whose aim is to make this city an international pole and a world point of reference for this art restoration. The goal of this project – which is called 'Florence, City of Art Restoration' (CSPS, 2001) and has institutions such as the Opificio delle Pietre Dure as partners – is to broaden and improve the already-present competencies of local workers by bringing out their stronger points: the centuries-old experience and tradition on the one hand, and the city's penchant for scientific research and technological innovation on the other.

Conclusions

Can Florence be considered a creative city? The first question advanced at the beginning of this chapter might be answered in the affirmative, but only with caution, since the information available can offer only a general picture of the phenomenon. Several studies and researches carried out on this topic by different authors, with different units of analysis and diversified methodologies, show Florence to have a significant quotient of creativity. Its value emerges by comparison with the general Italian situation, whose performance is not so good when put alongside those of other European countries and of the United States (Florida and Tinagli, 2004, 2005). Florence offers itself as the creative pole for its province, as it can attract more than half of the province's creative firms. At the same time, 'new entries' approach in the shape of small creative towns that certainly make the scene richer.

The second question this study intended to answer was: what kind of creativity is typical of Florence? This city is certainly complex and multifarious; the creative industries are quite relevant and they are mainly positioned in the traditional sectors rather than those propelled by the new economy. However, Florence's creativity does not stop at tradition, but can be found in processes of redevelopment of mature sectors, such as handicraft and art restoration, or the classical sectors of the 'made in Italy'. This is the reason why, to better comprehend the full creative capacity of Florence, the vision of the creative economy's contribution must be widened and adapted to our specific contexts.

This contribution to the investigation of the relationship between culture, creativity and local development is meant to suggest a new, wider notion of creativity for HC places, opening up to suggestions coming from the cultural cluster and the district approaches. Another proposition offered is the cross-fertilisation between the concepts of creative habitat and district creative atmosphere.

The economic enhancement of culture can take place in our cities of art through forms of cultural clustering and cultural districtualisation. However, culture is not only a production factor but also a source of innovation since it can revitalise new *filières* and new professions, thus becoming a creative engine itself. The human factor is what combines new resources and activities, and what changes yesterday's cultural industries into tomorrow's creative industries.

The impending question then becomes another: is Florence, as an HC place, also a creative place capable of attracting the creative class? Furthermore, are these places, rich with cultural, artistic, human and environmental resources, capable of generating wealth and economic development for the future?

Notes

This research was financed with funds of the Project ex-60 per cent 'Creative industries in cities of art' (University of Florence); and the PRIN/2005 Project no. 20051370540, 'Territorial benchmarking: analytical tools applied to performance and competitiveness in local production systems' (national coordinator: University of Milan; partners: Universities of Florence, L'Aquila, Lecce and Brescia). I wish to thank Barbara Nencioni for data collection; and also Dafna Schwartz, Tommaso Cinti, Francesco Capone and two anonymous referees for their helpful comments.

1 Local contexts (cities and regions) with a cultural vocation have long ago taken up a strategic dimension in political and economic planning, as shown by the last general policy actions taken by the European Community, such as the *Agenda 2000* (European Union, 1999) and the first, fifth and sixth framework programmes (European Commission, 1998, 1999, 2001).
2 Among the artistic resources belong the set of artistic assets and works of art in the strict sense (e.g. monuments, architectural complexes, works of art, buildings, archaeological sites); 'cultural resources' refers to that set of activities, behaviours, habits and customs of life that make one place different from any other (e.g. universities and research centres, typical arts and crafts, contextual knowledge, events and manifestations, or the neighbourhood 'atmosphere'); among 'human resources' fall those expressly ascribable to human capital (e.g. artists, writers, scientists, artisans); and 'environmental resources' refers to typical elements of the urban, natural and environmental landscape (e.g. urban morphology, ornamental gardens, parks, streets, squares, neighbourhoods, characteristic flora and fauna).
3 The processes of cultural districtualisation are: increase in the division of labour based on Cultural, Artistic, Human, and Environmental Heritage; relationships between specialised productive skills and a general core of cultural needs; building up of local labour markets (teams); presence of specific integrators of different specialisations (for instance, Florence museums); routines of district socialisation applied by individual and collective agents; institutionalisation of formal and

informal successful routines; development of knowledge and know-how within the districts; development of a strong sense of belonging; the presence of new entrepreneurs and entrepreneurship; and others still to be identified (Lazzeretti, 2001).

4 These three factors have already been considered separately as resources for the economic development of a place. According to the 'theory of human capital' (Lucas, 1998), regional development is channelled by the concentration of highly educated, qualified and productive people, so far as firms manage to create a synergy with them, instead of a mere supplier–client relationship ('talent'). The conventional approach (Glaeser, 2000) explains regional development as the result of the existence of firms able to set off technological evolution ('technology'). The 'theory of social capital' (Putnam, 2000) finds the reasons of local economic development in a community's ability to interact and be the source of social trust and affluence ('tolerance'). In Florida's model these three elements, previously considered on a case-by-case basis, are integrated and explained in their interaction as factors of local economic growth.

5 This figure is given by the rate of units employed in the high-tech industries (hardware and physical products, software and services, telecommunications and audio-visual) compared to total employment.

6 The index of innovation capacity is given by the number of requests for patent rights per thousand inhabitants.

7 Tolerance is measured by the index of diversity (incidence of foreigners in the total population, and variety of ethnic groups), and by the indexes of integration of immigrants (graduated foreigners, mixed marriages, incidence and educational level of foreign children) and homosexuals (attitudes towards gays and lesbians) (Florida and Tinagli, 2005).

8 'The index is based on the number of writers, designers, musicians, actors and directors, painters and sculptors, photographers, and dancers. Regions in which these 'Bohemians' are over-represented possess a milieu that favours openness to creativity and artistic expression' (Florida and Gates, 2001, p. 2).

9 The Opificio delle Pietre Dure is a public institution and a world leader for restoration and for training restoration workers with a long-established tradition, as it was founded in the sixteenth century.

10 'With this definition, creative industries constitute a very large portion of capitalist economies – design, fashion, film, multimedia, software, publishing, advertising, arts and so on (about 15 sectors)' (Wu, 2005, p. 2).

11 The *CIMD* covers advertising, architecture, the art and antiques market, crafts, design, designer fashion, film and video, interactive leisure software, music, the performing arts, publishing, software and computer services, and television and radio.

12 Forty-four municipalities belong to the province of Florence: Bagno a Ripoli, Barberino di Mugello, Barberino Val d'Elsa, Borgo San Lorenzo, Calenzano, Campi Bisenzio, Capraia e Limite, Castelfiorentino, Cerreto Guidi, Certaldo, Dicomano, Empoli, Fiesole, Figline Valdarno, Firenze, Firenzuola, Fucecchio, Gambassi Terme, Greve in Chianti, Impruneta, Incisa in Val d'Arno, Lastra a Signa, Londa, Marradi, Montaione, Montelupo Fiorentino, Montespertoli, Palazzuolo sul Senio, Pelago, Pontassieve, Reggello, Rignano sull'Arno, Rufina, San Casciano in Val di Pesa, San Godenzo, San Piero a Sieve, Scandicci, Scarperia, Sesto Fiorentino, Signa, Tavarnelle Val di Pesa, Vaglia, Vicchio and Vinci.

13 A review of the use of location quotients can be found at an Italian level in Sforzi (1997) and at an international level in De Propris (2005). In detail, Drejer and Vinding (2005) apply location quotients to the knowledge-intensive services; Pratt (1997) and Bassett *et al.* (2002) to the cultural sector in the United Kingdom; and finally, García *et al.* (2003) to the Spanish sector.

14 Two different indicators were in fact calculated: the first measures the incidence and the second the percentage. In the present analysis, only the first index has been considered, and is calculated as the ratio (HS firms in the city/city industry)/(HS firms in Italy/Italian industry).

15 Publishing (economic and legal) consultancy, architecture, advertising, cinema, radio and television, and cultural activities. These sub-sectors were identified by Caves (2000).

16 Taken together, the nine local labour systems of Milan, Turin, Trento, Padua, Trieste, Parma, Bologna, Rome and Florence account for a 90 per cent of total employees of creative industries in Italy.

17 Interestingly, other researchers obtained similar results. For instance, the cultural industries of the United Kingdom in 1996 accounted for 4.5 per cent of overall employment (Pratt, 1997).

18 We must note that, since the creative industries belong to different sectors, data on creative firms are also computed within the various sectors of Florence province.

References

Allen Consulting Group. (2001) *The economic contribution of Australia's copyright industries*. Sydney: Australian Copyright Council and Centre for Copyright Studies.

Amabile, T. M. (1988) A model of creativity and innovation in organizations. *Research in Organizational Behavior*, 10, pp. 123–167.

Amadasi, G. and Salvemini, S. (2005) *La città creativa. Una nuova geografia di Milano*. Egea and the Institute for Competitiveness and Prosperity.

Arts Council England (2003) *Creative industries policy*. Manchester: Arts Council England, North West.

Barron, F. and Harrington, D. M. (1981) *Creativity in context: updated to the social psychology of creativity*. Boulder, CO: Westview Press.

Bassett, K., Griffiths, R. and Smith, I. (2002) Cultural industries, cultural clusters and the city: the example of natural history film-making in Bristol. *Geoforum*, 33, pp. 165–177.

Becattini, G. (1989) Il distretto industriale come ambiente creativo. In F. Benedetti (ed.) *Mutazioni tecnologiche e condizionamenti internazionali*. Milan: Franco Angeli.

Becattini, G. (2004) *Industrial districts: a new approach to industrial change*. Cheltenham, UK: Edward Elgar.

Bellandi, M. (1992) The incentives to decentralized industrial creativity in local systems of small firms. *Revue industrielle*, 59, pp. 99–110.

Belussi, F. and Sedita, R. S. (2005) The economics of intangible: some theoretical bases in networks of creativity with a focus on cultural, design, and science based industries. Paper presented at the Summer Conference on Dynamics of Industry and Innovation: Organizations, Networks and Systems, DRUID, 27–29 June, Copenhagen.

Benhamou, F. (2004) *L'economia della cultura*. Bologna: Il Mulino.

Bodo, C. and Spada, C. (eds) (2004) *Rapporto sull'economia della cultura in Italia 1990–2000*. Bologna: Il Mulino.

Brooks, A. C. and Kushner, R. J. (2001) Cultural districts and urban development. *International Journal of Arts Management*, 3 (2), pp. 4–15.

Capone, F. (2006) Identification and analysis of cultural creative systems in Italy (1991–2001). *Proceedings of the XIVth International Conference of the ACEI*, 6–9 July, Vienna.

Caves, R. (2000) *Creative industries: contracts between art and commerce*, Cambridge, MA: Harvard University Press.

CCIAA (2004) *Banca dati Stock View. Statistiche economiche territoriali*, September, Florence: Chamber of Commerce.

Chong, D. (2002) *Art management*. London: Routledge.

Cooke, P. (2005) Culture, clusters, districts and quarters: some reflections on the scale question. *Sviluppo Locale*, XI (26), 7–31.

CSPS (Comitato Scientifico del Piano Strategico) (2001) *Progettare Firenze. Materiali per il piano strategico dell'area metropolitana*. Florence: Edizioni Comune Network.

DCMS (Department of Media, Culture and Sport) (1998) *The creative industries mapping document 1998*, London: HMSO.

DCMS (Department of Media, Culture and Sport) (2001) *The creative industries mapping document 2002*, London: HMSO.

De Propris, L. (2005) Mapping local production systems in the UK: methodology and application. *Regional Studies*, 39 (2), pp. 197–211.

Dei Ottati, G. (2005) El efecto districto: algunos aspectos conceptuales de sus ventajas competitivas. *Economia Industrial*, 359, pp. 73–81.

Demel, K., Falk, R., Harauer, R., Landsteiner, G., Leo, H. and Ratzenbck V. (2004) *An analysis of the economic potential of the creative industries in Vienna*. Kulturdokumentation, Mediacult, Vienna: Wifo.

Drejer, I. and Vinding, A. L. (2005) Location and collaboration: manufacturing firms. Use of knowledge intensive services in product innovation. *European Planning Studies*, 13 (6), pp. 879–898.

Dziembowska-Kowalska, J. and Funck, R. H. (2000) Cultural activities as a location factor in European competition between regions: concepts and some evidence. *Regional Science*, 34, pp. 1–12.

European Commission (1998) *First European Community Framework Programme in Support of Culture (2000–2004)*. Luxembourg: Office for Official Publications of the European Communities.

European Commission (1999) *Le Cinquième programme-cadre*. Les programmes de recherche de l'Union Européenne 1998–2002. Luxembourg: Office for Official Publications of the European Communities.

European Commission (2001) *Exploitation and development of the job potential in the cultural sector in the age of digitalisation*. DG Employment and Social Affairs.

European Union (1999) *Agenda 2000*. Brussels.

Evans, G. (2003) Hard branding the cultural city: from Prado to Prada. *International Journal of Urban and Regional Research*, 27 (2), pp. 417–440.

Evrard, Y. and Colbert, F. (2000) Arts management: a new discipline entering the millennium? *International Journal of Arts Management*, 2 (2), pp. 4–13.

Florida, R. (2002) *The rise of creative class*. New York: Basic Books.

Florida, R. and Gates, G. (2001) *Technology and tolerance: the importance of diversity to high-technology growth*. Brookings Institution Survey Series. Washington, DC: Brookings Institution.

Florida, R. and Lee, S. Y. (2001) Innovation, human capital, and diversity. Presentation given at the APPAM 2001 Research Conference, November, Washington, DC.

Florida, R. and Tinagli, I. (2004) *Europe in the creative age*. London: Demos.

Florida, R. and Tinagli, I. (2005) *L'Italia nell'era creativa*. Milan: Creative Group Europe.

Frost Kumpf, H. A. (1998) *Cultural district: the arts as a strategy for revitalizing our cities*. Washington, DC: Institute for Community Development and the Arts, Americans for the Arts.

García, M., Fernández, Y. and Zofío, J. (2003) The economic dimension of the culture and leisure industry in Spain: national, sectoral and regional analysis. *Journal of Cultural Economics*, 27 (1), pp. 9–30.

Gertler, M., Florida, R., Gates, G. and Vinodrai, T. (2002). *Competing on creativity: placing Ontario's cities in North American context*. Toronto: Ontario Ministry of Enterprise, Opportunity and Innovation.

Glaeser, E. L. (1999) *The future of urban research: nonmarket interactions*. Washington, DC: Brookings Institution.

Glaeser, E. L. (2000) The new economics of urban and regional growth. In G. Clark, M. Gertler and M. Feldman (eds). *The Oxford handbook of economic geography*. Oxford: Oxford University Press.

Gottlieb, P. (1995) Residential amenities, firm location, and economic development. *Urban Studies*, 32, pp. 1413–1436.

Greffe, X. (2003) *La gestione del patrimonio culturale*. Milan: Franco Angeli.

Howkins, J. (2001) *The creative economy: how people make money from ideas*. London: Penguin.

ISTAT (2002) *Classificazione delle attività economiche*. ATECO 2002. Rome: ISTAT.

Jacobs, J. (1984) *Cities and the wealth of nations*. New York: Random House.

Jeffcutt, P. and Pratt, A. C. (2002) Managing creativity in the cultural industries. *Creativity and Innovation Management*, 11 (4), pp. 225–233.

Kotlin, J. (2000) *The new geography*. New York: Random House.

Krackhart, D. and Porter, L. W. (1985) When friends leave: a structural analysis of relationship between turnover and stayer attitude. *Administrative Science Quarterly*, 30, pp. 242–261.

Kunzmann, K. R. (1994) Kultur, ein Standortfaktor? In R. Funck (ed.) *Regionale Identität im offenen Europa*. Heidenheimer Schriften zur Regionalwissenschaft August. Lösch in memoriam 12, Heidenheim an der Brenz: Kulturamt.

Lazzeretti, L. (2001) I processi di distrettualizzazione culturale delle città d'arte: il cluster del restauro artistico a Firenze. *Sviluppo Locale*, 8 (18), pp. 61–85.

Lazzeretti, L. (2003) City of art as high culture local system and cultural districtualization processes: the cluster of art restoration in Florence. *International Journal of Urban and Regional Research*, 27 (3), pp. 635–648.

Lazzeretti, L. (ed.) (2004) *Art cities, cultural districts and museums*. Florence: Firenze University Press.

Lazzeretti, L. (2005) Note sul modello di distrettualizzazione culturale dei luoghi ad alta intensità culturale, *Sviluppo Locale*, XI (26), forthcoming.

Lazzeretti, L. and Cinti, T. (2001) La valorizzazione economica del patrimonio artistico delle città d'arte: il restauro artistico a Firenze. Florence: Firenze University Press.

Lazzeretti, L. and Nencioni, B. (2005) Cultural and creative industries in 'high cultural place': the case of the art city of Florence. *Proceedings of the XXIst conference of the Regional Studies Association*, 28–31 May, Aalborg, Denmark.

Leslie, D. (2005) Creative cities? *Geoforum*, 36, pp. 403 405.

Lloyd, N. and Clark, T. N. (2001) The city as entertainment machine. In K. F. Gatham (vol. ed.) *Research in urban sociology*. vol. 6, *Critical perspectives on urban redevelopment*. Oxford: JAI/Elsevier, pp. 357–378.

Lucas, R. E. Jr (1998) On the mechanics of economic development. *Journal of Monetary Economics*, 22, pp. 1–42.

McCain, R. A. (2004) Cognitive economics, creativity, and the economics of the arts. Paper presented at the XIIIth Conference of the Association for Cultural Economics International, June, Chicago.

Marshall, A. (1920) *Principles of economics* (8th ed.). London: Macmillan.

Mommaas, H. (2004) Cultural clusters and the post-industrial city: towards the remapping of urban cultural policy. *Urban Studies*, 41 (3), pp. 507–532.

Mood, A. M., Graybill, F. A. and Boes, D. C. (1988) *Introduzione alla statistica*. Milan: McGraw-Hill.

Mossetto, G. (1992) *L'economia delle città d'arti: Modelli di sviluppo a confronto, politiche e strumenti di intervento*. Milan: Etas libri.

OECD (2005) *Culture and local development*. Paris: OECD.

Pierce, N. (2002). Lively economies require lively cities. *Washington Post*, 9 June.

Pilotti, L. and Rinaldin, M. (2004) Culture and arts as knowledge resources towards sustainability for identity of nations. *Finanza Marketing e Produzione*, 22 (1), pp. 5–37.

Porter, M. E. (1998) *On competition*. Boston: Harvard Business School Press.

Pratt, A. (1997) The cultural industries production system: a case study of employment change in Britain, 1984–91. *Environment and Planning A*, 29 (11), pp. 1953–1974.

Putnam, R. (2000) *Bowling alone: the collapse and revival of American community*. New York: Simon & Schuster.

Sacco, P. and Pedrini, S. (2003) Il distretto culturale: mito o opportunità? *Il Risparmio*, 51 (3), pp. 101–155.

Santagata, W. (2005) Cultural districts, clusters and economic development. In V. Ginsburgh and D. Throsby (eds) *Handbook on the economics of art and culture*. Amsterdam: Elsevier.

Scott, A. J. (1988) *New industrial spaces: flexible production organisation and regional development in North America and Western Europe*. London: Pion.

Scott, A. J. (2000) *The cultural economy of cities*. London: Sage.

Scott, A. J., Agnew, J., Soja, E. and Storper, M. (2001) Global city regions. In A. J. Scott (ed.) *Global city regions*. Oxford: Oxford University Press.

Sforzi, F. (1997) *I sistemi locali in Italia*. Rome: ISTAT.

Simonton, D. (1999) *Origin of genius: Darwinian perspectives on creativity*. Oxford: Oxford University Press.

Spencer, G. (2006) Cognitive diversity and creative advantage in city-regions. Paper presented at the DRUID Winter Conference, January, Alborg, Denmark.

Throsby, D. (2001) *Economics and culture*. Cambridge: Cambridge University Press.

Towse, R. (ed.) (2002) *A handbook of cultural economics*. Cheltenham, UK: Edward Elgar.

Towse, R. (2004) Towards an economics of creativity? In *Proceedings of the workshop Creative Industries: A Measure for Urban Development?*, 20 March, Vienna.

Trullèn, J. and Boix, R. (2001) Economia della conoscenza e reti di città: città creative nell'era della conoscenza. *Sviluppo Locale*, 8 (18), pp. 41–61.

UNESCO (2003) *The international creative sector: its dimensions, dynamics and audience development*. Austin, TX: UNESCO.

Woodman, R. W., Sawyer, J. E. and Griffin, R. W. (1993) Toward a theory of organizational creativity. *Academy of Management Review*, 18, pp. 293–321.

Wu, Weiping (2005) Dynamic cities and creative clusters. World Bank Policy Research Working Paper no. 3509.

Wynne, D. (ed.) (1992) *The culture industry: The arts in urban regeneration.* Aldershot, UK: Avebury.

Wyszomirsky, M. (2004) Defining and developing creative sector initiatives. In *Proceedings of the workshop Creative Industries: A Measure for Urban Development?*, 20 March, Vienna.

Zanni, L. and Bianchi, P. (2004) The construction of interindustry networks as value creation strategies in the cultural districts: synergies between museum and fashion business according to Italian and foreign experience. In B. Sibilio Parri (ed.) *Creare e valorizzare i distretti industriali.* Milan: Franco Angeli.

Zukin, S. (1995) *The cultures of cities.* Oxford: Blackwell.

10 Reflections on innovative alliances involving technological science and the creative industry: a case study involving the Roskilde region and Musicon Valley

Birgitte Rasmussen and Jens-Peter Lynov

The Roskilde region and Musicon Valley

Back in 1998 the city of Roskilde in Denmark celebrated its millennium. At that time a visionary think-tank was appointed with the aim of looking ahead and focusing on the future development of the city and the region. To describe the region's strong points and characteristics, the think tank identified three key areas: (1) music; (2) knowledge; and (3) water (i.e. inlets and springs). Out of that came the first vague idea about a "musicon valley", which represents the vision of developing the Roskilde region in terms of the topics music and knowledge with the ambition "to create an international power centre for the creative industries in the region".

The idea of developing a creative industries cluster was concretised with the establishment of the Musicon Valley organisation in 2001 (www. musiconvalley.dk). Musicon Valley builds on the region's strong points:

- Roskilde offers a range of cultural activities, especially music events, including the Roskilde Festival (attracting an audience of about 70,000 people each year), the regional music club Gimle, the International Franz Schubert Society of Denmark, the School of Sacred Music, folk music festivals and jazz festivals.
- Roskilde is a centre of education where about 20,000 young people presently attend various short or medium-level and higher education programmes.
- The region has a high concentration of knowledge and research institutions (Roskilde University Centre, Risø National Laboratory, the CAT Science Park, the National Environmental Research Institute, the Danish Meat Research Institute).
- Roskilde is a historical town, with its famous cathedral and various popular museums (e.g. the Viking Ship Museum, the Lejre Experimental Centre) located in beautiful surroundings. Roskilde Fjord can be experienced on board a Viking ship or on a modern cruise vessel.

At the municipality level, the city council has formulated an ambitious plan for developing Roskilde and the region as a musical power centre (Roskilde Kommune, 2005).

In this chapter we describe and discuss the development of the Musicon Valley initiative seen from the perspective of Risø National Laboratory, a government research institution and one of the key institutions in the region. Our aim is to focus on the incentives, challenges and dynamics of cooperation between the creative industries and technological science. We look at Risø as a player in regional development. Moreover, we want to discuss how these two aspects can encourage a government research institution to bring its role and rationale into better agreement with emerging societal expectations and demands.

Musicon Valley growth environment

The Musicon Valley organisation is supported by the municipality of Roskilde, the county of Roskilde and the Danish Ministry of Science, Technology and Innovation. Furthermore, Danish companies contribute with competencies, skills and manpower.

In 2002 the Musicon Valley initiators invited Risø to join the event. The immediate reaction was "Risø has no interest in joining such a rock circus", a statement that illustrates some of the challenges and barriers in this kind of cooperation and development. At first glance it may be hard to see the perspectives of cooperation with a new and extremely untraditional partner, and it is of utmost importance to identify the person(s) inside the organisation who can see the possibilities, who has/have the enthusiasm to join the cooperation with an open mind, and who is/are willing to work for it.

However, Risø decided to join the initiative for two main reasons:

- We are interested in contributing to regional development.
- We realised that technology has an essential role to play in the creative industries, as indicated in Figure 10.1. The outside parts of Figure 10.1 reflect the various supporting and underlying services and functions necessary for arranging and performing cultural events.

Risø's contribution has mainly been within the frame of Musicon Valley Growth Environment (in Danish: Musicon Valley Vækstmiljø), a triennial self-contained activity under the Musicon Valley umbrella organisation addressing education, technology and market related to light and sound technologies. The activities are organised in four main fields, as illustrated in Figure 10.2. The other participants are Roskilde University Centre, the Roskilde Business College, the Roskilde Technical College, DPA Microphones, DPA Soundco, Seelite, ComTech, DELTA, Martin Professionals, CAT Innovation, the Roskilde Festival and DR Productions.

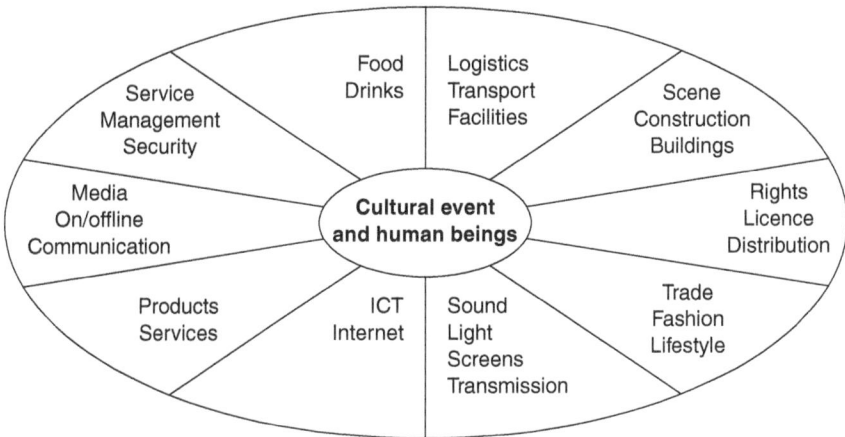

Figure 10.1 Supporting and underlying services and functions necessary for arranging and performing cultural events

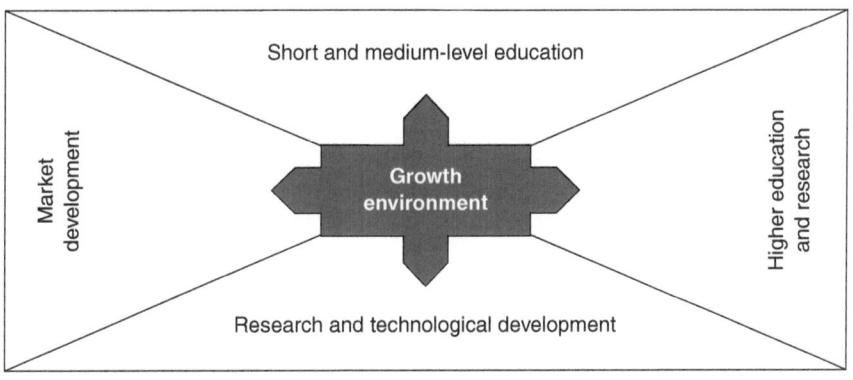

Figure 10.2 Musicon Valley Growth Environment

The Musicon Valley Growth Environment is partly financed by the Ministry of Science, Technology and Innovation, and was selected as one of 11 regional growth environments in Denmark on the basis of the following criteria:

- establishing knowledge and learning networks that can support regional development within strong regional occupational fields
- focusing research and education on regional occupational needs and applications
- increasing the role of research and education institutions in regional occupational development by closer cooperation between regional enterprises together with research and education institutions

- developing new post-school and higher education programmes in accordance with industrial and commercial demands
- ensuring improved anchoring and more dynamic interaction between technological developers and regional development within industry and commerce.

Setting the scene

Risø is a government research institution with about 700 employees under the Danish Ministry of Science, Technology and Innovation. Risø contributes to the development of environmentally acceptable methods for energy technologies, industrial production and bio-production. Risø collaborates with universities, research institutes, technological institutes and the industrial sector on a national, a European and an international basis.

From our point of view, a government research institution such as Risø has to face at least four main challenges that can also be seen as four incentives for identifying new research areas together with new commercial and scientific constellations, networks and partners. These are:

- science's new role in society, with science being asked to demonstrate the societal benefit of research investments
- the demand for successful innovation and product development in a globalised economy
- the economic and social transformation in the United States and Europe from an industrial to a creative economy
- the focus on research institutes as dynamos in regional knowledge-based economies.

New mission for national laboratories: science's new role in society

Risø was established in 1958 and, like many of Europe's governmental mission-oriented research laboratories founded in the years following World War II, had mission statements related to defence or nuclear research. When nuclear energy became less popular in many West European countries during the 1980s and after the fall of the Berlin Wall in 1989, the European governments' strategic interest in "nuclear research" and "defence" was no longer an important rationale behind mission-oriented research. Partly as a consequence of this, government R&D expenditure fell dramatically during the 1980s and the first half of the 1990s (European Commission, 1997).

During the 1990s, low economic growth, unemployment and lack of technological innovation compared with Japan and the United States came high on the political agenda in Europe (CEC, 1993, 1995). Today, government policies on science and technology are often defined according to this agenda. The prime characteristics of current and future developments in the research world are the ever-increasing foci on application and capitalisation of research ("entrepreneurial" science) and on convergence between public and private research. The boundaries and relationships between research carried out in the public sector (universities, research institutes, etc.) and research performed in the private sector are in a state of flux. Public and private organisations are taking on tasks that were formerly the province of other sectors, and shaping these relations is increasingly a subject of research and technology management at different levels (Leydesdorff, 2000). University–industry–government relations can be considered a triple helix of evolving networks of communication and cooperation. The triple helix model argues that a knowledge infrastructure is generated in terms of overlapping institutional spheres, each taking the role of the other and with hybrid organisations emerging at the interfaces. For many countries and regions the common objective is to realise an innovative environment consisting of university spin-off firms, trilateral initiatives and strategic alliances among firms, governmental research laboratories and academic research groups (Etzkowitz and Leydesdorff, 2000).

It seems as if a new social contract between science and society is under development. Under the prevailing contract, science is expected to produce reliable knowledge (i.e. in areas such as defence and nuclear power), provided merely that science communicates its discoveries to society. The contract under development must ensure that scientific knowledge is socially robust, and that its production is seen by society to be both transparent and participatory (Gibbons, 1999).

Universities all over Europe and the rest of the world have difficulties in redefining their role in this new paradigm (Etzkowitz *et al.*, 2000). Still, the universities' role is linked with the production of new scientists and technicians, and the universities' impact on society is assured through graduates' employment in academia, industry and government. The impact of government laboratories is more difficult to define. Managers of mission-oriented national laboratories have faced a myriad of challenges, owing to the shift of today's competitive environment. Science research organisations, which operate at the cutting edge of creative innovation, require organisational designs that are capable of supporting this growing trend. The advent of an increasingly sophisticated and demanding clientele has confronted many knowledge-based organisations with the need to become more innovative and better at generating customised solutions. There is an ongoing debate as to how a traditional organisation could best be redesigned to meet the demands of this complex and dynamic environment (Simpson and Powell, 1999).

Innovation processes

Words such as "innovation" and "invention" are often used interchangeably, but, while closely related, these notions are not the same (Kolodovski, 2005) (see Figure 10.3).

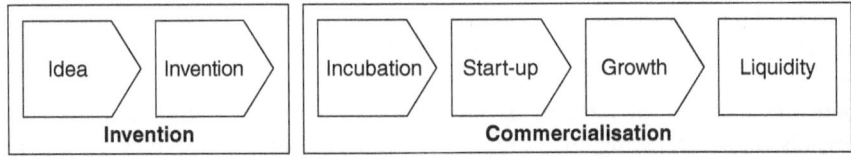

Figure 10.3 Innovation = Invention + Commercialisation. Source: Kolodovski, 2005

Invention is the creation of a new product or process. Commercialisation is the conversion of the invention into widespread use. The commercialisation part typically takes ten to a hundred times more time and resources than the invention part. Innovation can be presented as six distinct steps (see Figure 10.2):

- idea: identifying an interesting problem to solve, or discovering new technological capabilities
- invention: matching a problem and technical capabilities to create new solutions
- incubation: research and planning how to launch the invention on to the market
- start-up: establishing an organisation and generating a business plan
- growth: building up the organisation and working for growing sales to reach a profitability level
- liquidity: receiving dividends from the company, or selling shares to other investors.

From our point of view it is important to see innovation activities and processes as dynamic interactions and interplays in a large institutional set-up (see Figure 10.4). In this approach the focus is on relations, learning processes and actions between different actors and areas of knowledge. Further, innovation activities are not considered unified and linear, but rather as interactive processes involving huge numbers of actors.

New agenda: creative content of products and services

Referring to Florida and Tinagli (2004), the United States and Europe are going through a period of sweeping economic and social transformation – from an industrial to a creative economy. This transformation is fundamentally based on human intelligence, knowledge and creativity. Creativity is the motive power of economic growth. Today, between 25 and more than 30 per cent of the workers in the advanced industrial nations are employed in the creative sector of the economy (science and engineering, R&D, technology-based industries, arts, music, culture, aesthetics, architecture, design, etc.).

Experience and creative content are becoming still more important elements

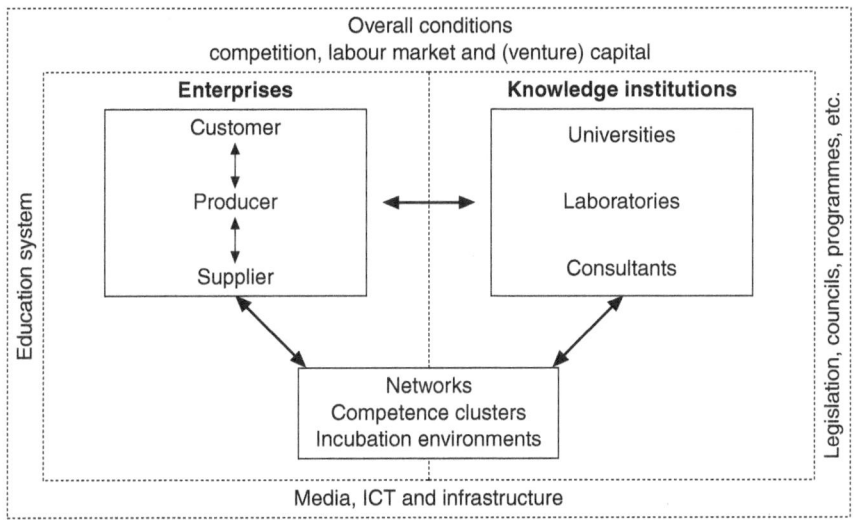

Figure 10.4 Innovation system. Source: Andersen, 2004

of any product or service as the determining competitive edge. Knowledge-intensive industrial manufacturers acknowledge that the immaterial dimensions of both their products and their brand/organisation are becoming increasingly important to address. Referring to Mandag Morgen (2002), there is an increasing demand to balance the increased use of technology (high tech) with human facets (high touch). Developers of products and services should focus not only on efficiency and rational solutions but also on high-touch dimensions such as art, intellectual fellowship and creativity.

The regional dimension of the knowledge-based economy

Regions are expected to have a core role in the development of the European Research Area. They can be a pivotal factor in driving economic growth through, for example, the development of regional innovation strategies, local-level partnerships and clusters of related enterprises and researchers. Dynamic regions can contribute to turning Europe into the most competitive knowledge-based economy in the world by 2010, a goal set by the March 2000 Lisbon European Council. The European Research Area concept implies that efforts should be coordinated effectively at different administrative and organisational layers: at European, national, regional or even local level. In this way, measures would not only be mutually consistent but would be better adapted to the potential of the regions themselves (CEC, 2001). This finding highlights the importance of developing science and technology for all policy fields by the European Commission together with national and regional actors. To be successful, the European Research Area requires coherent development of research in close dialogue with societal actors affected by these policies (Europa Kommissionen, 2002).

In regional development, Regional Innovation Systems are key elements of innovation policy for the innovativeness and competitiveness of firms and regions. These systems can be defined as interacting knowledge generation and exploitation subsystems linked to global, national and other regional systems (Asheim *et al.*, 2006). According to Cooke and Memedovic (2003), the key dimensions of a regionalised innovation system are (1) processes and policies supporting education and knowledge transfer; (2) arrangements for the governance of innovation; (3) the level of investment; and (4) the type of firms and their degree of linkage and communication (networks, partnerships, etc.). Asheim *et al.* introduce the concept "construction of advantage", saying that in the future it will not be sufficient to rely on competitive advantage being automatically created through co-location; rather, the advantage needs to be more consciously and proactively constructed in a way that takes into account sectoral and regional specificities. This points to a new and more dynamic role for the public sector (including universities and research institutes) in cooperation with private-sector partners.

As suggested by Lagendijk and Cornford (2000), the regional development industry needs concepts, notions, theories and models that can help organisations to undertake the task of developing regional economies. A huge number of conferences, seminars, etc. have already been prepared to facilitate circulation of ideas to regional development industries and players. In short, the regional development industry resembles the description of Mode II knowledge given by Gibbons *et al.* (1994). While Mode I knowledge is disciplinary based, hierarchical, science oriented and based on the linear model of knowledge flows, Mode II knowledge is interdisciplinary, heterogeneous, organisationally transient, more socially robust and reflexive.

Lessons learned

Knowledge and knowledge transfer

Knowledge transfer involves two actions: transmission (sending or presenting knowledge to a potential recipient), and absorption by that person or group (Davenport and Prusak, 1998). If knowledge is not absorbed, it has not really been transferred. Knowledge transfer has a large impact on changes in behaviour and generation of new ideas, and, consequently, innovation is inseparably bound up with knowledge transfer. Also, the Musicon Valley initiative has to face the fact that considering and evaluating the means and processes required to support and encourage knowledge transfer between the different actors is a big challenge and task. An essential question has therefore been how to establish meeting places and agendas that will appeal to scientific communities, artists and creative environments as well as the corporate sector and its manufacturers.

There is no tradition of contact between the creative content providers, traditional industry, service sectors and research institutions. In establishing meeting places and agendas it is crucial to be aware that knowledge transfer across institutional and disciplinary boundaries can involve tacit, explicit and cultural knowledge to varying degrees, and that many factors can be barriers to knowledge transfer; examples are differences in culture and vocabulary, lack of time and meeting places, and lack of absorptive capacity in recipients.

The creative content providers rarely get in touch with technological research institutions, or with social and economic disciplines such as management, or enterprise dynamics. This finding is supported by observations, presented by a Danish expert group, that many companies within the creative economy need to strengthen their knowledge and competencies regarding development and optimisation of the company's operations and processes. It is often difficult to transfer traditional explanation models related to management, organisation, innovation, market, etc. to creative economy companies, and as a result there is a need to develop knowledge that targets the specific conditions of these companies (Ministeriet for Videnskab, Teknologi og Udvikling, 2005).

One of the key questions is the incentive for interaction and cooperation between science and the creative industries. Bowker and Star (1999) have introduced the concept "boundary objects", which are objects that both inhabit several communities of practice and satisfy the informal requirements of each of them. Boundary objects so to speak represent something of common interest to meet around, and they have to be plastic enough to adapt to the local needs and constraints of the several parties employing them, yet robust enough to maintain a common identity across sites.

From the perspective of a technological research institution, the challenge of identifying boundary objects is at least twofold. It is easier said than done for technological scientists to express and explain how their skills and competencies can be used by the creative industries. The creative industry domain is a completely new and unexplored area for a research institution such as Risø, and it is very hard to see where, why and how our experience and competencies can contribute to the development of the creative sector in a meaningful way. Furthermore, the actors in the creative sector are less trained in the exercise of transferring working life experience and difficulties into structured problem-oriented projects. Figure 10.5 illustrates the usual understanding and view on natural scientists and people involved with cultural events such as the Roskilde Festival. Of course these drawings are caricatures; on the one hand they reflect the prejudices between the different actors, and on the other they represent a realistic picture of the difficulties in finding common interests and working fields. It has not been a straightforward process to find the links between the scientific world and the world of cultural events.

Structured dialogue addressing practical problems is of the utmost importance in identifying common interests and fields where the different competencies and capacities can be brought together into new initiatives. We have

What is the link between

The polite answer: **not obvious !!**

Figure 10.5 A key question for the Musicon Valley Growth Environment

chosen what could be called a user–driven innovation approach – that is, innovations derive from new functional requirements posed by professional users.

During the first years of the Musicon Valley initiative, a few concrete technological projects have been identified either as part of the Musicon Valley Growth Environment or in the form of contacts established through other Musicon Valley arrangements and activities:

- *Wireless transmission of signals (audio, light and picture) that is protected against illegal copying.* The users of light and sound technology have formulated that it is a hard and resource-demanding job to install buried cables for transmission of signals. Technically, this kind of wireless transmission is possible, but until now it has been troublesome to get from the invention phase to the commercialisation phase. One explanation is that the problem was raised by the technology users, not by the producers and developers.
- *Windshield systems for microphones with a water-repellent surface.* The Danish company DPA Microphones is interested in developing a product that gives new possibilities for using microphones in wet conditions, such as when it rains. A cooperative project has been established between producers and researchers in order to develop the technology for surface treatment.
- *Development of LED 3 W white diode source of light to replace conventional 15–20 W glow lamps.* The project is aimed at developing a high-quality LED (light-emitting diode) lamp. Novel micro- and nanostructured optical elements are being developed for efficient colour mixing and light control. The project is a collaborative one between Risø and the Danish industrial partners NESA, RGC-Lamps and Nordlux. A new project is continuing and extending this work, and moreover includes development of new lamps for this new generation of innovative light

sources. This work is done in cooperation with the Danish companies Asger BC Lys and Louis Poulsen Lighting. Both projects are supported by ELFOR, Dansk Eldistribution.

Cross-institutional and interdisciplinary collaboration

Cross-institutional and interdisciplinary collaboration can be an attractive and necessary approach to find societally robust solutions to demands and needs in society. The knowledge required for societal development cannot solely come from the offerings of any single discipline (Kahn and Prager, 1994). According to Gibbons (1999), expertise has to bring together knowledge that is itself distributed, contextualised and heterogeneous; it cannot arise at one specific site, or out of the views of one scientific discipline or group of highly respected researchers. Rather, it must emerge from the bringing together of the many different "knowledge dimensions" involved. Its authority depends on the way in which such a collective group is linked, often in a self-organised way. According to Klein (2000), research has become increasingly interdisciplinary during recent decades, and new social and cognitive forms have already altered the academic landscape, new practices have emerged and disciplinary relations have realigned. However, such arrangements often face great obstacles to make the collaboration profitable and successful. Setting up networks and projects by mixing different kinds of working fields, competences and skills, for example, also means that different requirements with respect to documentation of knowledge and competences, funding, success criteria, planning horizons, etc. will be taken into account when joint projects are in the pipeline.

In the perspective of the boundary objects concept, central aspects are differences in merits and working routines. Scientific work is characterised by processes with long time horizons, 10–15 years or even more. In the scientific world the most prominent success criterion is documentation of scientific work through publication of articles in peer-reviewed international journals. Other success criteria are patents and industrial cooperation. The process from initiation of a project and generation of results through article writing to acceptance and publication in a journal may take several years. The situation is quite different for cultural event organisers: they often have to manage very short planning horizons, as they have to offer events with popular or new artists. As an example, for the Roskilde Festival the success criteria are happy festival participants and volunteers, contented performers and musicians, pleased sponsors and politicians, and balance in economy.

Advertising and marketing are other fields where scientific institutions and cultural event organisers have different traditions and behaviour. Also in this field, the different view on documentation plays a role. Scientists are not used to public exposure of preliminary results; they strongly prefer to disclose only reliable and substantiated results. For the event organisers on the other hand, advertisement and marketing are essential means by which to get into contact with their customers. In particular, the event organisers' way of behaving in connection with advertisement and obtaining publicity concerning Musicon Valley arrangements has been a disclosure of project ideas in their early stage of development. The instructive process has dealt with getting experience in handling the general requirement that research institutions need to be more visible to the surrounding society. If an institution such as Risø is going to maintain its level of activity, we have to make improvements in respect of visibility and communication to society as well as to the Danish government.

The Musicon Valley Growth Environment has many points in common with the concepts contained within the actor-network theory (originally developed by Latour, Cannon and Law). According to Stalder (1997), in the dynamic development of networks three phases can be distinguished:

- *Emergence:* Networks are put into place by actors. Networks allow actors to translate their objectives, be it conscious human choice or prescription of an object, to other actors and add the other actors' power to their own.
- *Development:* A network can develop in two different directions, towards convergence or towards divergence of its actors.
- *Stabilisation:* Networks that are not able to stabilise themselves to a certain degree disappear from the scene. An actor-network thrives for stabilisation because none of the entities which make it up would exist without that network in that form.

This suggestion can be supplemented by observations reported by Kahn and Prager (1994) indicating that all networks have four common stages in their organisational development:

- *Listening across the gulf:* Early meetings of networks show a familiar pattern of behaviour: successive pronouncements more or less on the subject, heard with varying degrees of attention and comprehension by their listeners. The underlying task at this early stage is the search for a common theme that is specific enough to attract members intellectually, but general enough to give them room for exploration in their own terms.
- *Conceptual translation:* The underlying task is to develop a common language, a prerequisite for collaborative work. The result is a shared conceptual vocabulary, smaller and less specialised than the vocabulary of any single discipline, but enabling each member to assimilate the work of other disciplines to his or her own.
- *Onset of collaboration:* The third stage involves activities of consultation, marked by a high degree of mutual tolerance, an eagerness to help and a willingness to be helped. In some networks this level of integration leads quickly to the major collaborative efforts; in others the process is more gradual.

- *Joint projects:* Some networks tend to reach this stage in their second year, some later.

The stages observed by Kahn and Prager look recognisable from a Musicon Valley perspective. The Musicon Valley Growth Environment has now been running for about three years and its development stage is somewhere between stage 2 and stage 3. We have had several meetings and arrangements in order to obtain a common language ("dictionary building") and to establish a common frame of reference. We have identified common interests and possibilities of collaboration, but we have also identified substantial discrepancies concerning the overall understanding of the Musicon Valley Growth Environment. To some extent there has been a divergence among some of the actors, and it is not likely that the network will stabilise in its present form.

As part of the activities performed in the Musicon Valley Growth Environment, a pilot study was carried out addressing the conditions for the formation of a competence cluster among Danish actors working with sound and light technologies within the live event sector. The study concluded that at present no incipient cluster formation can be identified directly. This conclusion was largely drawn on the observation that three very different understandings and interests exist among the respondents regarding the overall idea of and possibilities for cooperation within the frame of the Growth Environment (Rasmussen and Skjerning, 2005):

- a broad covering approach addressing the creative industries with emphasis on events or products where the increase in value is established through some kind of staging or production of a clear-cut distinction between the sound technology sector and the light technology sector motivated in the physical and technological differences between sound and light technologies
- a clear-cut distinction between technology producers, professional users and end users motivated in conflicting economic interests and large cultural differences.

Innovation processes and understanding

Innovation processes demand a considerable amount of resources – human resources as well as capital. There is no guarantee of success, as innovation processes are very complex and depend on a huge amount of driving forces, interests, competencies, etc.

Looking at the innovation process proposed by Kolodovski (see Figure 10.3), the activities within the Musicon Valley Growth Environment have mainly been focused on the first step – that is, getting ideas and identifying interesting problems to solve, or discovering new technological capabilities. To some extent the activities have been concentrated on step 2 – that is, matching problems and technical capabilities to create new solutions. The most resource-demanding steps – that is, the commercialisation steps – have only been addressed to a minor extent.

The Musicon Valley Growth Environment has been very focused on building networks and on the assumption that placing people in new constellations confronting them with each other's viewpoints and competencies will create new ideas and that this is the hard core of an innovation process. However, in terms of innovation theory this assumption is not well justified. Ideas are necessary, but are only the beginning of a hard and resource-demanding process.

Regional anchoring

The Musicon Valley organisation has been launched and supported by the municipality and the county of Roskilde. At local and regional levels, expectations of the Musicon Valley initiative have been rather high. However, at the municipality level there has not been a clear understanding of the huge amount of resources needed to accomplish the innovation processes. The financial support from the municipality has been rather low and allocated only on an annual basis; this kind of short-term financial support is not in accordance with the rather long time horizons needed for innovation through networks.

Future work and perspectives

The relations, arrangements and cooperation in the Musicon Valley Growth Environment have been an eye-opener at Risø. We have learned new ways of being visible outside the scientific world. The public relations activities have been of great value for Risø. It has been overwhelming and a positive experience that the media have shown a noteworthy and considerable interest in Musicon Valley and in Risø's participation. In this chapter we would like to emphasise the coverage in *Nature* of Risø's participation in the Roskilde Festival under the heading "Science rocks" (Smaglik, 2003). Inside the organisation this type of coverage is of significant importance in order to obtain acceptance and recognition of the efforts spent on the Musicon Valley Growth Environment. Furthermore, we hope that different kinds of coverage will play a role in creating an image of Risø as being in the forefront with respect to creativity and technology and that the institution can in that way be a central player in the development of the region.

Funding for the Musicon Valley Growth Environment from the Ministry of Science, Technology and Innovation has been allocated for a three-year period. In terms of the time horizons for the development of new networks and joint projects, three years is probably too short a time frame. The Musicon Valley Growth Environment experienced difficulties in reaching common interests within the domain of light and sound technologies, and in building up new constellations between the event industry and the techno-logical science sector. It is important to have more than one shot, and Risø therefore wanted to explore the possibilities in another technological domain; materials science was chosen as the new domain, as this is a core competence field at Risø.

In 2004, Risø arranged a conference entitled "Materials and Innovation in the Creative Industries". Surprisingly, the conference attracted about 130 participants and was covered by the press. The follow-up on this interest has been a set of initiatives to establish closer contact and cooperation between designers and scientists. A series of meetings under the headline "12 designers meet 12 scientists" were arranged, resulting in identification of areas of common interests and also initiation of new concrete projects.

Risø is a technological research institution that must be capable of com-peting internationally; otherwise, the institution will not meet the require-ments for getting funding. The strong priority given to the international level may conflict with involvement and contribution at the regional scale. It is of interest to Risø to be a central player in the regional development and to be located in a strong region in Denmark in order to attract competent staff. Therefore, Risø has to strike a balance between regional, national and inter-national interests and perspectives.

Musicon Valley Growth Environment has been a part of the Musicon Valley initiative. It has now been running for approximately three years and it may be worthwhile to stop and reflect on how the entire initiative impacted on the regional development of Roskilde. First of all, a period of three years is a very short time in which to evaluate the results, in view of the time horizons normally seen in relation to innovation and regional develop-ment processes. Therefore, we can only discuss observations of the initial stage of a development process.

The most conspicuous observation is that Musicon Valley has been able to fulfil the role of being a catalyst for new ideas and initiatives. Examples of ideas or initiatives supported or initiated by Musicon Valley are the following:

- Several seminars and conferences have been organised, addressing differ-ent aspects of cultural development and creative industries. The arrange-ments included highly qualified presentations and they have contributed to building up a common reference and knowledge platform. Further, the programmes normally contained long coffee breaks, and in that way the seminars and conferences also functioned as informal meeting places providing room for free-flowing conversations.
- Musicon Valley has been represented in activities at the Roskilde Festival promoting and demonstrating technology and science to a broad young audience. These activities attracted attention at governmental level as one year the Minister for Research participated in the opening ceremony, expressing his support for the cooperation and initiatives within Musicon Valley.

- The visions behind the Musicon Valley initiative have been integrated in the strategies at municipality level, especially the strategy addressing the region as a musical power centre. The establishment of the Musicon Valley organisation has contributed to the trustworthiness of the strategy and helped it find political favour. Further, some of the actors from the Musicon Valley sphere have been involved in the development of a band academy in Roskilde for rhythmic music.
- BACKSTAGE is a cross-disciplinary knowledge forum originally initiated by Musicon Valley and today hosted by the Roskilde Festival. The intention is to strengthen the event trade as an actor in the global knowledge-based experience economy.
- Education has been one of the keywords. At Roskilde University a new course in performance design has been offered, addressing live performances and cultural events. Roskilde Business College has introduced a new course in performance management.
- "Danish Sound Design" is a new audio network in Denmark. Musicon Valley activities contributing to the development of the network have been a background report describing and analysing the Danish audio sector (Musicon Valley Vækstmiljø, 2006), together with the organisation of meetings and conferences.
- An ambitious idea that has been on the local agenda for more than five years is the establishment of a rock museum in Roskilde. The idea emerged from the Danish rock music environment, and Roskilde was selected as host for the museum owing to the Roskilde Festival's international position in rock music. Musicon Valley has been involved in the development work. At present the idea is still alive but there is still lot of work to be done.
- The combination of music, culture and knowledge in the vision of Musicon Valley has recently inspired local actors and politicians to consider the possibilities of an annual winter festival in Roskilde, with spirit and knowledge as the overall festival themes.

References

Andersen, M. M. (2004) An innovation system approach to eco-innovation: aligning policy rationales, Paper given at the conference "The greening of policies: interlinkages and policy integration", Berlin, 3–4 December 2004.

Asheim, B., Coenen, L., Moodysson, J. and Vang, J. (2006) Constructing knowledge-based regional advantage: implications for regional development, Paper given at "Innovation Pressure, International ProACT Conference: Rethinking Competitiveness, Policy and the Society in a Globalised Economy", 15–17 March, Tampere, Finland, www.proact2006.fi/.

Bowker, G. C. and Star, S. L. (1999) *Sorting Things Out: Classification and Its Consequences*, Cambridge, MA: MIT Press.

CEC (1993) *Growth, Competitiveness, Employment: Challenges and ways forward into the 21st century*. White Paper. Bulletin of the European Communities, Supplement 6/93.

CEC (1995) *Green Paper on Innovation*. Bulletin of the European Communities, Supplement 5/95.

CEC (2001) *Communication from the Commission: The regional dimension of the European research area*. COM (2001) 549 final.

Cooke, P. and Memedovic, O. (2003) *Strategies for regional innovation systems: Learning Transfer and Applications*, UNIDO (United Nations Industrial Development Organisation).

Davenport, H. and Prusak, L. (1998) *Working Knowledge: How Organizations Manage What They Know*, Boston: Harvard Business School Press.

Etzkowitz, H. and Leydesdorff, L. (2000) The dynamics of innovation: from National Systems and "Mode 2" to a triple helix of university–industry–government relations, *Research Policy*, 29, pp. 109–123.

Etzkowitz, H., Webster, A., Gebhardt, C. and Terra, B. R. C. (2000) The future of the university and the university of the future: evolution of ivory tower to entrepreneurial paradigm, *Research Policy*, 29, pp. 313–350.

European Commission (1997) *Second European Report on S&T Indicators*. ECSC–EC–EAEC, Brussels/Luxembourg.

Europa Kommissionen (2002) *Praktisk vejledning i regionalt fremsyn i Danmark*, ed. B. H. Jørgensen, I. Miles, M. Keenan, G. Clar and C. Svanfeldt, EUR 20478.

Florida, R. and Tinagli, I. (2004) *Europe in the creative age*, Pittsburgh Carnegie Mellon Software Industry Center.

Gibbons, M. (1999) Science's new social contract with society, *Nature*, 402, C81–C84.

Gibbons, M., Limoges, C., Nowotny, H., Schartzman, S., Scott, P. and Trow, M. (1994) *The new production of knowledge*, London: Sage.

Klein, J. T. (2000) A conceptual vocabulary of interdisciplinary science, in P. Weingart and N. Stehr (eds) *Practising interdisciplinarity*, Toronto: University of Toronto.

Kolodovski, A. (2005) *Innovation process* (prepared for Risø National Laboratory).

Lagendijk, A. and Cornford, J. (2000) Regional institutions and knowledge: tracking new forms of regional development policy, *Geoforum*, 31, pp. 209–218.

Leydesdorff, L. (2000) The triple helix: an evolutionary model of innovations, *Research Policy*, 29, pp. 243–255.

Kahn, R. L. and Prager, D. J. (1994) Opinion: interdisciplinary collaborations are a scientific and social imperative, *The Scientist*, 8 (14): 12.

Mandag Morgen (2002) *Næste generations udviklingspolitik: mellem kultur, erhverv og kompetencer*.

Ministeriet for Videnskab, Teknologi og Udvikling (2005) *Det innovative humaniora og samfundsvidenskab: oplæg til en forskningspolitisk handlingsplan*, rapport fra en ekspertgruppe.

Musicon Valley Vækstmiljø (2006) *Dansk lydteknologi, en global styrkeposition: kortlægning af danske virksomheder og forskning som bygger på og anvender lydteknologi*, ed. Flemming Madsen and Jan Voetman.

Rasmussen, B. and Skjerning, A. (2005) *Musicon Valley Vækstmiljø. Pilotstudie af forudsætningerne for udvikling af en kompetenceklynge blandt danske aktører relateret til live-event teknologi med særligt fokus på lys og lyd*, Risø-R-1489(DA).

Roskilde Kommune (2005) *Musikbyen Roskilde*.

Simpson, B. and Powell, M. (1999) Designing research organisations for science innovation, *Long Range Planning*, 32 (4), pp. 441–451.

Smaglik, P. (2003) Science rocks, *Nature*, 424, 10 July, p. 233.

Stalder, F. (1997). *Actor-network-theory and communication networks: Toward convergence*, University of Toronto, http://felix.openflows.org/html/Network_Theory.html.

Part III

Knowledge transfer, R&D outsourcing, open innovation

11 Research, knowledge and open innovation: spatial impacts upon organisation of knowledge-intensive industry clusters

Philip Cooke

Introduction

This chapter reports the elaboration and testing of a new theoretical approach to understanding economic development. The focus is on microeconomic geography, informed by theoretical insights such as those of Krugman (1995) on spatial monopoly, Penrose (1959/1995) on the knowledge capabilities of firms and their networks, and Chesbrough (2003) on 'open innovation'. In brief, the theory framework suggests that the following has been superseded as a Western industrial organisation model. When firms exploited administrative scale advantages to perform most business functions in-house, among the key functions so performed were those involving knowledge exploration and exploitation (Chandler, 1990; March, 1991). This involved the combination of research and development (R&D), and administrative command and control to enable research to be transformed into commercialised knowledge as products and services, including innovations in both. This often led to co-location of such functions in geographic proximity, then later some decentralisation of facilities within corporate expansion.

However, during the 1990s it became clear that a significant shift in location of R&D by large firms was occurring through outsourcing to smaller, specialist research-intensive firms and public research organisations (PROs). These research suppliers were often themselves co-located not in metropolitan centres but, for example, in university towns. Although Chesbrough (2003) noted the phenomenon of what he termed 'open innovation', he failed to offer any analysis of its economic geography, something a recent test also fails to do (Laursen and Salter, 2004). This is an important lacuna since it leaves undecided which are the beneficiary regions, what specific knowledge assets they combine and whether 'economic governance' (Goodwin *et al.*, 2002) may assist in broadening the spatial domain of such knowledge advantage. For example, efforts to construct regional advantage in scientific research for economic development are common but only infrequently successful (Fuchs and Shapira, 2004; Wink, 2004).

Hence, the chapter seeks to explore future business and governance expectations regarding knowledge outsourcing by large corporations, where the leading and future global locations with regional advantage, or even spatial monopoly, regarding open innovation and how immutable any such patterns are likely to be. Briefly, the research reports on global knowledge flows among R&D customers and suppliers discovered in UK survey research in two highly research-intensive industries, namely information and communication technologies (ICT) and biopharmaceuticals. These results, in turn, facilitate a more rounded picture than hitherto of the economic and geographical evolution of firms in these industries regarding research and other knowledge management strategies. A few steps forward are made in this regard in this chapter. These tests of new thinking, with their intent to compare and contrast intent are intended to offer a modest advance to knowledge and practice in the fields of economic geography and regional science policy (see, for example, Cooke, 2004a).

The background evidence for spatially focused open innovation

This research began with a suite of papers specialising in understanding spatial dynamics of bioscience (Cooke, 2001, 2002, 2003, 2004b). These discovered a remarkable inversion of the 'scale' thesis so beloved of economic geographers who see globalisation as a linear, hierarchical and totalising process in which 'only size matters' (Brenner, 2001; Bunnell and Coe, 2001; Mackinnon et al., 2002; Bathelt, 2003). Brenner (2001) refers to this view of the world, somewhat suffocatingly, as 'scalar enveloping'. The stark finding from the five years' bioregional research, since supplemented for a further five to ten years through the award by the UK Economic and Social Research Council of a Centre for the Social and Economic Analysis of Genomics (CESAGen) to the present author and colleagues at Cardiff University is that in bioscience, scale of the kind propounded by the above advocates has very little purchase on contemporary reality. *Au contraire*, increasingly it is the case that small actors such as 'star' scientists, their research teams and academic entrepreneurial firms, often collaborating with their peers in centres of research excellence in a few bioregions around the world, generate most of the leading-edge knowledge upon which so-called big pharma relies for its continued profitable existence. This chapter explores the same also for ICT.

So much are R&D outsourcing and open innovation predominating in biotechnology that a widely cited industry contention is that an R&D outsourcing share of 30 per cent in 2003 will translate into one of 50 per cent in 2010 (CHA, 2004). Buckley (2005) reported the head of fabric and homecare products research at Procter and Gamble (not a pharmaceuticals company; rather, a consumer products firm, known in the main for toothpaste and washing powder) as saying:

In 1970, only 5% of global patents were issued to small entrepreneurs. Today that figure is around one-third and rising. My biggest competitor today is a person with an idea. P&G is no longer closed to ideas from outside, as it was until about 2000 … it is difficult for P&G to generate enough big ideas internally to fuel significant growth. The company estimates some 1.5 million scientists in academia or industry around the world have expertise relevant to P&G. So why not tap them?

Unfortunately, it is not yet known where, geographically, these 1.5 million academics and industry personnel are located. Perhaps an astute graduate student will turn the question into a thesis project. However, it may be argued that not all sectors are equally affected by 'open innovation' and that in the chemicals industry, for example, scale still 'envelops' key functions.

Nevertheless, on the basis of a limited amount of documentary research on agro-food chemicals it can be seen that key firms are increasingly becoming somewhat like pharmaceuticals firms in some respects, and evidence of agro-food biotechnology clustering is plentiful (Ryan and Phillips, 2004). Of particular interest have been instances of divested or privatised chemicals companies such as Syngenta (former ICI and Novartis), Avecia (former ICI), Dow Agroscience (from Dow Chemicals) and DSM (former Dutch State Mining) gradually transforming from a tradition in chemicals towards biological products and services. As they do so, they become more inclined to outsource R&D and practise 'open innovation'. Contrariwise, firms like BASF from Germany withdrew from biologics in 2001 while Monsanto and Bayer Cropscience have curtailed genetically modified (GMO) crop research. Little is known about research strategies in agro-food firms such as Nestlé or Unilever except that they have a traditional inclination not to outsource R&D significantly (Valentin and Lund-Jensen, 2003). Nevertheless, some fifteen agro-food chemicals concentrations involving smaller knowledge-intensive producing, research businesses and PROs exist globally, of which two-thirds are in Europe and North America, with the remainder in Australia.

Chesbrough (2003) writes mostly about the ICT industry where open innovation involving smart SMEs and university research centres of excellence is shown to be capturing an ever-larger share of industrial R&D, at least in the United States, as Table 11.1 reveals. Evidence from the automotive industry is further provided by Schamp *et al.* (2004), where the key outsourcing of knowledge, at least in Germany, has been that of design and supply chain management, which has been largely taken over by engineering consultancy firms. So much have such firms grown in some cases that even manufacturing supplier firms like Bernardi have been acquired by consultants. Interestingly, the 'cluster' of such consultancy businesses centres upon Germany's business services region around Frankfurt, rather than in automotive heartlands like Stuttgart, Munich or Cologne. Finally, research on knowledge-intensive business services such as management consultancy and software shows that they play the central role in transferring the research knowledge from centres of excellence to firms that enables the latter to remain innovative (Aslesen, 2004). These findings on the role of consultants suggest that knowledge transfer from exploration stage (research) to exploitation stage (commercialisation) is by no means straightforward and is mediated by an examination stage (value assessment or evaluation) at which some knowledge is discarded and only that with clear applicability retained. This further suggests that the much written about but seldom in-detail researched question as to how tacit or implicit knowledge is translated into explicit or codified knowledge is assisted by research revealing the nowadays crucial importance of third-party intermediaries in knowledge transfer. As Table 11.2 indicates, this function can be captured by the intermediary concept of *complicit* knowledge. Complicit knowledge is possessed by a third party, probably with a background in the knowledge-base of the tacit knowledge holder, who therefore belongs to the same 'epistemic community'. However, the complicit knowledge will also extend professionally into the explicit knowledge epistemic community, probably in business of some kind. Thus, the intermediary is complicit in two epistemic communities, a scientific (or maybe a symbolic, i.e. creative arts) and a commercial one.[1] In Table 11.2 the three categories of knowledge are set against three spatial categories, the first being a *knowledge domain*, or a space where exploration knowledge originating endogenously and exogenously, and possibly capable of being recombined into a commercial innovation is concentrated. The second refers to the *knowledge capabilities* such a domain accumulates through the recruitment and retention of talent available to work in research or in the commercialisation of such innovations. Finally, supporting the knowledge transformation process are its institutions, which through network relationships interact in

Table 11.1 Percentage of US industrial R&D by size of enterprise

Company size	1981	1989	1999
<1,000 employees	4.4	9.2	22.5
1,000–4,999	6.1	7.6	13.6
5,000–9,999	5.8	5.5	9.0
10,000–24,999	13.1	10.0	13.6
25,000+	70.7	67.7	41.3

Source: National Science Foundation, cited in Chesbrough (2003).

Table 11.2 Knowledge: from implicit domains to regional innovation systems

	Implicit	*Complicit*	*Explicit*
Knowledge domain	Invention	Translator	Appropriation
Knowledge capability	Talent	Research	Technique
Innovation system	Institutions	Networks	Interactions

support of regional or, for clusters, localised innovation. Networks, for example, are a paradigm case of institutional complicit knowledge intermediation. Regions with such knowledge domains, capabilities and innovation systems experience increasing returns to scale and become knowledge quasi-monopolies that are not necessarily easy for firms or other researchers to gain entry to. Thus, against the argument of Maskell and Lorenzen (2004) that, for example, clusters are simply markets – for if they were, why cluster? – the argument here is that they monopolistically exercise a shared interest in protecting themselves against too much competition. Adam Smith spoke of the conspiracy against the layman represented by firm association as follows: 'People of the same trade seldom meet together, even for merriment and diversion, but the conversation ends in a conspiracy against the public, or in some contrivance to raise prices' (Smith, 1776, bk 1, ch. 10, pt 2).

In the research that follows, the results touch many dimensions of this R&D outsourcing and 'open innovation' economic geography, but only for biotechnology and ICT. The main focus is on collaborative and cooperative relations between firms, some of which are in clusters, many of which are not. An attempt is made to show how clusters are associated with more competitive business results, a factor in stimulating 'increasing returns', spatial knowledge (quasi-) monopoly, and the rationale for R&D outsourcing to such locations. It further shows clustering rationales differ between the two sectors studied.

Research questions

Hence, theory suggests that larger corporations increasingly outsource their R&D and expect to continue so to do (i.e. Chesbrough, 2003 for ICT, biotechnology and household care goods; Schamp *et al.*, 2004 for automotives). Target suppliers exist in locations where numerous 'regional knowledge capabilities' from exploration (research), examination (testing, trialling) to exploitation (commercialisation) expertise are available. What are the economic geographical implications of this? For example, what implications are there for such issues as geographical and functional or organisational proximity? And are there significant differences between the respective R&D outsourcing practices and strategies of, on the one hand, biotechnology and ICT firms? We shall tackle this question in the empirical sections of this chapter.

This may also help us make progress regarding whether these R&D outsourcing and in-contracting movements are secular or cyclical. In other words, to what extent are currently observable practices and strategies regarding this immutable? Moreover, how far are they confined to the United States and what expectations are there of increased outsourcing of research in other countries and in these sectors globally? In other words, is there reversibility in the trend to R&D outsourcing that resulted in a US decline from 78 per cent of industrial R&D being conducted in-house by firms employing over 10,000 in 1989 to 55 per cent in 1999, while it rose from 9 per cent to 23 per cent in firms employing under 1,000 during the same period (Chesbrough, 2003)? What are foreseeable main impulses for continuation or any change in trend?

In what ways has 'economic governance' assisted in constructing advantage for regions displaying accomplished knowledge capabilities relevant to the target industry sectors? This was investigated from the company viewpoint and from that of agents of economic governance in line with the general research approach. Skills formation, recruitment and retention of 'talent' for these activities was clearly a crucial variable. Thus, the extent to which distinct economic governance strategies, including also regional research, science and technology support policies, were adopted in furtherance of constructed advantage and whether such approaches are capable of emulation for less accomplished settings are matters of key importance to the offsetting of developmental barriers occasioned by 'asymmetric knowledge and information' (Akerlof, 1970). In the theoretical approach under examination it is proposed that spatially 'asymmetric knowledge' of many kinds, but particularly that regarding innovation governance (Penrose, 1959/1995), is a prime cause of regional economic disparities. As such, understanding of knowledge deficits is of fundamental importance to the improvement of economic development theory and practice. This research addresses this in questioning UK biotechnology and ICT firms and finds economic governance to be of minor importance in firm transactions.

Framework, methods of analysis, rationale, data and information

A number of concepts have been adduced thus far, and at this point it is important to demonstrate how they frame the research. This process may be assisted by first listing and defining the main ones that orientate the proposed research:

- *Asymmetric knowledge:* after Akerlof (1970), whereby specific knowledge expertise is cumulative and path dependent, and attracts increasing returns to scale (Krugman, 1995), tending towards spatial knowledge monopoly.
- *Knowledge capabilities:* after Penrose (1959/1995), firms (and their host regions) grow competitively by practising dynamic organisational and knowledge networking expertise (see also Teece and Pisano, 1994).
- *Open innovation:* after Chesbrough (2003), the practice of contractual R&D outsourcing from corporations to PROs and specialist firms. Locations with asymmetric knowledge capabilities also yield up localised knowledge spillovers from PRO conventions of 'open science' that outweigh possible knowledge 'leakages' but add to regional knowledge monopoly, further benefiting open innovators.

- *Constructed advantage:* after Foray and Freeman (1993), whereby economic governance raises systemic regional innovation capabilities from attention to the interplay between economic, governance, knowledge and environmental advantages, thereby improving the likelihood of attracting and retaining 'talent' and regional competitiveness.

Whereas science is often stable and paradigmatic, not least for purposes of pedagogy, research may be destabilising and even iconoclastic (Latour, 1998). Therein lies the attraction for entrepreneurship from new knowledge that is capable of being commercialised in new markets that give expectations of high returns on investment. Accordingly, among the knowledge capabilities often seen to coexist with research are those associated with venture capital, patent lawyers, specialist consultants and other intermediaries involved in knowledge transfer.

In Figure 11.1 the four main concepts interact in an evolutionary spiral with three intermediate outcomes, resulting in a generative growth process (and in the negative triggering decline or adjustment, where, for example, initial asymmetric knowledge endowments are surpassed or become redundant). In turn, Figure 11.1 leads to three hypotheses that relate to the research questions outlined in the previous section and which the proposed research aims to test. These are as follows:

- *Hypothesis 1*: Firms experiencing knowledge asymmetries that reduce the effectiveness of in-house research capabilities are more likely to engage in 'open innovation' with research supplier firms and/or PROs in knowledge-capable regions.

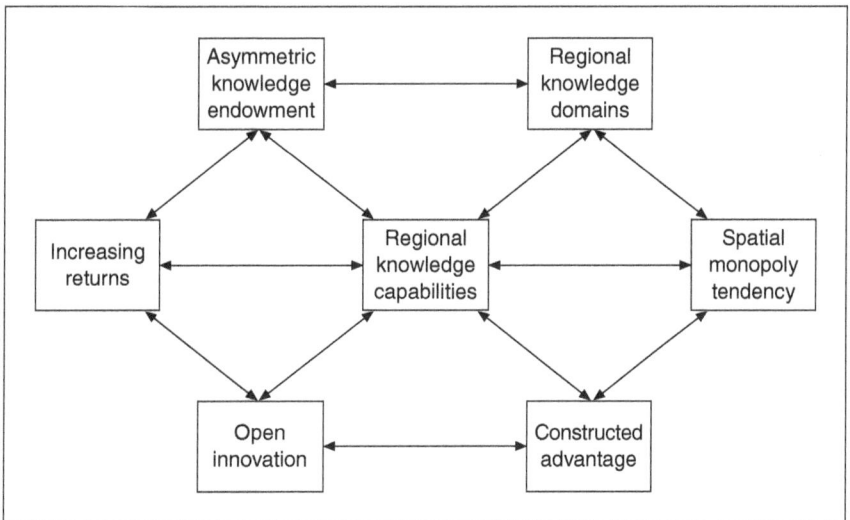

Figure 11.1 Generative growth processes attracted by regional knowledge capabilities

- *Hypothesis 2*: Customers choosing open innovation set in motion increasing returns processes that generate both growth and diversification of research outsourcing activities. These are likely to lead to specific qualities of knowledge capability in specific regions according to the spatial knowledge domain, which evolves over time, possibly entailing certain kinds of spatial knowledge monopoly.
- *Hypothesis 3*: Knowledge-capable regions are more likely to have benefited from economic governance practices and strategies further to construct advantage regarding such elements as 'talent' recruitment and retention and regional science policy. Such economic governance agencies are likely to have high benchmarking interaction with 'learning regions' aspiring to construct advantage by overcoming knowledge asymmetries.

The rationale of the research was thus to assess the extent of open innovation in knowledge-capable regions as an important new element in regional economic growth. Moreover, such growth is hypothesised to occur also in locations that need not be metropolitan or global cities. Shifts of this kind, to the extent they are generic, are important for theory and policy. This is so since if economic activity can shift to growth locations in which public research infrastructure is a pronounced element of a region's constructed advantage, this holds out growth prospects for less favoured locations, with, for example, medical schools, universities or other kinds of public research organisation (PRO). What remains obscure at present, albeit subject to hypothesis as discussed above, is the extent to which tendencies to spatial monopoly attend the operation of increasing returns forces. In Krugman's modelling work he hypothesised a zero-sum outcome from what he himself admitted was a simplistic two-location contest between jurisdictions. However, examination of the evolution of global 'bioregions' suggests that many can be sustained in relative proximity and at a distance from bioscience, particularly pharmaceuticals production (Cooke, 2004b). Thus, despite official scepticism regarding widespread regional aspirations to evolve knowledge-intensive 'clusters' in growth industries, there are instances reported of their emergence in unlikely places (Fuchs and Shapira, 2004; Wink, 2004; Groot *et al.*, 2004). Is this also true for ICT in the United Kingdom compared to the United States? Hence, this research seeks to elaborate theory and test it at least partially by reference to the two target sectors to augment complementary knowledge of its evolution in US ICT and household care industries as shown by Chesbrough (2003), and in automotives (Schamp *et al.*, 2004).

Empirical findings from a UK genomics biotechnology clustering survey

Research into the microeconomics of the UK biotechnology sector's structure, innovation characteristics and spatial distribution has been conducted at

CESAGen by the present author and associates during 2003–2004. In this contribution a few key findings are reported which relate to the themes under discussion. It is important to note that the firms sampled excluded those active neither in drug development nor in genomics. Regarding genomics, most UK biotechnology firms are also active in genomics research with a view to developing genomics-based medicines. With respect to the production of platform technologies, diagnostics and devices, some genomics biotechnology firms are also active in such production lines, warranting inclusion in the sample, though their genomics activities were those primarily inquired about in the survey. Thus, the sample size was, at 156 firms, less than the universe. The response rate was 15 per cent.

Regarding the age of sampled firms, this is shown in Table 11.3. It reveals that UK genomics biotechnology firms surged in rate of formation during two economic boom periods in the 1980s and late 1990s. The latter date coincided with the publication of the human genome but perhaps more importantly, with the stock market boom driven by technology business equities, including biotechnology. This is revealed as the more important driver, since, post-boom, the rate of new genomics firm formation in the United Kingdom has been far slower. Clearly, biotechnology firm formation is as sensitive to general business cyclicality as is the rate of new chemical entity (NCE) production, which is shown to be cyclical (Table 11.4), peaking relatively recently in 1996 in the United States. The turnover of the UK genomics biotechnology sector grew from $1.8 billion in 1999 to $3.4 billion in 2003, with mean turnover per firm rising from $45 million to $66 million over the same period.

The microeconomic indicators for the CESAGen respondents shown in Table 11.5 reveal that our respondents were smaller than the industry average, at least in terms of turnover, and the mean employment size confirms that respondents were smaller than the industry average. Nevertheless, other indicators show that the responding firms have a healthy turnover, fall squarely in the small firm size category, are moderately high exporters of their output, and spend, at 21 per cent more than, for example, the high pharmaceuticals average of 18 per cent of turnover on R&D. Foremost, they are shown to be significant patent holders, and, importantly, nearly half had registered new patents during the year prior to the survey in 2004.

Table 11.3 Date of establishment of UK biotechnology firms conducting genomics

Date of establishment	Frequency	Percentage
Before 1980	4	3
1980–1989	62	39
1990–1995	17	11
1996–2000	56	36
After 2001	17	11

Source: CESAGen Genomics Survey.

Table 11.4 Declining pharmaceutical productivity over time

Date	US NCE approvals	R&D expenditure, 2001 ($ billions)
1963	19	2
1967	20	2
1971	18	3
1975	17	3
1979	16	3
1983	16	4
1987	19	5
1991	25	8
1995	25	15
1996	45	17
1997	37	18
1998	23	22
1999	32	25
2001	23	30

Source: BIGT, 2004.

Table 11.5 Economic and innovation indicators for UK genomics biotechnology firms, 2003

Indicator	Quantity
Mean employment	44
Mean turnover	$30 million
Mean exports/turnover	24%
Mean R&D expenditure/turnover	21%
Firms with new patents	47%
Mean patents per firm	4.5

Source: CESAGen Genomics Survey.

Table 11.6 Business and innovation partnering by UK genomics biotechnology firms

Indicator	Percentage
Collaboration with firms/institutes	77
Clustering spatially with collaborators	78
Cooperating specifically on Innovation	70
Cooperating on innovation in home region	18
Cooperating on innovation in the UK	23
Cooperating on innovation in the EU	18
Cooperating on innovation globally	28

Source: CESAGen Genomics Survey.

Finally, of key interest is assessing the extent to which the results echo the thesis that knowledge-driven firms occupy specific 'spatial knowledge domains'. They are hypothesised to share geographical coexistence with similar firms and knowledge sources, notably research laboratories or centres of research excellence. This is rather than, for example, locating near to client

Table 11.7 Scaling for proximity by UK genomics biotechnology firms

Proximity factor	Range	Mean	Standard deviation
University research	1–4	3.25	0.851
Genomics services	2–4	3.15	0.745
Business environment	2–4	3.05	0.759
Qualified workforce	2–4	3.05	0.826
Regional agency/grants	1–4	3.05	0.826
Other public research	2–4	2.90	0.968
Collaborators/competitors	1–4	2.80	0.951
Suppliers	1–4	2.70	0.801
Private research	1–4	2.60	0.940
Technology transfer	1–4	2.55	1.050
Customers	1–4	2.40	0.995

Source: CESAGen Genomics Survey.

firms such as pharmaceuticals companies. In making judgements regarding this, Table 11.6 is indicative. It reveals that a large majority of UK genomics firms not only collaborate on business-related matters, but cluster geographically to engage in such activities. However, with specific regard to *innovation* (as distinct from other interactions such as research, joint patenting, purchasing or supplying, and other more informal collaboration) – the act of *commercialising* new knowledge in the form of a product or service new to the firm or new to the market – firms behave distinctively. Firms innovate in partnership with other actors in their region, mainly their cluster, to an equivalent amount that they innovate in partnership with actors in the European Union. Moreover, they innovate more than either of those categories with partners in the United Kingdom more generally, and finally their *innovation* partner is most likely to be outside Europe, currently the United States or to a lesser extent Asia in most cases.

Hence, we learn that geographical proximity is relatively *unimportant* for genomics firms transforming new knowledge into new products or services or conducting cooperative *innovation*. So why do genomics biotechnology firms in the United Kingdom cluster so frequently? The answer lies in Table 11.7, which reports on the scaling importance ranked from 1–4 accorded by firms to proximity for accessing a variety of services. It is clear from this result, and confirmed in results shown in Table 11.7, that the most important reason given by UK genomics firms for locating in proximity is access to university research, followed by specific services expertise required in genomics. Of slightly less value than these factors are three of equivalence: the proximate business environment, covering such things as legal, financial and other business-related services; the presence of an appropriately skilled workforce, not wholly university trained, since technical and ICT skills are also important; and regional development agencies with grant-giving capacity. Hence, exploration or basic research knowledge and finance may be said to be the key business enhancements arising from proximity. Of these, university research scales highest. Since, in the United Kingdom, all key genomics universities pre-date genomics and most pre-date biotechnology, it is reasonable to infer that for a science-driven industry such as genomics, the presence of university genomics research is responsible for the 'spatial knowledge domain' and 'spatial knowledge capabilities' that assist formation of the regional cluster. This is significantly more important than proximity to other kinds of R&D or business transaction relations. Regarding R&D specifically, Table 11.8 shows that the *regional* university ranks high but not highest, which is university research links elsewhere in the United Kingdom, and in third position R&D links in North America. 'Collaborator' is defined broadly, to include market and non-market exchange. Perhaps surprising are the low R&D links from UK genomics cluster firms to collaborators in the European Union but the relatively higher numbers of R&D collaborators who are customers for the research, suppliers of inputs for the research, or competitors of the R&D-practising firm. Most of the latter collaborations arise through EU Framework Programmes for Science, Technology and Innovation, in which a few universities mobilise larger networks of SMEs. An interesting sidelight on this is that genomics firms either have few direct competitors in their region, or they do not collaborate with them significantly if they are present. Contrariwise, they collaborate substantially with competitors elsewhere. The fact that 30 per cent of firms have collaborations with competitors in the United Kingdom may suggest that the former interpretation is more likely. If so, it casts an interesting sidelight on the 'spatial knowledge capabilities' thesis. That is, it indicates that genomics firms have no desire to conduct R&D with local competitors because they already know its likely content, as a result of 'open science' and localised knowledge spillovers among firms competing in highly specific local niches (Owen-Smith and Powell, 2004; Caniëls and Romijn, 2006).

Figure 11.2 shows how this operates in the United Kingdom's Cambridge cluster. Both proximate specialist firms and university research are networked in joint genomics projects. This applies also to micro-firms in, for example, the Babraham Bioincubator. But noticeable also are the strong partnership links with distant 'big pharma', normally the ultimate customer partners for

Table 11.8 Economic geography of R&D collaborators of UK genomics biotechnology firms

Collaborator	Region (%)	UK (%)	EU (%)	North America (%)	Asia (%)	Rest of world (%)
University	30	44	4	17	9	0
Consultant	22	35	4	13	4	0
Supplier	17	30	30	30	0	0
Other R&D	13	37	7	11	5	0
Customer	5	35	30	35	13	9
Competitor	4	30	26	22	9	4

Source: CESAGen Genomics Survey.

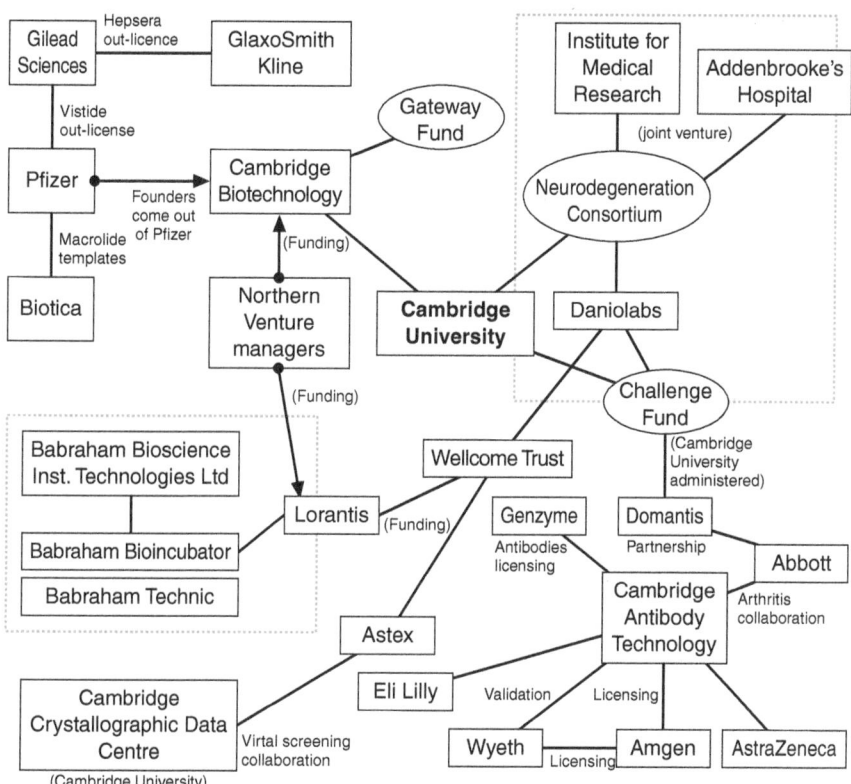

Figure 11.2 Sample of proximate and distant networking in the Cambridge genomics cluster. Source: CESAGen Genomics Survey

genomics drugs being researched in world-leading firms like Cambridge Antibody Technologies. It is noticeable that regional hospitals, venture capitalists and bioimaging suppliers also contribute to Cambridge's specific knowledge domains in neurology and inflammation. Moreover, not all firms are university spin-outs; one is a former Pfizer employee start-up. Thus, absence of distant spillovers means that firms form collaborator relations with 'distant networks' (Fontes, 2006) to augment R&D knowledge for themselves. These occur broadly equally in the European Union and North America, as well as more extensively in the home country. This leads logically on to the next section of this contribution, which seeks to throw light upon where and with whom such networks tend to be, at least with respect to leading genomists engaging in the key joint R&D output of co-publication.

Thus, in relation to the theory and hypothesis testing discussed in the previous section, these results confirm the following. First, UK genomics SMEs (which is nearly equivalent to saying UK biotechnology companies, since most perform genomics research and exploitation activities) rely heavily on outsourced R&D contracts from large pharmaceuticals firms. These contracts are also made for *exploration* rather than *examination* (e.g. clinical trials) or *exploitation* (e.g. innovation, commercialisation) research. These two forces, as shown in Figure 11.2, form a close set of network interactions in geographical proximity because of the concentration of regional knowledge capabilities, even in spatial knowledge domains. Thus, Cambridge's lead in monoclonal antibody research since 1975, when Milstein and Köhler performed their Nobel Prize-winning research discovering 'Mabs' at the Medical Research Council Molecular Biology Laboratory, carries through to the strong R&D in-sourcing performance of a Cambridge firm like Cambridge Antibody Technologies (at the heart of a global R&D network in Figure 11.2). Hence, hypothesis 1 is well supported by this evidence. Hypothesis 2 is also supported by virtue of the same 'spatial knowledge domains' thesis. Hypothesis 3 gains weak support for economic governance and 'talent', which nevertheless score less importantly than the kind of specialist genomics research and services (non-governmental) that firms find attractive and enabling highly valued collaboration in the Cambridge cluster. At this point, therefore, the theory of 'regional knowledge capabilities' gains strong support, more regarding globally scarce and geographically proximate *research* capabilities than globally less scarce but nevertheless *functionally* proximate (big pharma) *innovation* capabilities when products and services are being readied for the market.

Empirical findings from a UK ICT clustering survey

Precisely the same methodology was elaborated for the study of reasons why firms cluster or do not cluster in ICT as for genomics biotechnology. The key difference is that the universe and sample size of firms surveyed was vastly bigger for the former postal questionnaire survey than the latter. Basically, sectors were grouped according to computing, telecommunications and software, and the first two further into hardware and services. Reference to Table 11.9 gives a comparison for ICT with data on longevity of firms in biotechnology (Table 11.3). It shows, unsurprisingly, that the earliest establishment date of ICT firms is often earlier than for biotechnology firms, but

Table 11.9 Date of establishment of UK survey respondent ICT firms

Date of establishment	Frequency	Percentage	(UK biotechs)
Before 1947	8	3	0
1948–1970	16	6	0
1971–1980	24	9	3
1981–1990	69	26	39
1991–1999	124	47	47
2000–2003	23	9	11

Source: ESRC ICT Collective Learning Survey.

that a similar bunching of firm formation among survey respondents in ICT occurred in the 1980s and, particularly, in the 1990s. Similarly, founding activity after 2000, when the technology stock markets nosedived worldwide, is comparably meagre. More surprisingly, for the 1990s, is the exact similarity at 47 per cent in the percentage of all ICT and biotechnology formed during that decade. Thus, for both, very differently composed, sectors, just under half the respondent firm populations arose during the 1990s. Moreover, proportionately more biotechnology firms were founded than ICTs in the 1980s. This suggests that arguments that the biotechnology firm formation cycle is different from other sectors because it is science rather than market driven are unfounded, at least by comparison with ICT. Of course, science plays an important role in ICT too, but lead times from discovery to innovation are generally much shorter, as are timescales from firm formation to flotation on stock markets. Moreover, testing and trialling while regulated for safety reasons worldwide is less complex and drawn out because much of it is computerised, as with structural calculation of finite elements algorithms.

In this sector a proportion of firms are larger organisations. In the sample, respondent firms splitting 11 per cent large firms, 89 per cent small and medium enterprises (SMEs). Regarding the corporate headquarters of respondents, the majority, as we have seen, being SMEs, 37 per cent were UK domiciled, 33 per cent North American, 16 per cent European, 7 per cent Asian and 7 per cent domiciled elsewhere.

With respect to key performance indicators of UK ICT respondent firms, the data in Table 11.10 provide this and facilitate comparison with the UK biotechnology firms. Clearly, there are differences, notably the greater mean size, although still well within the SME category, of ICT compared to biotechnology firms. But biotechnology firms sampled are more value-adding – productive – per employee, as their mean turnover is almost as large. Both sectors spend a significant share of turnover on R&D with biotechnology, true to its strongly science-driven character, outstripping ICT, although the ICT ratio is also rather high. As indicators of intellectual property regimes in the two sectors, our results conform to the conventional wisdom in the literature, namely that ICT tends to be a low patent-holding sector, while, again,

Table 11.10 Economic and innovation indicators for UK ICT and biotechnology firms, 2003

Indicator	ICT	Biotechnology
Mean employment	115	44
Mean turnover	$35 million	$30 million
Mean exports/turnover	40%	24%
Mean R&D expenditure/turnover	17%	21%
Firms with new patents	11%	47%
Mean patents per firm	8.0	4.5

Source: ESRC ICT Collective Learning Survey.

biotechnology's science-driven nature and heavier R&D mean that nearly half had new patents within the year before filing the questionnaire. However, the mean number of patents held per firm tells a different story, which is that the average ICT firm holds nearly twice as many patents over the longer term as compared with the average for the respondent biotechnology firms. This is probably due to two factors: the greater size, in terms of employee numbers, of ICT firms; and their greater longevity. But it suggests strongly a different regime between sectors, where ICT firms may exploit patents over a lengthy period of time, while biotechnology firms exploit theirs swiftly and, because of the complex nature of discovery in this sector, in patent 'families'.

Moving on, Table 11.11 begins analysis of central importance to understanding the distinctive roles of proximity and collaboration, the heart of the matter of determining whether and why clustering among firms occurs habitually and similar for more distant networking. Recall that a strong finding of the UK genomics biotechnology research was that firms cluster for *research* but utilise *distant networks* (often with large pharmaceuticals firms) for *innovation*. The comparison in Table 11.11 is in some ways remarkably similar for these distinctive sectors, and in other ways remarkably different. First, the similarities: strikingly, *collaboration* in general is remarkably and similarly high for both at 80 per cent for UK ICT and 77 per cent for biotechnology. Comparably, specifically collaborating on *innovation* is almost identically high among respondents: 73 per cent of ICT firms doing collaborative innovation, 70 per cent of those in biotechnology.

Cooperating on *innovation* in the home region is comparably low, as it is within the European Union and with global collaborations at between 18 per cent and 30 per cent respectively. We summarise this as meaning that both fields are highly collaborative sectors with respect to innovation, with varying amounts of this occurring locally and globally. However, collaborating nationally is much more common in UK ICT than biotechnology. Notably, clustering with *collaborators* is also much less frequent for ICT than biotechnology. Thus, on key *proximity* indicators for ICT we have a picture of a large majority of firms collaborating on *innovation* with other domestic but not particularly local firms and organisations. Contrariwise, biotechnology

Table 11.11 Business and innovation partnering by UK ICT and biotechnology firms

Indicator	Percentage ICT	Biotechnology
Collaboration with firms/institutes	80	77
Clustering spatially with collaborators	27	78
Cooperating specifically on Innovation	73	70
Cooperating on innovation in home region	27	18
Cooperating on innovation in the UK	57	23
Cooperating on innovation in the EU	20	18
Cooperating on innovation globally	30	28

Source: ESRC ICT Collective Learning Survey.

firms are nearly equivalently collaborating on innovation non-domestically. Furthermore, it will be recalled that biotechnology firms collaborate strongly on R&D locally. We now need to see if the situation is similar or different for UK ICT firms.

If we move to the issue of *proximity* in the ICT firm rationale for inter-action, the picture is different between ICT and biotechnology in this regard. It will be recalled, as Table 11.12 summarises, that for biotechnology, prox-imity to universities for accessing *research* was the first imperative, whereas proximity was relatively less important for innovation. This frequently took place in partnership with distant network actors, frequently – as Figure 11.2 showed illustratively – with US or other transnational corporations. For ICT, the proximity rankings for business interaction are given in Table 11.12. These data show that universities are ranked medium as 'proximity partners'.

Most strikingly, 'customers', ranked lowest in biotechnology, rank highest for ICT, and other public research, such as that conducted in non-university laboratories, is ranked very low by ICT but is of medium influence in terms of proximity drivers by biotechnology firms. Thus, a picture is relatively easily and correctly formed of ICT and biotechnology as having polar oppos-ite rationales for proximate interaction in research and innovation. Whereas biotechnology firms cluster around universities and, to a lesser extent, other public laboratories for research knowledge and related interactions, mean-while interacting distantly with customers, many of which are pharmaceuti-cals transnationals, ICT firms prefer to cluster close to customer firms, keeping research at a distance. This is an original finding for both industries and tells us much about their nature, and the differences between them. First, both collaborate intensively but ICT more nationally than either locally or globally, as in the case of biotechnology. Second, ICT is more market than science focused in its proximity practices, a sign that innovation is more important and swifter than in biotechnology. Third, and of policy relevance,

Table 11.12 Scaling for proximity by UK genomics biotechnology firms

Proximity factor	ICT mean	Biotechnology mean
University research	3.09 (6)	3.25
Services	3.69 (3)	3.15
Business environment	3.50 (4)	3.05
Qualified workforce	3.69 (2)	3.05
Regional agency/grants	2.49 (8)	3.05
Other public research	1.96 (10)	2.90
Collaborators/competitors	2.89 (7)	2.80
Suppliers	3.44 (5)	2.70
Private research	2.11 (9)	2.60
Technology transfer	1.67 (11)	2.55
Customers	3.91 (1)	2.40

Source: ESRC ICT Collective Learning Survey.

a region is well advised to have localised ICT multinational customers to help promote its nascent ICT cluster, while for biotechnology this is relatively unimportant, and proximity to an accomplished medical or other biosciences research capability is of greater importance for cluster building.

Finally, we can to a considerable extent compare the economic geography of R&D collaborations by ICT and biotechnology firms. The nature of the data deployed makes comparison in a single table impossible, but Table 11.13 summarises the position for UK ICT. Recall that the main lineaments of such collaborative economic geography for biotechnology were as follows. First, UK biotechnology's favoured R&D collaborator was UK universit(ies), followed mostly by UK 'other (public) R&D', consultants and customers. Competitors and suppliers in the UK were as popular as the best-scoring collaborator in the host region. This was the regional university, followed by regional consultancy, then supplier, public R&D, while regional customers and competitors were negligible R&D collaborators. Indeed, customers anywhere globally were of more importance (Table 11.13). For ICT, the picture of R&D collaboration is significantly more national in orientation but also more regularly regional and much less global than for biotechnology, for most kinds of R&D collaborator, as Table 11.13 shows.

Here, it is clear that most UK ICT collaboration in R&D occurs nationally, with the host region some way behind, but much more engaged, except for customer–collaboration interaction, for most variables than the non-national level. A partial exception to this is that 'suppliers' are relatively important to R&D collaboration in both the European Union and North America, as indeed are customers. Thus, a picture forms of UK ICT firms much engaged in transatlantic supply chains bolstered by UK and regional R&D collaborations with a wide range of support actors, especially universities. Hence, while R&D is less a factor in proximate location for UK ICT firms, especially compared to the proximity force of innovation and market partners, UK and regional R&D is more important for R&D collaboration than that from abroad, including North America, which is a nexus of R&D collaboration of minor significance. Thus, in terms of the thesis advanced at the outset of this chapter that clusters gather for different reasons but that both ICT and biotechnology clustering in the United Kingdom, driven as

Table 11.13 Economic geography of R&D collaborators of UK ICT firms

Collaborator	Region (%)	UK (%)	EU (%)	North America (%)	Asia (%)	Rest of world (%)
University	34	45	11	5	2	2
Consultant	33	56	5	3	1	2
Supplier	18	50	18	7	4	3
Other R&D	24	61	3	3	0	0
Customer	25	40	16	11	4	5
Competitor	23	39	18	8	6	6

they are by different imperatives – research for biotechnology, innovation for ICT – are intimately involved in interacting collaboratively with customer firms with which they engage for purposes of conducting 'open innovation' and/or 'R&D outsourcing' kinds of collaboration. Further, these firms value proximity in this regard: to repeat, with national and regional consultants, customers and universities for ICT firms and with national and regional universities, but more transatlantic customers and suppliers, for biotechnology firms. Hence, a further elaboration is a greater valuation by the latter of *functional* proximity as compared with *geographical* for innovation through distant networks.

Concluding remarks

This chapter has proposed theoretical elaboration and has mobilised evidence that give confidence in the following key observations. First, and in theoretical terms, an advance in our understanding of the persistence and conceivable reinforcement of an asymmetric economic geography of prosperity and accomplishment, in an evolving and intensifying knowledge economy, science driven and otherwise technologically sophisticated economic activity gives rise to demands upon industry organisation that reinforce collaborative activity among smaller knowledge-intensive businesses, on the one hand, and between smaller smart firms and university laboratories towards customer (and supplier) firms, many of which can, on the other hand, be shown to be large or even transnational corporations. This is important and original support for the thesis that *regional knowledge capabilities* increasingly determine the distribution of growth regions, currently favouring those that gain increasing returns from asymmetric knowledge distribution that assists in the construction of regional advantage in terms of talent recruitment and retention, spatial knowledge quasi-monopolies, and 'R&D outsourcing' or 'open innovation'. In UK ICT and biotechnology such features are pronounced, with key bioregional capabilities attracting these advantages to clusters like Cambridge and Oxford, while for ICT, London and its satellites in the M25 and M4 corridors are the dominant market-led magnet.

Second is the important question – unanswered until now – as to whether UK firms in the biotechnology and ICT industries perform better in collaborative proximity (a synonym for 'clusters') than not. In particular, it is interesting and important to separate collaborator performance from general performance. Convincing evidence for this can only at this stage of the analysis be offered for ICT, hence the indications are in the present section rather than the main body of the chapter. The key indicator data are presented in Table 11.14. They show that on most indicators of economic performance, collaborating UK ICT firms' mean performance is generally better than the mean scores in the respondent group as a whole, consisting of both collaborators and non-collaborators. Thus, collaborators have superior performance regarding market share; superior performance regarding innovation (more

Table 11.14 Collaboration and performance of UK ICT firms

Indicator (%)	N	Total (mean)		Collaborators (mean)	
		2000	2003	2000	2003
Market share (high/v. high)	25	–	23.2	–	32.1
New products (last 3 years)					
>5	154	–	72.0	–	66.7
5–20	46	–	21.5	–	27.2
R&D as percentage of turnover	115	19.9	16.6	16.1	17.7
New patents	81	10.6	10.6	15.2	17.2
Employees (number)	181	84	115	143	215
Turnover ($£$ million)	165	13.1	17.9	19.5	25.5

Source: ESRC ICT Collective Learning Survey.

new products or services in the preceding three years), except for the category generating fewest innovations; higher R&D as a share of turnover in 2003 (but not 2000); more new patents in 2000 and 2003; more employees per firm; and greater turnover in both 2000 and 2003. Hence, it is not a perfect competitive advantage but a reasonably convincing one. Accordingly, it is not unreasonable to propose that ICT firms engaging in collaborative activity with others are more capable on the R&D and patenting input side of the innovation relation and they benefit on the output side with greater market share.

A step further towards clarifying the tantalising issue of collaboration, performance and proximity, at least for UK ICT firms, is provided in the final table (Table 11.15) of this chapter. Questions were answered by firms on what advantage, if any, spatial proximity gave to their cognitive or knowledge capabilities. Furthermore, as a cross-check, firms were asked about the extent to which *innovation* activities were conducted in a local cluster, the definition 'closely proximate collaboration' being provided to define such a context. The results from asking these two questions reveal that collaborators favour spatial proximity to a greater extent than the respondent group as a

Table 11.15 Proximity, cognitive and innovation advantages for UK ICT firms

Proximity indicator	Total respondent share (%)	Collaborator respondent share (%)
Facilitates knowledge exchange	12.6	18.2
Swifter, clearer knowledge exch.	10.0	13.9
Reduces interaction cost	9.2	11.3
Facilitates informal communication	12.4	17.6
Reduces uncertainty	8.0	12.8
Facilitating collective learning	6.8	11.1
Innovation cooperation in cluster	22.8	27.3

Source: ESRC ICT Collective Learning Survey.

whole, although it must be said that respondents answering these questions were low in number, with even those stressing clustering cooperation for innovation a minority compared to those conducting such activities with intra-firm cooperation, and intra- or even extra-UK innovation cooperation. This is, of course, totally consistent with Table 11.6 regarding clustering in ICT.

Finally, this research has tended to find support for the superiority of collaboration in respect of a variety of performance indicators, and clustered cooperation for innovation being supported more by the collaborating part of the firm sample than the respondent group as a whole. This broadly applies in ICT and biotechnology, but, as we have seen, less regarding clustering for innovation activity by ICT than by biotechnology firms, and much more for research interactions by biotechnology than ICT firms. So, does this evidence lend support to the group of concepts that, in this chapter, connect asymmetric knowledge, R&D outsourcing and 'open innovation' whereby knowledgeable clusters attract collaborative interaction concerning research and knowledge exchange and cooperative innovation for commercialisation? This is difficult to be certain about. Clearly, in the United Kingdom's biotechnology sector there is considerable asymmetry in the location of, for the most part, research-driven clustering. The fact that innovation attracts distant networks of cooperation for this sector is consistent with the characteristic feature of biotechnology, which is that R&D and innovation consistently involve externalised inter-firm and firm–laboratory interactions. To that extent biotechnology conforms fairly well, judged from these UK data, to the standard narrative. ICT shares some of this character but in an inverse manner. Research is less of a cluster driver than innovation activity, but the latter is not as pronounced as supply-chain innovation stretching globally and intra-firm interactions, the latter being partly a function of differing firm size between the two sample. The one thing that appears to be almost transparent, especially in the ICT data, is the superiority for firm performance of collaborative knowledge exchange and innovation activity over stand-alone competition, even for the large, dominating firms in biotechnology, though possibly somewhat less for ICT firms, a few of which made it into the UK ICT respondent group.

Acknowledgements

This research was funded by ESRC through the Centre for Economic and Social Analysis of Genomics (CESAGen) and by an ESRC research contract 'Collective learning in knowledge economies: milieu or market?', award no. RES-000-23-0192. The sponsors are thanked for these supports. David Knight and Carla De Laurentis are Research Associates at CESAGen and the Centre for Advanced Studies, Cardiff University, respectively funded by these sources, and their assistance in generating data analysed here is gratefully acknowledged.

Note

1 A concrete example of this occurs in the following. An Israeli medical doctor specialising in gastro-intestinal medicine was frequently told by patients that endoscopy (the use of a camera in a tube for internal patient diagnosis) was painful to throat and oesophagus. He thus had implicit knowledge of a problem. Research on nano-cameras, for example that led Bogdan Dragnea at Indiana University in Bloomington to get an image of what goes on inside living cells and a greater understanding of how viruses work, is also occurring in Israel, some of it at the behest of the intelligence service. Therefore, implicit and limited explicit intelligence knowledge of nano-camera technology existed in Israel. The doctor mentioned the endoscopy problem to an entrepreneur acquaintance formerly in the Israeli army rocketry service, who knew of the existence of nano-cameras for guidance purposes. So, he had complicit knowledge. The idea of a camera in a pill arose from the conversation, and a commercial product is now in prospect from GivenImage, the entrepreneur's firm.

References

Akerlof, G. (1970) The market for 'lemons': qualitative uncertainty and the market mechanism, *Quarterly Journal of Economics*, 84, pp. 488–500.

Aslesen, H. (2004) Knowledge intensive business services and regional development: consultancy in city regions in Norway, in P. Cooke and A. Piccaluga (eds) *Regional Economies as Knowledge Laboratories*, Cheltenham, UK, Edward Elgar.

Bathelt, H. (2003) Growth regimes in spatial perspective 1: Innovation, institutions and social systems, *Progress in Human Geography*, 27, pp. 789–804.

Bioscience Innovation and Growth Team (BIGT) (2004) Bioscience 2015: *Improving National Health, Increasing National Wealth*, London: BIA/DTI/DoH.

Brenner, N. (2001) The limits to scale? Methodological reflections on scalar structuration, *Progress in Human Geography*, 25, pp. 591–614.

Buckley, N. (2005) Procter's gamble on outside ideas has paid off, *Financial Times*, 14 January, p. 11.

Bunnell, T. and Coe, N. (2001) Spaces and scales of innovation, *Progress in Human Geography*, 25, pp. 569–589.

Caniëls, M. and Romijn, H. (2006) Localised knowledge spillovers: the key to innovativeness in industrial clusters?, in P. Cooke and A. Piccaluga (eds) *Regional Development in the Knowledge Economy*, London, Routledge.

CHA (2004) *Successful Outsourcing of Pharmaceutical R&D: Trends and Strategies*, Waltham, MA, Cambridge Healthtech Advisors.

Chandler, A. (1990) *Scale and Scope*, Boston, Harvard/Belknap.

Chesbrough, H. (2003) *Open Innovation*, Boston, Harvard Business School Press.

Cooke, P. (2001) Biotechnology clusters in the UK: lessons from localisation in the commercialisation of science, *Small Business Economics*, 17, pp. 43–59.

Cooke, P. (2002) Biotechnology clusters as regional, sectoral innovation systems, *International Regional Science Review*, 25, pp. 8–37.

Cooke, P. (2003) The evolution of biotechnology in three continents: Schumpeterian or Penrosian?, *European Planning Studies*, 11, pp. 757–764.

Cooke, P. (2004a) Biosciences and the rise of regional science policy, *Science and Public Policy*, 31, pp. 185–197.

Cooke, P. (2004b) Globalisation of bioregions: the rise of knowledge capability, receptivity and diversity, Regional Industrial Research Report 44, Cardiff, Centre for Advanced Studies.

Fontes, M. (2006) Knowledge access at distance: strategies and practices of new biotechnology firms in emerging locations, in P. Cooke and A. Piccaluga (eds) *Regional Development in the Knowledge Economy*, London, Routledge.

Foray, D. and Freeman, C. (1993) *Technology and the Wealth of Nations: The Dynamics of Constructed Advantage*, London, Pinter.

Fuchs, G. and Shapira, P. (eds) (2004) *Rethinking Regional Innovation and Change*, Berlin, Springer.

Goodwin, M., Jones, M., Jones, R., Pett, K. and Simpson, G. (2002) Devolution and economic governance in the UK: uneven geographies, uneven capacities?, *Local Economy*, 17, pp. 200–215.

Groot, H., Nijkamp, P. and Stough, R. (eds) (2004) *Entrepreneurship and Regional Economic Development*, Cheltenham, UK, Edward Elgar.

Krugman, P. (1995) *Development, Geography and Economic Theory*, Cambridge, MA, MIT Press.

Latour, B. (1998) From the world of science to the world of research, *Science*, 280, pp. 208–209.

Laursen, K. and Salter, A. (2004) Searching high and low: what types of firms use universities as a source of innovation? *Research Policy*, 33, pp. 1201–1215.

Mackinnon, D., Cumbers, A. and Chapman, K. (2002) Learning, innovation and regional development: a critical appraisal of recent debates, *Progress in Human Geography*, 26, pp. 293–311.

March, J. (1991) Exploration and exploitation in organisational learning, *Organisation Sciences*, 2, pp. 71–87.

Maskell, P. and Lorenzen, M. (2004) The cluster as market organization, *Urban Studies*, 41 (5/6), pp. 975–93.

Owen-Smith, J. and Powell, W. (2004) Knowledge networks as channels and conduits: the effects spillovers in the Boston biotechnology community, *Organization Science*, 15, pp. 5–21.

Penrose, E. (1959/1995) *The Theory of the Growth of the Firm*, Oxford, Oxford University Press.

Ryan, C. and Phillips, P. (2004) Knowledge management in advanced technology industries: an examination of international agricultural biotechnology clusters, *Environment and Planning C: Government and Policy*, 22, pp. 217–232.

Schamp, E., Rentmeister, B. and Lo, V. (2004) Dimensions of proximity in knowledge-based networks: the cases of banking and automobile design, *European Planning Studies*, 12, pp. 607–624.

Smith, A. (1776) *The Wealth of Nations*, New York, The Modern Library, 1994.

Teece, D. and Pisano, G. (1994) The dynamic capabilities of firms: an introduction, *Industrial and Corporate Change*, 3, pp. 537–556.

Valentin, F. and Lund-Jensen, R. (2003) Science-driven discontinuities and the organisation of distributed innovations: the case of biotechnology in food processing technologies, Paper presented to DRUID Summer Conference 'Creating, Sharing and Transferring Knowledge', Copenhagen, 12–14 June.

Wink, R. (ed.) (2004) *Academic–Business Links*, Basingstoke, UK, Palgrave.

12 The outsourcing of knowledge production and its implications for regional path dependence

Alex Burfitt and Chris Collinge

Introduction

An important change in the organisation of knowledge production across economies nationally and internationally over recent years has been the growth of R&D outsourcing. But while this development has been acknowledged within the literature, it has not received the level of attention it deserves, given the interest there is in knowledge economies. So while there has been some examination of the pressures that lead firms to undertake outsourcing, little at all has been published about the growth of specialist R&D services firms to undertake this work, the routes through which these firms have emerged, and their role in regional innovation systems (but see Howells, 1997, 1999; Koschatzsky, 2004; Readman and Hales, 2000). The first purpose of the present chapter is therefore to address this deficit by examining the evidence generated through a regional case study of the R&D services sector. But the reorganisation of R&D has implications for the process of regional development, a process that is often conceived within evolutionary economics in terms of path dependence, and so the second and more general purpose here is to explore – albeit tentatively at this stage – the implications of this reorganisation for notions of path dependence.

The chapter begins by placing R&D within the wider context of knowledge production, and considers how this process has been conceptualised within economic geography, including its relationship to the informal process of localised learning and its interpretation in terms of path dependence. We then present secondary data showing trends in the organisation and performance of R&D, including the growth of outsourcing, and identify factors that have been put forward to explain the emergence of an R&D services sector. The sub-region of Leicestershire in the East Midlands of England contains a significant concentration of employment in R&D services, one that has gone largely unremarked. On the basis of the findings from a survey of R&D services firms in this sub-region conducted in 2003 we describe the characteristics and activities of the R&D services sector there, identify its likely determinants or origins, and suggest some implications for local economic development. Finally, the case study is used to raise some more general questions regarding the notions of path dependence and lock-in, and to suggest how these questions might be addressed by distinguishing different components of continuity and change.

Knowledge production and R&D

In this section we consider the ways in which knowledge production – through both formal R&D and informal or localised learning – is conceptualised within economic geography, and address in this context the significance of regional path dependence. We also review recent thinking on the mechanisms and drivers of R&D outsourcing, and on the resulting development of an R&D services sector.

Localised learning and path dependence

As a focus for the formal exploitation of knowledge within industry, R&D is seen as an important part of the wider knowledge economy and, with its outsourcing, of the development of knowledge-intensive business services. For Arrow, knowledge was a public good that could not be produced efficiently in competitive markets, given its non-excludability, indivisibility, non-appropriability and non-tradeability (Arrow, 1969). It was therefore the responsibility of public agencies such as universities to generate most new knowledge through formal R&D activities. More recently, however, it has been argued that knowledge *does* have a degree of natural appropriability and excludability, and that a major source of knowledge is localised learning between complementary and perhaps cooperating firms (Loasby, 1999; Antonelli, 1999). On this basis it has been recognised that there may be a 'knowledge trade-off' whereby the excludability provided by intellectual property rights, while giving firms the incentive to produce knowledge by preventing uncontrolled leakage, may diminish the beneficial spillover of knowledge into a public pool and hence the efficiency of the economy as a whole (Shankerman and Scotchmer, 2001)

There has been considerable interest over recent decades in the region as an important setting for economic development. Many authors have sought to identify the characteristics of successful regions, and have focused upon localised networks and knowledge interactions within regional systems of innovation (e.g. Braczyk *et al.*, 1997). From this point of view the collective learning process within a community of firms is thought to depend upon trust and spatial proximity, factors that contribute to learning through the formation of new spin-out enterprises, the level of inter-firm interaction, and flows of skilled personnel between firms (Keeble *et al.*, 1999; Maskell *et al.*, 1998). Generally speaking, the literature has also placed increasing emphasis upon the social and institutional context of agglomeration (upon 'associational economies' or 'innovative milieux'), and it is generally for their contributions to localised learning that these contexts are considered (Camagni,

1999; Cooke and Morgan, 1998). Despite the strong emphasis there has been upon localised learning, it is still recognised, however, that R&D is a necessary condition for successful innovation. The literature has, for example, shown that R&D has a major impact upon productivity, and that for every increase in R&D spending the output increases by significantly more (OECD, 2001; Coe and Helpman, 1993). Indeed, one of the main hypotheses has been that formal R&D activities reinforce spillovers through the inter-firm or interpersonal network relations they establish, and that the specialised nature of much R&D activity necessitates a cooperative relationship with information pooling between participants (e.g. Jaffe *et al.*, 1993; Mowery and Rosenberg, 1999; Cooke *et al.*, 2003). The outsourcing of R&D might on this basis be hypothesised to contribute to the creation of a more innovative milieu.

Consideration of the dynamics of the relationship between R&D and other forms of learning within an innovation system leads into the wider historical question of regional development. There is of course an extensive literature within evolutionary economics and, increasingly, economic geography that accounts for the reproduction of economic characteristics from one time period to the next in terms of path dependence and indeed lock-in. Discussion of path dependence is frequently couched in technological terms, arguing that past technologies (product and process) embedded in particular sectors endogenously enable and constrain future technological trajectories (Mowery and Rosenberg, 1999). But this notion is also deployed within discussions of innovative milieux, in which firms are embedded in regional institutional and cultural contexts (such as the Third Italy) that govern future developments by guiding local collective learning (Camagni, 1991; Hassink, 2005, p. 530). Several mechanisms are likely to be invoked here – including the occupational mobility of human capital conveying knowledge from one firm to another, and interpersonal networks which convey information – factors that are themselves conditioned by the techno-industrial specialisations of the area (Boschma and Lambooy, 1999, p. 415; Lawson, 1999). But there is a tendency within economic geography to invoke 'path dependence' in a relatively unexamined way, without separating out the components of continuity and change that are involved, or explaining why any particular combination of these is reached (see Kenney and Von Burg, 1999, p. 99; Hassink, 2005, p. 523; Meyer-Stamer, 1998). These are issues to which we return after the case study presentation below.

Outsourcing R&D functions

Research and development is hard to define and correspondingly difficult to identify or quantify. However, the standard definition of R&D is contained within the OECD's 'Frascati Manual', which states that 'Research and experimental development comprise creative work undertaken on a systematic basis in order to increase the stock of knowledge, including knowledge of man, culture and society, and the use of knowledge to devise new applications' (OECD, 1994, p. 7). This definition highlights both the creation of new knowledge and the use of existing knowledge to create new products and processes. A crucial factor in differentiating R&D from other activities lies in the reason why it is being performed, and in particular that it involves 'the presence of an appreciable element of novelty and the resolution of scientific and/or technological uncertainty' (ibid., p. 8).

Recent analysis of R&D activity has argued that firms, including multinational corporations, can no longer sustain an adequate level of innovation solely on the basis of in-house research. Although other responses to this pressure are available, R&D has nonetheless been adopted as a strategy in many cases. Indeed, the experience in R&D is not dissimilar to that of other services, which are being decentralised and outsourced as large companies narrow their focus in a post-Fordist manner. There has, therefore, been an expansion of R&D externalisation and outsourcing over recent years in developed economies. So while outsourced R&D appears relatively minor in proportion to the totality of R&D in the UK economy, it still had a value of £2 billion in 2001, and in this context represents a significant area of economic activity. Furthermore, the proportion of contracted out or extramural research in advanced economies has grown significantly since the 1970s, with a doubling of this in real terms in the United Kingdom between 1985 and 1995 (Whittington, 1990). The share of total business enterprise R&D in the United Kingdom that is contracted out has grown from 5.5 per cent in 1985 to 10 per cent in 1995 (Howells, 1997, p. 7; 1999). There is at the same time evidence to suggest increased internationalisation of R&D outsourcing, with some transnational corporations (TNCs) subcontracting R&D functions to local firms in developing economies (UNCTAD, 2005).

Push factors towards the externalisation of R&D would include the increased complexity, cost, and riskiness of the research process – as more difficult problems are tackled and as consumer tastes become more sophisticated – and the need to spread this risk and draw upon a wider pool of talent by combining the in-house with the outsourced R&D activities. Collaboration is encouraged because many technologies are now required for each new product, and different disciplines must be combined in problem solving (Howells, 1999). In a drive to reduce the innovation cycle time in different industries there is a pressure to improve the interface between basic research and the developmental process that leads directly into innovation. Another reason given for externalising, however, may be that large firms have sought to outsource their more routine work while retaining higher-skilled activities in-house. It may be that a 'core'/'periphery' workforce is now appearing in the R&D sector, with the former working on the less routine tasks and the latter on the more routine ones, although it cannot necessarily be concluded that this division coincides with the in-house/outsourced distinctions.

As regards 'pull' factors, specialist external research agencies provide an opportunity for firms to scan for technological challenges and opportunities in the wider economy, and provide for staff within R&D departments to learn. Howells (1997, 1999) suggests that the use of R&D services firms is likely to be based upon the pursuit of cost reductions, and speed of delivery, as well as access to specialist expertise within large firms. Ringe (1992) in his survey of UK firms found access to specialist expertise to be the most frequently cited reason for using outsourced R&D, together with the access to specialist equipment, additional research labour, timescales, and cost controls. In addition, the expansion of the R&D services sector provides scale economies in R&D that many firms can never achieve on their own. It can, however, be argued that there are corresponding disadvantages from externalising R&D, including the possible loss of learning opportunities and the sacrifice of core knowledge competences (Byatt, 1979).

Growth of an R&D services sector

Taken together, the trends described above have led to the emergence or expansion of a contract research market in which R&D activity is traded and in which there is now a significant cadre of specialist client and contracting firms (Howells, 1997, p. 3; 1999; Readman and Hales, 2000). The capital intensity of R&D has increased as not only automation but also computer-aided modelling and design have come on stream. Indeed, in this context it should be noted that some activities regarded as R&D (such as screening, testing, and analysing samples) can be automated and are now often undertaken by robots and other automated devices. CAD and CAM have improved the interface between the designer and the manufacturer, as have Laboratory Information Management Systems (LIMS). But this kind of automation need not be associated with routinisation and a loss of skill, as in the case of 'discovery research services' provided to pharmaceutical and chemical companies by molecular designers who search for novel drugs and chemical compounds (Howells, 1997, p. 13; 1999). In any case, this increased capital intensity may be conducive to the establishment or expansion of specialist R&D firms serving small businesses that could not afford themselves to set up the facilities they need to use.

It is not unreasonable to assume that the presence of a specialist R&D sector within a locality can itself make a significant contribution to the fortunes of that area by contributing to a high value added, skill-intensive milieu in tune with the knowledge economy. Indeed, it is argued by some that the expansion of R&D services firms can make a significant contribution to the innovation infrastructure of a nation or region by contributing to the flow of R&D into SMEs in particular. But as Howells (1997) points out, there is a policy challenge here in terms of linking SMEs to the R&D services market, and question marks remain (especially after the study outlined here) as to whether local SMEs benefit from these firms. Most R&D services firms prefer, or find themselves in, contractual relations with large firms that have bigger R&D budgets. SMEs may lack not only the resources but also the expertise required to deal effectively with subcontracting R&D companies: 'the simple existence of a large number of contract research and technology organisations in a national or regional economy does not imply that local small firms will be adequately provided for in relation to scientific and technical services' (Howells, 1997, p. 19; and see Cohen and Levinthal, 1990).

The R&D services sector in Leicestershire

This section provides an account of the development and activities of the R&D services sector in Leicestershire, a sub-region of the East Midlands region of England in the United Kingdom. The area contains a significant concentration of employment in R&D services, one that has gone largely unremarked. Equally, the processes underlying the development of this important agglomeration have not been studied. This section therefore explores the path-dependent processes underlying the growth of the sector in the context of ongoing regional economic restructuring. As in many European regions, this involved a rapid decline in the region's traditional manufacturing base and the growth of a new services economy. The ties between previous economic activities and the development of the R&D service sector are strong, however, and the growth of this sector has been shaped by previous industrial arrangements and the wider regional innovation system that underpinned them. Indeed, to some extent the new R&D services sector represents a reconstitution and re-embodiment of knowledge production and diffusion processes that were taking place within the old manufacturing base and its regional innovation system (RIS) – a process driven by the vertical disaggregation of the firm base and the associated rise of outsourcing as a means of servicing businesses' knowledge requirements.

Methodology

The data for this chapter are based on a study of the R&D services sector in Leicestershire (Burfitt and Collinge, 2004). These firms have been defined as those falling within Standard Industrial Classification (SIC) code 73.1, 'research and experimental development on natural sciences and engineering'. The focus of the study is therefore on the private firm component of the contract research community and excludes other organisations such as public laboratories, universities and bridging institutions (Readman and Hales, 2000). It does, however, include one applied industrial research and technology organisation (AIRTO) that now functions as a private 'not for profit' company.

Leicestershire is a sub-region covering the former county of Leicestershire and consists of the two local authority areas of Leicester City Council and Leicestershire County Council. The study involved analysis of secondary data on the R&D sector in Leicestershire and the East Midlands, principally the Annual Business Inquiry run by the Office for National Statistics and provided through NOMIS. The bulk of the data, however, was generated through a survey of firms in the R&D sector in Leicestershire. The survey used a database prepared from Dun & Bradstreet, Companies House, Leicestershire County Council, and the Loughborough Advanced Technology Initiative. A total population of 59 eligible firms was identified, and these firms were surveyed in a programme of telephone interviews in spring 2003. Some 35 firms responded, a rate of 59 per cent. The survey work was complemented by in-depth case studies of five firms.

R&D services employment

The R&D services sector employed 101,049 workers in Great Britain[1] in 2002, some 0.39 per cent of national employment (see Table 12.1). These activities were most strongly concentrated in the South East and Eastern regions of the country. The East Midlands, in contrast, is not a significant region in aggregate terms for this sector, and the sector's share of total regional employment is significantly below the national rate. Table 12.1, however, demonstrates that the geographical distribution of R&D services employment in the region is distinctly uneven. In particular, it is clear that Leicestershire dominates regional employment in the sector (60.2 per cent) and contains a significant concentration of employment (location quotient of 1.44).

Since the mid-1990s the national R&D service sector has grown in line with the national economy, and consequently its share of total national employment has remained relatively stable. But in contrast both to the national picture and to the regional one, where the share of employment in this sector has fallen over this period, R&D service employment has grown rapidly in Leicestershire in this period, increasing from 0.42 per cent of total employment in 0.78 per cent in 2000, although falling back somewhat by 2002.

The growth of the R&D services sector in Leicestershire has occurred in the context of significant economic restructuring. From 1995 to 2002, Great

Table 12.1 R&D service employment in the East Midlands by sub-region, 2002

	R&D services employment	Share of total sub-regional employment (%)	Location quotient
Derbyshire	210	0.06	0.14
Leicestershire	2,214	0.57	1.44
Lincolnshire	223	0.09	0.24
Northamptonshire	111	0.04	0.10
Nottinghamshire	915	0.21	0.53
East Midlands	3,675	0.21	0.54
Great Britain	101,049	0.39	–

Source: NOMIS.

Britain's share of employment in manufacturing fell from 17.6 per cent to 13.4 per cent. This was paralleled by even greater reductions in Leicestershire (28.3 per cent to 21.1 per cent), where manufacturing traditions were strong. Service activities have expanded to replace this lost employment in most instances. In particular, business services have grown nationally in this period from 13.3 per cent to 15.3 per cent of total employment, while in Leicestershire they have increased from 9.3 per cent to 11.6 per cent (a rate of 24 per cent). R&D services form a highly successful component of this sector, with a growth rate in the sub-region of 35 per cent over this period.

Sector profile

The following sections explore the core characteristics, processes of formation, and activities of firms in the sub-region.

Firm characteristics

The recent growth of the R&D services sector in the sub-region is reflected in the age of the firm base: some 68.4 per cent of firms have been established since 1990. This profile can be contrasted with the findings of an earlier study on the high-tech manufacturing sector in the sub-region, where 72 per cent of firms had been established prior to 1990 (Bentley *et al.*, 2002). The survey also indicated that R&D services firms predominantly had their roots in the locality, with 84 per cent originally having been first established in the sub-region. Overall, firms could be characterised as being small to medium-sized, independent (only 6 per cent had their headquarters outside the region), and occupying single sites. There were two significant exceptions to this, however: the Production Engineering Research Association (PERA) and Advantica. Both these firms are large employers in terms of the sector (employing over 250 workers each), and both have their roots in the public sector. PERA operates as an AIRTO, while Advantica is the R&D component spun out of the privatised public utility company British Gas.

Mode of market entry

Building on Howells (1997), firms were classified from the survey findings according to their mode of establishment. Three main forms of development were identified: new firm start-ups, spin-outs from business or higher education, and those with their roots in the public sector (e.g. privatised institutions or private companies belonging to public institutions). Table 12.2 demonstrates that new firm formation was the predominant mode of market entry, with 48 per cent of firms having been established in this manner. Spin-outs were also important, with 40 per cent of firms originating from this source, with higher education institutions (HEIs) being the predominant contributor. This mode of market entry has become increasingly important

Table 12.2 Mode of market entry of R&D services firms

Mode	Source	Percentage of respondents
New firms	Ex-employees of local firms	23
	HE-related employees	9
	Management buy-outs	6
	Other	10
Spin-outs	Business	14
	HEIs	26
Government linked	Privatisation or AIRTO	11

Source: Burfitt and Collinge (2004).

since the mid-1990s, a reflection of the new commercial environment for HEIs nationally (Potts, 2002). Finally, some 11 per cent of firms had their roots in the public sector through privatisation or through the creation of government-sponsored innovation support bodies such as PERA.

Firm activities

The provision of contracted R&D services is at the heart of the activities of these firms, with 69 per cent of firms providing these services. While these services were frequently centred on the technical aspects of R&D, it was also clear that a number of firms offered ancillary R&D services, including R&D project management services, R&D staff recruitment, training for R&D, and training for R&D management. Many firms also provide a range of accompanying advanced producer services aimed at supporting R&D and underpinning innovation such as design services (69 per cent) and the provision of technical data (40 per cent). However, in addition to these R&D-centred activities, many firms also undertake activities such as testing, equipment maintenance, and manufacturing. Therefore, while there are a number of firms that provide R&D services only, the bulk of firms undertake a range of activities centred on a core of R&D work.

The notion of a passive sector undertaking routinised contractual work for innovative manufacturers (see Howells, 1997) was not consistent with the activities of survey respondents. In particular, there was evidence that certain firms had adopted a more proactive approach to secure contracts. A number of firms led or participated in networks undertaking EU or national government funded research. Firms also sought to appropriate knowledge generated in their contractual research; some 40 per cent held patents, 34 per cent had copyrights, 18 per cent had trademarks, and 13 per cent had registered designs. Firms therefore not only received fees for undertaking contractual research but also sought to generate intellectual property (IP). On occasion, ownership of IP was accepted in lieu of fees. Many firms attempted to capitalise on this IP independently, and 46 per cent of firms contracted out the manufacture under licence of products that had resulted from their R&D activities. There was also some evidence of other forms of IP capitalisation such as the creation of spin-out companies by R&D service firms.

Development of the Leicestershire R&D services sector

The R&D services sector in the sub-region is rooted in its business base and also in the surrounding institutional infrastructure provided by HEIs and other public organisations. The following sub-sections take each of these factors in turn in order to identify the path-dependent links between this seemingly new area of activity in the local economy and the sub-region's previous competencies.

Links to the existing industrial base

The sub-region has a strong manufacturing history (Bentley *et al.*, 2002). Despite recent de-industrialisation, this heritage continues to be reflected in a substantial over-representation of this activity as a share of employment relative to the national figure (21.1 per cent compared to 13.3 per cent in 2002). It has particular strengths in low-tech sectors such as textiles, clothing, paper products, and rubber and plastics. However, it also contains high levels of employment in certain medium- and high-tech activities, including mechanical and electrical engineering, precision instruments, and pharmaceuticals. The wider region also shares many of these strengths and has additional strengths in industrial equipment and transport equipment. Clearly, the stage of life cycle of a number of these industries has resulted in their concentration in the sub-region generating negative rather than the previous positive aspects of lock-in.

Table 12.3 demonstrates that the sub-region's R&D services sector relates closely to the engineering and pharmaceutical components of the regional

Table 12.3 Main sectoral markets of R&D services firms (percentage of respondents; multiple responses possible)

Main sectoral markets	Percentage of respondents
All manufacturing industries	17.6
Pharmaceuticals	17.6
Mechanical engineering	14.7
Electrical engineering	11.8
Motor vehicles	11.8
Energy	11.8
Higher education	5.8
Construction	5.8
Others	17.6

Source: Burfitt and Collinge (2004).

economy. It provides data on the key sectoral markets for survey respondents and illustrates that while some firms offered services across a range of manufacturing industries (17.6 per cent), many firms' activities were centred on a limited number of sectors. Pharmaceuticals and mechanical and electrical engineering sectors were identified as key sectoral markets by 44.1 per cent of firms. The wider regional motor vehicles cluster was also important to 11.8 per cent of respondents.

These findings support a generalised view of the development of the sector derived from the survey and case studies in which R&D service firms have emerged to provide materials engineering, structural design, and development services, and other specialist R&D services either directly to the region's aerospace, vehicles, motor sports, and marine sectors or to the network of specialised mechanical and electrical engineering firms that in turn underpin product innovation among the region's original equipment manufacturers (OEMs). A similar process has occurred with the pharmaceutical sector in the region, which supports an associated R&D contract firm base, much of which is located in the sub-region.

It is also the case that the historical sectoral mix within the area has provided it with a business culture attuned to the provision of technical knowledge for external clients. With respect to knowledge production and consumption for innovation, firms involved in the mechanical engineering and precision instruments sectors have been characterised as 'specialised suppliers' (Pavitt, 1984). These provide large-scale production firms in 'scale-intensive' activities with specialised knowledge and experience. The key focus of their activity is on the provision of product innovations for use in other sectors. They are often small, technologically specialised, and with a high degree of resource focused on product innovation. Firms in the electrical engineering and pharmaceutical sectors are viewed as 'science-based' firms with extensive in-house R&D resources. Owing to the fundamental importance of basic science within their activities, they are also associated with strong links with the university base.

The prevalence of both of these types of firms in the sub-region suggests that a portion of sub-regional manufacturing has been typified by a culture of R&D and knowledge-related networking. In this sense it is the competences and practices of these elements of the local economy (Lawson, 1999), competences that have increasingly become available given the absolute decline of the region's industrial base, rather than simply the demand they generate for technological knowledge, that have underpinned the development of the sub-regional R&D services sector.

The role of HEIs

Table 12.2 demonstrated the importance of the HE base as a source of R&D service firms. The sub-region is well represented with regard to the HE sector, which contains three separate institutions and 56,745 students in

2003/2004, giving it a density of 63 students per 1,000 head of population, the highest for any of the East Midlands' sub-regions and well above the regional figure of 40 or the national figure of 32.[2]

The sub-regional HEI sector is also attuned to the historical structure of the regional economy. De Montfort University, for instance, focuses on regional strengths such as textiles, fashion, and design, and its School of Art and Design is credited with underpinning the production of a large pool of design workers feeding the 'Leicester design cluster' (Comedia, 2001). Given the decline in the regional textile base, it and other HEIs are increasingly orienting design courses towards engineering design, a process with clear implications for the R&D service sector. The region's engineering heritage is also reflected in the sub-regional HEI base, and Loughborough University contains the largest engineering faculty in the country (HEFCE, 2001), while Leicester University has a nationally important Space Research Centre. Finally, the region's pharmaceutical strengths are echoed in the presence of a major medical school at Leicester University.

The survey and case studies indicated that this closely aligned knowledge base had contributed to the development of the R&D services sector in a number of ways. First, HEIs have provided a steady and appropriate supply of labour to meet the needs of local industrial employers. This labour pool is of course as much available to R&D service firms as it is to the manufacturing base. In fact, the decline of the regional engineering base has increased the supply of labour available to R&D service firms, as the fall in manufacturing jobs has not been matched by a reduction in student provision: the number of students on engineering courses in the United Kingdom remained stable from 1996 to 2003 while the number of engineering-related jobs (SIC codes 28–35) fell by 23 per cent.

In addition to graduates, staff employed in the sub-regional HEI sector are available to the R&D services sector either as employees or as entrepreneurs. Some 9 per cent of firms in the survey had been established not as formal HE spin-outs but by individuals previously employed in these institutions. These informal start-up activities have recently been augmented by the adoption of a highly proactive approach to spin-out activities. This reflects a new national commitment to develop HEIs as bodies capable of directly boosting economic development through product commercialisation and firm creation. This agenda has been pursued vigorously in the sub-region, particularly at Loughborough University, and some 26 per cent of sub-regional R&D services firms have their roots in the university spin-out process. To some extent this represents a repositioning of the HEI base within the regional innovation system, as it has moved from a primary role of supplying firm inputs in the form of labour and basic research, to one in which it contributes directly to the size of the firm base through its spin-out activities, themselves based on a new focus on applied research.

The role of public research organisations

The sub-regional R&D services sector has also benefited from the presence of PERA, an AIRTO located in the area in 1964. Originally founded by the government as a research and technology organisation to support the area's engineering base, it is now a private 'not for profit' company. PERA provides R&D and technical services to private clients, but the bulk of its activity is centred on the delivery of a range of Department of Trade and Industry initiatives. While it is a major direct provider of R&D services, perhaps its most significant recent contribution to the R&D services sector has been the creation of a substantial pool of R&D labour; given the normally small size of the R&D services sector firms, any firm employing over 250 workers represents a major concentration of labour. In particular, PERA has underpinned the supply of a large number of highly skilled R&D entrepreneurs who have subsequently gone on to establish independent firms in the sub-region. In both the survey and the case studies, PERA was identified as an organisation that current owners of independent R&D service firms had worked for previously. Consequently, much in the way that the role of HEIs in the sub-region has changed, while PERA's direct importance within the regional innovation system has declined in line with the engineering base, it now makes a fundamental contribution to the development of the 'new' R&D services sector as staff members have left and established these firms.

Trajectory of continuity and change

The links between the R&D services sector and the sub-regional economy are strongly apparent. However, the current mode of development for the sector is distinctly non-local. Indeed, the transference of R&D activity from an integrated regional engineering base to an expanded R&D services sector is also associated with a new development trajectory with regard to these competences (Boschma and Lambooy, 2002), one focused on the provision of highly specialised technological services to large firms operating in national and international markets.

Geographical markets

Evidence of the different trajectory of the R&D services sector compared to the manufacturing base from which it has emerged is provided in Table 12.4. It compares the geographical market shares of firms in the sub-regional R&D services sector with those in the high-tech manufacturing base.[3] The implication from the table is that, in general, and taking into account firm size, sub-regional R&D services firms are more strongly oriented towards national and international markets than those in the high-tech manufacturing sector. In contrast, regional markets, while not their dominant market, are more important for high-tech manufacturing firms.

Table 12.4 Sales by geographical market among Leicestershire R&D and high-tech manufacturing firms, by firm turnover

Firm turnover	Sector	Market location (%)		
		East Midlands	UK	International
<£100k	R&D (2003)	25	62	12
	High-tech (2002)	40	54	6
£100k–£1m	R&D (2003)	24	53	25
	High-tech (2002)	23	58	19
>£1m	R&D (2003)	12	43	46
	High-tech (2002)	25	45	31

Sources: Bentley *et al.* (2002); Burfitt and Collinge (2004).

Other data from the survey go some way to explaining the focus on extra-regional markets among R&D services firms. Some 71 per cent of firms identified multinational companies as customers in comparison to only 13 per cent that provide services to SMEs. This reflects the view that the specialised nature of R&D services means that SMEs are less likely to possess the resources or competences to be able to deal effectively with R&D contract firms. In contrast, the resource bases and absorptive capacity of larger firms, particularly multinationals, make them more suitable customers. Consequently, the specialisation at the heart of the R&D services sector necessarily drives a switch towards non-local markets, or to ones densely populated by major corporations.

Innovation partners

A further method through which to explore the development trajectory of the R&D service sector is to examine the geography of networks that firms utilise for innovation. Table 12.5 illustrates geographical patterns of interaction by firm size from the survey. It demonstrates that, in general, national linkages predominate over regional ties, while international linkages are the

Table 12.5 Geographies of cooperation for innovation by firm turnover, 2001–2003

Respondent turnover	Cooperation with any organisation (%)		
	East Midlands	UK	International
<£100k	50	63	0
£100k–£1m	7	87	13
>£1m	0	89	33
All firms	23	80	14

Source: Burfitt and Collinge (2004).

least important. While this general pattern is in line with that found in similar surveys, the rate of interaction at the national level is marked (see Cooke *et al.*, 2000; Kaufmann and Tödtling, 2001). The table also illustrates the importance of firm size in shaping geographies of interaction for innovation (Tödtling and Kaufmann, 2001). In particular, regional ties are significantly more important to very small firms; some 50 per cent of these types of firm have this form of interaction. In contrast, the regional level is of almost no significance as a source of innovation partners among larger firms. Rather, national and international interactions predominate as firm size grows.

These patterns differ markedly to those generated in the previous study of the high-tech manufacturing sector in the sub-region[4] (Bentley *et al.*, 2002), where some 33 per cent of medium-sized firms and 47 per cent of the largest firms had regional innovation linkages of some kind. National linkages were more important to both these firm types (42 per cent and 53 per cent respectively), but this scale of interaction was significantly less important to high-tech manufacturing firms than to R&D service firms of similar sizes (87 per cent and 89 per cent respectively). Accounting for size, R&D service firms in the survey therefore participate in innovation networks that function disproportionately at the national level compared to high-tech manufacturing firms in the sub-region.

Components of path (in)dependence

The sub-region of Leicestershire has been able to develop a significant concentration of R&D service employment and activities over the past decade, a change that has coincided with a period of prolonged decline in sub-regional and regional manufacturing. The rapid growth of this sector identifies it as an especially vigorous part of the new service economy that has emerged in the area. The main commercial activity of this firm base is the provision of contract R&D services of various kinds, though it is clear that firms also undertake a range of more general innovation support activities. In many instances the objective of firms is not simply to obtain fees for contracted services but to generate and appropriate IP that can then be exploited independently.

The development of this sector in the sub-region can be linked to the general growth of knowledge-intensive services across Europe, including the United Kingdom and to the 'unprecedented demands for specialist expertise, at a time when private and public agencies have sought to reduce the overhead costs of employing such expertise in-house' (Wood, 2001, p. 188). However, we should be careful not to view the development of the R&D services sector in the sub-region as the inevitable expression of a series of general restructuring trends. While these processes can no doubt be taken into account, the development of R&D service firms in the sub-region is the specific outcome of a complex reorganisation of knowledge production processes that already existed in the area's established manufacturing-based economy. This reorganisation has in particular involved the redeployment of existing regional competences previously contained within (or focused upon) manufacturing activities, rather than the wholesale development of new activities. Several components of continuity and change can be identified.

Competences

The activities of many pre-existing manufacturing firms in the sub-region were themselves R&D-intense, and involved the provision of technological knowledge to support innovation in their own customers (such as OEMs in the regional vehicle and aerospace industries). The sub-regional economy therefore has a long-established culture of the provision of R&D services, albeit one that was submerged within the manufacturing base. New R&D services firms are clearly dependent in this regard upon the competences and culture of the old regional and sub-regional industrial structure. Indeed, the decline of this manufacturing context implies that these competences have increasingly been released and made *available* to form R&D services firms. But the establishment of an R&D services sector relies upon the formation of a new generation of small enterprises in which these competences can be put to work, enterprises that involve novel business models and forms, including significant use of intellectual property rights (IPR). At the same time, it has required the expansion of outsourcing efficiencies and customers to make use of these competences. That is to say, the production of this continuity therefore relies upon the appearance of these and other key discontinuities.

Markets and networks

The newly emerged R&D services sector is influenced, in terms of the market it serves, by the previous and continuing industrial structure of the sub-regional and regional economies – finding customers particularly from among the mechanical and electrical engineering sectors. But despite this continuity, compared to those sections of the manufacturing base from which they have emerged, R&D services firms are significantly more focused upon national and international markets, and are substantially more likely to participate in national than regional innovation networks. This transformation reflects the highly specialised nature of R&D service activity, where interaction with local SMEs is regarded as unusual and the primary customers are major corporations functioning in national and international markets. We could say that to preserve their sectoral or technical market they have had to sacrifice their traditional geographical market and look further afield for their customers.

Innovative milieu

The emergence of the R&D services sector is also related to the activities of non-firm elements of the regional innovation system. So, HEIs and a major quasi-public facility that had previously provided graduate labour, basic R&D, and applied R&D services to the manufacturing base now provide these same inputs to R&D service firms. However, in a departure from their previous role, these institutions now contribute directly to the size of the firm base through spin-out activities and the provision of highly skilled entrepreneurs able to capitalise on the increased demand for R&D services. Although this shift appears subtle, these additional activities are markedly different from those they had previously undertaken, while remaining rooted in the long-standing research competences of each type of organisation.

Path dependence or independence?

The development of the R&D services sector has been prompted by the decline of the existing industrial base, the release of a strong culture of R&D activity, the emergence of an enlarged market for technological knowledge, and a shift in the activities of non-firm components of the regional innovation system that are themselves densely located in the sub-region. In the process, knowledge production activities previously undertaken in the manufacturing sector have *migrated* to the service base, while their reorientation from regional to national and international markets represents a clear breakout from the cycle of long-standing decline into which the regional manufacturing base is locked.

It could be argued that against a background of sustained manufacturing decline the recent emergence of a specialist R&D services sector in Leicestershire represents a departure from path dependence – indeed, a modest but significant introduction of novelty that breaks with the past. But at the same time it could also be argued that this sector shows a striking continuity with the past, in which important competences have been retained and simply repackaged for a new market context. So which is it – continuity or discontinuity? Perhaps the simplest answer we can give is that the case study illustrates the process of path dependence *through* path independence – through the introduction of novel entrepreneurial roles on the part of long-established institutions, new business forms and models that put old R&D competences to work, and new geographical markets that permit the pursuit of old sectoral markets; novelties that permit traditional strengths to survive into the future.

Conclusions

A significant trend in the production of knowledge has been the growth of R&D outsourcing, although little attention has been given within the literature so far to the growth and behaviour of the R&D services sector. In this chapter we have attempted to redress this deficit by presenting the findings from a study of R&D services firms within the Leicestershire sub-region of the United Kingdom. Great stress has been placed within economic geography over recent years upon the importance of endogenous or localised learning, in which regional proximity is an important feature. This emphasis is perhaps understandable given the prior tendency to externalise innovation and knowledge, and indeed to treat these in reified terms as the formal product of educational and research institutions. But formal R&D remains very important to the knowledge production process, and, as some scholars have observed, there is – or can be – strong complementarity between formal and informal learning, between formal R&D on the one hand and localised learning and proximity on the other. The case study presented in this chapter shows, however, that in some circumstances market pressures lead to the 'clustering' of R&D activities at the national level and the *detachment* of these activities from many local linkages involving localised learning – that in order to harness local skills, R&D may need to detach itself from many local networks in order to benefit from those operating at wider geographical scales. The outsourcing of R&D is undoubtedly one of the most important reorganisations that is currently taking place within knowledge production, but the evidence suggests that this is by no means guaranteed to reinforce the importance of regional proximity in localised learning and may in some sectors produce the opposite effect, strengthening the significance of social or network proximities at national or international levels.

Notes

1 Great Britain includes England, Scotland, and Wales but excludes Northern Ireland, where the Annual Business Inquiry is operated separately.
2 The national figure is for 2001.
3 This study examined the 'high-tech' manufacturing sector in the sub-region of Leicestershire. This sector was defined at the two-digit level as those falling in SIC codes 24, 30, 31, 32, 33, 34, 35, and 73. Some 37 structured telephone interviews were undertaken in 2002.
4 Lists of types of possible collaborators contained in the two separate surveys were highly similar but not identical.

References

Antonelli, C. (1999) The evolution of the industrial organisation of the production of knowledge, *Cambridge Journal of Economics*, 23, pp. 243–260.

Arrow, K. (1969) Classificatory notes on the production and transmission of technical knowledge, *American Economic Review*, 59, pp. 29–35.

Bentley, G., Burfitt, A., Collinge, C., Gibney, J., and MacNeill, S. (2002) *High Tech Manufacturing in Leicestershire*, Report for Leicestershire County Council, CURS, University of Birmingham.

Boschma, R. and Lambooy, J. (1999) Evolutionary economics and economic geography, *Journal of Evolutionary Economics*, vol. 9, pp. 411–429.

Boschma, R. and Lambooy, J. (2002) Knowledge, management structure and economic co-ordination: dynamics and industrial districts, *Growth and Change*, 33, pp. 291–311.

Braczyk, H., Cooke, P., and Heidenreich, M. (eds) (1997) *Regional innovation systems*, UCL Press, London.

Burfitt, A. and Collinge, C. (2004) *The Leicestershire R&D services sector*, Report for Leicestershire Intelligence, CURS, University of Birmingham.

Byatt, I. (1979) *The British electrical industry 1875 to 1914*, Oxford, Oxford University Press.

Camagni, R (ed.) (1991) *Innovation networks: spatial perspectives*, London, Belhaven Press.

Camagni, R. (1999) The city as a milieu: applying GREMI's approach to urban evolution, *Revue d'Economie Régionale et Urbaine*, 3, pp. 591–606.

Coe, D. and Helpman, E. (1993) Internal R&D spillovers, NBER Working Paper no. 4444.

Cohen, W. and Levinthal, D (1990) Absorptive capacity: a new perspective on learning and innovation, *Administrative Science Quarterly*, 35, pp. 25–32.

Comedia (2001) *Creative Industries Study: Phase 1, Final Report*, Stroud, UK, Comedia.

Cooke, P. and Morgan K. (1998) *The associational economy: firms, regions and innovation*, Oxford University Press, Oxford.

Cooke, P., Boekholt, P., and Tödtling, F. (2000) *The governance of innovation in Europe: regional perspectives on global competitiveness*, London, Pinter.

Cooke, P., Roper, S., and Wylie, P. (2003) 'The golden thread of innovation' and Northern Ireland's evolving regional innovation system, *Regional Studies*, 37 (4), pp. 365–379.

Hassink, R. (2005) How to unlock regional economies from path dependency? From learning region to learning cluster, *European Planning Studies*, 13 (4), pp. 521–535.

HEFCE (2001) *The regional mission: the regional contribution of HE – the East Midlands, innovation through diversity*, Higher Education Funding Council For England.

Howells, J. (1997) Research and technology outsourcing, CRIC Discussion Paper no. 6, University of Manchester and UMIST.

Howells, J. (1999) Research and technology outsourcing and innovation systems: an exploratory approach, *Industry and Innovation*, 6 (1), pp. 111–129.

Jaffe, A., Trajtenberg, M. and Henderson, R. (1993) Geographic localisation of knowledge spillovers as evidenced by patent citations, NBER Working Paper no. 3993.

Kaufmann, A. and Tödtling, F. (2001) Science–industry interaction in the process of innovation: the importance of boundary-crossing between systems, *Research Policy*, 30, pp. 791–804.

Keeble, D., Lawson, C., Moore, B., and Wilkinson, F. (1999) Collective learning processes, networking and 'institutional thickness' in the Cambridge region, *Regional Studies*, 33, pp. 319–331.

Kenney, M. and Von Burg, U. (1999) Technology, entrepreneurship and path dependence: industrial clustering in Silicon Valley and Route 128, *Industrial and Corporate Change*, 8 (1), pp. 67–103.

Koschatzky, K. (2004) The role of R&D services in managing regional knowledge generation: a regional differentiation, in Karlsson, C., Flensburg, P., and Hörte, S.-A. (eds) *Knowledge spillovers and knowledge management*, Cheltenham, UK, Edward Elgar.

Lawson, C. (1999) Towards a competence theory of the region, *Cambridge Journal of Economics*, 23, pp. 151–166.

Loasby, B. (1999) *Knowledge institutions and evolution in economics.*, London, Routledge.

Maskell, P., Eskelinen, H., Hannibalsson, I., Malmberg, A., and Varne, E. (1998)

Competitiveness, localised learning and regional development: specialisation and prosperity in small, open economies, London, Routledge.

Meyer-Stamer, J. (1998) Path dependence in regional development: persistence and change in three industrial clusters in Santa Catarina, Brazil, *World Development*, 26 (8), pp. 1495–1511.

Mowery, D. and Rosenberg, N. (1999) *Paths of innovation: technical change in 20th century America*, Cambridge, Cambridge University Press.

OECD (1994) *A summary of the Frascati Manual 1993*, Paris, OECD.

OECD (2001) R&D and productivity growth: a panel data analysis of 16 OECD countries, in science, technology industry, Working Paper 2001/3.

Pavitt, K. (1984) Sectoral patterns of technical change: towards a taxonomy and a theory, *Research Policy*, 13, pp. 343–373.

Potts, G. (2002) Regional policy and the 'regionalisation' of university–industry links: a view from the English regions, *European Planning Studies*, 10 (8), pp. 987–1012.

Readman, J. and Hales, M. (2000) *RTOs in the service economy: country report – UK*, CENTRIM, University of Brighton.

Ringe, M. (1992) *The contract research business in the United Kingdom: the European dimension*, Report to the Directorate General for Telecommunications, Information Industries and Innovation, Commission of the European Communities, Luxembourg.

Shankerman, M. and Scotchmer, S. (2001) Damages and injunctions in protecting intellectual property, *Rand Journal of Economics*, 32, pp. 199–220.

Tödtling, F. and Kaufmann, A. (2001) The role of the region for innovation activities of SMEs, *European Urban and Regional Studies*, 8 (3), pp. 203–215.

UNCTAD (2005) *World investment report: transnational corporations and the internationalization of R&D*, New York: United Nations.

Whittington, R. (1990) The changing structure of R&D: from centralisation to fragmentation, in Loveridge, and Pitt, M. (eds) *The strategic management of technological innovation*, London, Wiley.

Wood, P. (2001) The United Kingdom: knowledge-intensive services and a restructuring economy, in Wood, P. (ed.) *Consultancy and innovation: the business service revolution in Europe*, London, Routledge.

13 Creativity and openness: outsourcing of knowledge-intensive services as a challenge for innovation systems in a metropolitan region

Rüdiger Wink

Introduction

The outsourcing and offshoring of production and services has been recognised as a major trend in the first decade of the twenty-first century (UNCTAD, 2004). Apart from arguments of cost savings, many outsourcing decisions by multinational companies (MNCs) are driven by the expectation of value added through access to external knowledge (Maskell *et al.*, 2005). Despite the observation of outsourcing as a common feature of most sectors, the argumentations, options, institutional designs and regional impact of such decisions differ across firms, regions and sectors. Within this chapter, the specificity of such a process is shown by investigating the relevance of regional characteristics that influence opportunities of MNCs to improve their creativity in the case of aeronautics as one sector requiring specified technological knowledge and cultures of interaction. The region analysed is the metropolitan region of Hamburg, which had the highest per-capita gross domestic product of any EU region in 2004 (€45,363). Within the past decade the aeronautics industry became the dominant manufacturing sector within the region, a result of Airbus's decision to locate its centre of excellence in cabin interior systems and final assembly for all single-aisle models within the region, and the decline of several other industries (Lublinski, 2003). As the aeronautics sector is generally confronted with huge structural changes (Zuliani *et al.*, 2003), sources of organisational, cultural and technological creativity become more evident and give rise to new threats and opportunities to the region.

The chapter is divided into three main parts. In the next section the main theoretical concepts for the investigation are presented and discussed. A special focus will be the relationship between openness and creativity, and its consequences for innovation systems. The second section deals with the specific challenges in the aeronautical sector based on new organisations of technology-based innovation and new spatial patterns of value-chain organisations. The third section provides empirical insights from the metropolitan region of Hamburg and its attempt to build a regional innovation system with linkages to knowledge outside the region. Finally, some conclusions are drawn, with a perspective on future research.

Creativity and openness: a theoretical view

The theoretical basis for the following investigation is rooted in the concepts of regional innovation systems (Braczyk *et al.*, 2004; Harmaakorpi and Melkas, 2005) and learning economies (Cappello and Faggian, 2005). Any knowledge is always based on the individual knowledge base, consisting of individual sets of information, cognitive filters to interpret and select new information, and capabilities to transfer existing information to different problems (Machlup, 1980). Changes of information are created by one's own or foreign experiences. Learning means in this context the capability to consciously or subconsciously connect these experiences with meaning, thus to transfer inflowing data from experiences into information (Rizzello, 2000). This mental model of experiences characterises the frames in which the stored experiences will be used in future situations (Denzau and North, 1994; Scharmer, 2001). As individual experience is limited, interaction is decisive for opportunities to learn and extend the knowledge base. Any interaction, however, is confronted by two major risks: (1) the risk of misperception of the message due to different cognitive patterns by the interacting partners; and (2) the risk of default due to a lack of mutual trust and secure expectations on the credibility of the interacting partner.

The risks of misperceptions are related to the cognitive context of the interaction. The individual cognitive patterns – based on genetically determined preconditions in the human brain and social experiences – determine how an individual interprets messages from interacting partners, connects these with already stored knowledge and decides on the actual meaning of this message for future contexts. If the interacting partners use different communication codes, they will come to different interpretations and conclusions reducing the relevance of the communication for the individual knowledge bases. Therefore, common communication codes serve as standards to reduce costs of misperceptions and cognitive translations (Wink, 2003). These standards have good network characteristics, as the individual benefit of every user is positively correlated to the number of users of this standard: the more individuals understand the code, the more options for communication are given.

The risk of default is caused by the asymmetrical distribution of information between the interacting partners typically discussed within the principal–agent framework (Hart and Holmström, 1987). Only the individual knows whether he or she correctly reveals the experiential knowledge. Therefore, every partner fears being exploited as long as he or she cannot actually prove whether the communication partners respond to the revelation of new experiences with reciprocal interaction. Two different problems occur: quality uncertainties, which means that the receiver actually does not know whether the data received are worth processing and whether the time needed to understand, interpret and apply the data will be wasted and lead to failure; and moral hazard, which includes the risk that a communication partner will be exploited by the other partners, if she is providing her best information but only receives worthless data (Blum and Müller, 2004). These fears can be reduced by common norms based on socio-cultural or legal rules to solve two institutional needs: an institution to reduce quality insecurities by credible signalling or screening; and an institution to overcome incentives for default by credible control and sanctions (Zaheer *et al.*, 1998; Nooteboom, 2002). Again, this requires certainty that all partners will comply with the norms.

Innovation systems offer solutions to the two basic problems of communication codes and institutions. They can prevent default by providing systemic linkages between actors and organisations with different experiential knowledge on the basis of common communication codes to overcome cognitive misperceptions, and common formal or informal institutional norms to increase trust between communication partners. The concrete design of innovation systems differs according to the specific requirements of the knowledge exchanged, the organisations and actors affected, and the historical background of the systemic linkages. A common feature of all innovation systems is that they try to enhance some kind of proximity between the nodes in the system, as proximity is seen as a key prerequisite to overcome both these barriers to learning. Proximity between partners makes it easier to communicate frequently and develop routines in using common communication codes. *Social and geographical* proximity helps to overcome risks of misperception and misinterpretation, as the communication partners have the chance to use frequent and repeated face-to-face (F2F) communication with continuous interaction to test whether the intended message has reached the sender (Bathelt *et al.*, 2004). This F2F communication is not restricted to specific professional events but also exists in private personal contacts (Dahl and Pedersen, 2004). Formal communication codes such as written language can readily be used if there is already a specific joint cognitive dimension (*cognitive proximity*), for example due to common professional or scientific backgrounds (Harhoff *et al.*, 2003). In these cases, publications and manuals are options for communication, although an additional temporary geographical proximity might be necessary to understand specific context conditions of the data provided.

Organisational and institutional proximity are means to build up specific and exclusive communication codes on the basis of formal and informal rules. By the term *organisational proximity* we mean shared formal relations ranging from relatively weak ties based on an inter-organisational contract (e.g. a joint venture) to strong hierarchical organisations with only a low level of autonomy for the individual (Boschma, 2005). Many concepts of knowledge management at the firm level look for necessary prerequisites for communication, including technological solutions, incentives for documentation and formalisation of non-formal experiential data, and opportunities for interaction and creation of codes by routines (Argyris and Schön, 1996; Nonaka *et al.*, 2000; Chen *et al.*, 2004). *Institutional proximity* refers to a more general set of formal or informal rules for individual behaviour (North, 1990). The stability of these institutions is again closely related to social and cognitive proximity, as they can support the effectiveness of interactions and options to sanctions against non-compliance with institutional rules (Coleman, 1986).

Institutions to prevent default are supported by *social proximity*, which creates trust through personal contacts (Nooteboom, 2002). Credibility is built up by personal reputation. Any non-compliance with the expectation of the communication partner will be sanctioned not only by loss of professional contacts but also by loss of personal contacts and social acceptance (Tura and Harmaakorpi, 2005). *Geographical proximity* might support this option by providing opportunities of social control via ongoing F2F contacts between different individuals spreading information on misbehaviour. Sanctions affect the relationship not only between sender and receiver but also that with other possible communication partners within the area (Gertler *et al.*, 2000). *Cognitive proximity* reduces the risks of quality uncertainties and moral hazard because of the lower level of asymmetries. The receivers of data are more able to identify sources of low quality, as they can stick to some formalised hints or can use their own experiential knowledge to test. Sanctions are extended to the loss of professional reputation. *Organisational proximity* might include specialists on examining new data before spreading them within the organisation (Harada, 2003). Sanctions cover the exclusion from the organisation with all its benefits and requests for compensation by the other members of the organisation (Foss, 1999). *Institutional proximity* contributes to the credibility of signalling and screening by securing these instruments with the help of either informal personal sanctions or external – public regulatory – sanctions. Similarly, institutional proximity helps prevent moral hazard by external incentives in contracts, for example shared risks of using data or obligations to compensate for any failure caused by wrong data (Tirole, 1999).

Creativity causes further challenges on learning codes and norms. Creativity means the novel interpretation of existing experiential knowledge or development of new experiences. Most studies refer to the individual skill and mentality requirements to initiate creativity (Florida, 2002; Amara *et al.*, 2005). In this chapter, however, the focus is directed to the structural prerequisites within networks and innovation systems. The difference to or even destruction of the incumbent knowledge by creative ideas causes further uncertainties (Aghion and Tirole, 1994), as communication codes have to be adjusted and new procedures have to be developed to confirm the new experiences and interpretations. Furthermore, creativity requires deviations from incumbent routines and expectations. The more homogeneous the cognitive patterns are and the more strongly the behaviour is predefined by norms, the lower the incentives for deviating behaviour are and the more difficult it is to assert deviations from the existing expectations. Consequently, many learning networks or knowledge clusters are confronted with the risk of lock-in and inertia, restricting the options for creative adjustment (Hassink, 2005). This lack of creativity in development contributes to the persistence of hierarchical patterns between regions: few regions are able to catch up or improve their position, because they are hindered by lock-in effects (Simmie, 2005). Openness of the existing codes and routines for other individuals and ideas can serve as a means to increase the opportunities for creative ideas to gain access to learning networks. This openness can be achieved via a high degree of flexibility of the codes and routines requiring trust in the existing structures to adjust (Nooteboom, 2002). It is the central question of this chapter, how regions can organise such flexible interactive structures within their innovation systems.

Many authors argue that innovation systems make use of gatekeepers having interactions with creative actors outside the network and integrating these experiences into the systems' interactions (Giulian and Bell, 2005; Bathelt *et al.*, 2004). But the challenge remains as to how these gatekeepers receive incentives to fulfil this function and how their input is related to the existing structures within the innovation system. These challenges are rather common for multinational organisations. They have to look for ways of linking decentralised communication partners from different parts of the world together and diffusing the knowledge exchanged among the decentralised units. Orlikowski (2002) provided a set of strategies to enable communication and learning processes in such diversified organisations based on empirical observation – that is, strategies to achieve openness within organisations. Again, these strategies follow different forms of proximity to overcome learning barriers:

- the emergence of a common identity – that is, the achievement of common cognitive frames and interpretation of new experiences (cognitive proximity)
- the intensification of F2F communication – that is, enhancing geographical proximity
- the use of standardisation – that is, improving institutional proximity by joint formal norms
- the creation of individual incentives for knowledge exchange – that is, creating some kind of cultural proximity by following joint objectives
- the promotion of opportunities to participate – that is, arranging organisational proximity with a wide range of possible participants.

These strategies, however, are restricted on a regional level by the framing conditions of the industry investigated. Therefore, we shall take a look in the next section at the specific challenges to proximity in the aeronautical sector.

Creativity and openness: framing conditions in the aeronautical sector

In the aeronautics sector there are several challenges to openness and creativity. One typical characteristic of the aeronautics sector is the specific role of the engineering sector. The development of new materials and new electronic solutions is heavily influenced by the natural sciences and engineering (Vincenti, 1990; Law, 1992). The integration into a complex system and implementation into industrial production require high-level engineering and design capabilities. As a result, the engineers need combinative knowledge to interact with scientists as well as to transfer new research expertise into concrete problem solutions (see de Vries, 2003, and Asheim, 2002, on different types of knowledge in sciences and engineering). Thus, engineering cultures have to be sufficiently open to be able to understand new scientific research solutions and sufficiently creative to be able to translate these results into technological specifications for a concrete industrial artefact. As the cognitive patterns within more abstract sciences and problem-oriented engineering contexts differ (Pitt, 2001), codes and norms have to be adjusted. Within the European consortium Airbus, an additional challenge is caused by the multinational structure of the firm, with shares of the work being distributed to locations in France, Germany, the United Kingdom and Spain (Thornton, 1995; Zuliani and Leriche, 2004). The management of the different interfaces requires openness to the different national specificities in engineering cultures causing specific cognitive patterns (see Wengenroth, 2000, on the different national engineering cultures).

Originally, these challenges were met internally within Airbus, but then the change of procurement strategies towards global modular sourcing led to an increased outsourcing of knowledge-intensive engineering services to integrated engineering system suppliers. These suppliers serve as nodes to translate and diffuse technological knowledge within the value chain, adjusting and standardising communication codes via technological specifications (Gann and Salter, 2000; Muller and Zenker, 2001; Grabher, 2004). This task requires creativity and openness towards new problem solutions.

For decades, Airbus was a company like no other multinational firm. With its specific legal constitution based on an agreement between four governments and the obligation to allocate shares of the work between the single locations according to national engagements, Airbus had to adjust to political requirements. Furthermore, the aeronautics sector has been influenced by the culture of military production and aerospace, where secrecy and national epistemic communities play a bigger role than in international markets. Airbus tried to make the best of its political role and obligations towards different national locations by defining a decentralised structure of 'centres of excellence' (CoE) in 2004 with their own responsibilities and decision-making processes. The CoE mainly refer to the operational parts. So far, six CoE have been defined:

- on wings at Filton and Broughton, UK
- on forward and aft fuselage at Nordenham, Varel, Bremen and Hamburg, Germany
- on nose and centre fuselage at Toulouse, Saint-Nazaire, Nantes and Méaulte, France
- on vertical tailplanes at Stade, Germany
- on pylon and nacelle at Saint Eloi, France
- on horizontal tailplanes at Getafe, Illescas and Puerto Real, Spain.

Additional CoE on cabin and customisation (Toulouse, Hamburg) and electronics (Toulouse) have been established.

This specialisation offers the opportunity of concentrating all knowledge on specific elements for aircraft innovation processes by still having the option of using competition between the locations as an innovation driver. For example, the wing producers in the United Kingdom have been challenged by the achievements of the German and Spanish locations in Stade and Getafe to increase the share of composites as an alternative to metals, because of their lower weight and greater flexibility. As a result of these challenges, new R&D investments have been attracted in the United Kingdom to overcome the deficits in composites. Specialisation also means the challenge to coordinate between the single CoE. As a consequence, Airbus introduced a 'concurrent engineering' programme to enable engineers at all locations to work simultaneously on joint projects. These joint and simultaneous activities help to manage interface problems as well as to develop common technological solutions for core systems of the aircraft. Further efforts have been made to improve the level of interaction between the different locations by staff exchange programmes and support for inter-regional cooperation between public authorities or private associations at the locations.

These processes, however, are still characterised by a relatively high level of integration within the firm. Looking at the references from other industrial sectors, Airbus consequently attempts to reduce this level of integration, wherever it makes sense. With the last two civil projects (A380, A350), Airbus tried to extend the strategy of dual and modular sourcing and combine this strategy with internationalisation. Traditionally, a high number of single components providers produced small shares within the aircraft value chain, with Airbus as the organisation responsible not only for defining technological requirements but also for developing new products in cooperation with R&D institutes and single suppliers. The suppliers look for geographical proximity to Airbus to get a better understanding of the actual needs of their customer and to build a social relationship – as they interpreted it – based on personal contacts to representatives of Airbus. All communication processes in these regional clusters, however, are focused on Airbus, with only weak ties between the single suppliers. This traditional supply chain management changed in the late 1990s. The number of suppliers was drastically reduced to only a few system suppliers. This process has already been realised, particularly in the segments of engineering service providers and at locations in the United Kingdom, while segments such as cabin interior systems needed additional time for adjustment. Suppliers have to offer integrative system products, participate in product development based on formal R&D investments and staff, and take part in the risk of selling the aircraft over a period of decades. Only large, integrated firms have the necessary financial resources and technological skills for these offers. As a consequence, a fierce process of consolidation in the aircraft supply sector started in the 1990s.

In contrast to automotive production, internationalisation in the aircraft markets is restricted by severe public formal quality standards to guarantee safety for the passengers. Furthermore, learning-curve effects are particularly important in the aircraft sector, with newcomers facing the problem of low productivity and quality. As a consequence, reallocation of production into low-cost countries is not as usual as in other industries, and is restricted to more standardised processes with a low share of aeronautics-specific technologies. As a general strategy, however, reallocations to other countries are welcome not only to reduce costs but, particularly, to increase sales markets, because many countries use local content requirements as trade policy instruments in the aircraft markets. Consequently, a large share of the Boeing 7E7 'Dreamliner' was produced in Japan, and a possible new assembly site for Airbus in China was also closely connected to sales of the A320. Further shares of local content requirements are realised by foreign direct investments of suppliers. To overcome the risks of internationalisation without losing its potential, Airbus used its demand power against the suppliers to encourage them to relocate production. Chosen destinations so far are, for example, Middle and Eastern European countries for German and French companies, North Africa for French and Spanish companies, and Asia and South Africa for British companies. The suppliers have to cope with the problem of meeting the formalised quality standards and the necessary transfer of expertise, which requires management skills that are not immediately available in many smaller component supply firms. For the recipients of the investments, this relocation means the opportunity to climb the quality ladder and acquire additional production shares with time (Cantwell, 1989).

The other perspective on internationalisation is the development of a premium quality strategy, which means to look for foreign locations to diversify or extend the knowledge base (Mol, 2005). Although aircraft production so far is mainly based on the knowledge of insiders – specialised engineers and mechanics, mostly produced by the domestic qualification system and controlled by domestic quality standards – new ideas, for example on new materials, new systemic integration, new customer devices, come from other sectors. A typical example is the integration of composites as an alternative material to metal. This material was first tested and introduced in the racing car market, as necessary funds for R&D investments, examination and application were available in these markets. Only after the experiences in the racing car market showed that these materials offer new qualities, such as a reduction of weight and greater flexibility in the event of crashes, was the attention of the big aircraft producers attracted. In earlier times, big industrial firms used to diversify themselves to build up an extended and diversified knowledge base (Garcia-Vega, 2006). Nowadays, they integrate the diversified knowledge from system suppliers, who are organised in international groups and active in different markets with key competences, and concentrate on the linkages between this knowledge and their core processes. The system suppliers for engineering and design services are organised in internationally decentralised structures and come close to the ideal of a transnational (network) company (Bartlett and Ghoshal, 1989; Harzing, 2000), as they have to be close to their main customers to find specific problem solutions, while at the same time connected to other segments of their firm to integrate the knowledge from other sectors or technological problems (Cantwell and Piscitello, 1999). The Airbus suppliers in this field are in the process of increasing the number of international locations in North America and Asia and diversifying their services along aeronautics, automotive and other industrial production systems.

Therefore, the relevance of geographical proximity as a means of organising knowledge interaction has changed. In the case of low-cost standardised segments, codification of knowledge is easier to achieve. As a consequence, quality norms are defined by Airbus or public authorities, and can also be used to assess the quality of products from foreign countries. Hence, the traditional assessments based on social proximity – personal contacts or trust – are substituted by more formal means of exchange such as certification processes (institutional proximity). In the case of premium-quality segments based on new integrative knowledge, cognitive and organisational proximity are enhanced by specialised system suppliers or the original equipment manufacturers (OEMs) themselves, which use a transnational (decentralised) network structure to exploit geographical proximity at different locations in the world but connect the single units by organisational rules and routines. These routines guarantee incentives for the single employees to exchange knowledge with the other members of the organisation and by frequent exchanges of knowledge, which can be supported by temporary geographical proximity such as company meetings or visits to labs in foreign countries to secure a common cognitive code for communication.

This change from integrated R&D and production systems to lean knowledge development and production based on modular sourcing to internationalised system suppliers, however, increases the complexity of knowledge governance. In an integrated system, knowledge is developed within the organisation, and the main task is to improve the facilities and incentives within an internal knowledge management system to make more explicit and codifiable for internal communication those elements of the knowledge that are process specific, stored within routines or based solely on individual capabilities and therefore 'tacit' (Orlikowski, 2002). Any cooperation with suppliers in this integrated system is clearly based on dependence of the suppliers on the definition of objectives, superior financial and market power, and system knowledge of the OEMs. With the modular knowledge sourcing system the asymmetry of knowledge interaction changes (Sanchez and Mahoney, 1996). Here, the OEMs depend on the specific interface capabilities of the system suppliers, which integrate knowledge from other fields and applications, but they need to have the complete system overview to anticipate the needs for compatibility between the different modules developed by system suppliers and the specific requests of their aircraft models in contrast to other products. In the automotive sector many of these complex system overview capabilities have been outsourced to engineering companies. In aircraft production, however, the OEMs still define most of the system characteristics of the model, but they recognise limits in governing the knowledge process.

Summing up, the market approach of innovation supports the increasing spatial diversification of activities. Reorganisations of value-chain systems and restructurings in big, integrated firms such as Airbus affect internationalisation processes and force system suppliers to build up integrative and system capabilities in fields where the OEMs do not observe strategic specificities of their sector and look for inputs from other technological fields. For emerging countries like India and China, these processes open up new opportunities to climb the quality ladder and develop their own specific innovation capabilities. These opportunities will further rise with growing purchasing power in the emerging economies. Customisation of new products according to the needs found in large sales markets will then make it necessary to build up culturally specific capacities for design and market exploitation in these countries. For incumbent aeronautical regions this spatial diversification brings severe challenges. Firms and research institutes need to develop leading-edge knowledge for the sector and to be capable of interacting with partners in other premium-segment (industrialised) regions in the world, while being able to manage low-cost relocations to emerging regions. Existing systemic linkages between firms, universities and research institutes in the regions have to be adjusted to integrate knowledge from other scientific disciplines as well as from other regions. The next section aims to describe these challenges for existing 'engineering cultures'.

Creativity and openness as challenges for engineering cultures?

When we speak of regional innovation systems, we are referring to systemic linkages between single innovation networks to enhance interaction of knowledge between the networks and to increase the innovative capacity of the whole system (Harmaakorpi and Melkas, 2005). These networks have relatively loose structures (compared to formal organisations) and are formed of heterogeneous groups (universities, firms, research units, services organisations, etc.). In the context of aeronautics, access to engineering knowledge plays a major role, as engineering still defines the basic technological paradigms and standards, while the engineers have to integrate knowledge from other (natural scientific) disciplines such as material sciences, information technology or electronics. Therefore, changes in the need for engineering knowledge and access to it critically affect the functionality of the innovation systems. In the fifth section we shall take a look at such changes in the aeronautical agglomeration of Hamburg.

With the decision to allocate the final assembly for single-aisle models to Hamburg and to establish a Competence Centre for Cabin Interior Systems in Hamburg, Airbus caused Hamburg to become increasingly attractive for aerospace suppliers. Historically, Hamburg has already been the location for a large aerospace company, which was finally integrated into the Airbus consortium. The sharp increase of work shares in Hamburg, however, led to increasing demand for aerospace engineers, a demand that could not be completely met by the regional labour market. A further challenge was caused by the development of a modular sourcing strategy by Airbus. Instead of relying on traditional in-house engineering, more and more orders were given to engineering system suppliers, which integrate different engineering services along interfaces. Consequently, this outsourcing process forced engineering companies to grow to be able to integrate different services and cover the risks of large long-term orders. As this integration was not restricted to work shares produced in Hamburg, transnational integration between Hamburg and the other major European Airbus locations had to be developed either by mergers and internal growth or by cooperation.

These changes require creative organisational solutions and openness to knowledge outside the region. Adjustments via the recruitment of engineers from other disciplines were limited, as the knowledge distance between the different engineering disciplines was assessed as being too wide. As a result, the typical advantage of a metropolitan region – the availability of a thick and diversified labour market – was not given in this segment, and openness of the regional markets and learning systems was needed. This is a relatively new phenomenon for the sector, as most companies in the aeronautical regions at Toulouse and Hamburg still report a very small proportion of employees coming from foreign countries. The integration of engineers from other countries not only requires the overcoming of language barriers and the organising of mobility but also requires cultural distances to be coped with. International differences in engineering cultures in general are rooted in the emergence of industries in the nineteenth century (Wengenroth, 2000). In the United Kingdom and United States, most of the engineering knowledge was created by empirical expertise, with only limited impact of scientific knowledge (Hall, 1974). Thus, knowledge transfer was based on communities of practice and face-to-face contacts, and attempts to codify knowledge were restricted until the end of the nineteenth century. These experiences still have an impact on education systems in the engineering science in these countries, which is relatively strongly focused on case studies and practical expertise.

In France, scientific schools were established relatively early, in the eighteenth century. The impact on knowledge emergence within industry, however, was limited, as most engineers were recruited by the state, and the small number of graduates recruited by industry were mainly appreciated as managers. Similarly, the emergence of technical schools in Germany had only a limited impact on industry, as most of the graduates went to the state (König, 1993). The most important challenge for industry in these countries was to catch up with UK industry by getting access to its – tacit – knowledge base, which was mainly done by the illegal recruitment of British workers, smuggling and reverse engineering (Ferguson, 1992). These difficulties brought a greater interest in codification of engineering knowledge to raise the efficiency of transferring imported knowledge. Consequently, new institutes of technology and engineering schools offered an increasing share of formal theory-driven knowledge within their courses in continental Europe after 1870. Furthermore, the role of the state and investment banks for the growing industry was more prominent in continental Europe than in the United Kingdom. This larger share of external stakeholders and shareholders made it essential for engineers to have a language with which they could communicate with the external world (Staudenmaier, 1985). This language could also be used for general proofs-of-principle, making it easier for investors to examine new ideas. As a result, theoretical knowledge became more relevant in continental European engineering and influenced the social status of engineers and their professional bodies in countries such as France, Italy, and Germany. Codification of engineering knowledge was introduced later in the United Kingdom and United States, and still plays a smaller role than in continental Europe. These historical cultural roots still explain differences in curricula and heuristics of engineering knowledge production between the continental European countries and the United Kingdom and United States.

These historical differences in cognitive perspectives, codification needs and social status also affect organisational structures. German engineers achieved a relatively high social status and common identity by cooperating in associations (Karl *et al.*, 2004). Therefore, engineers were not only integrated within industrial companies, but they organised themselves also independently within smaller service firms and partnerships. This strong focus on independence was underlined by exclusive communication codes between specialised engineers, close connections to networks based on leading universities and institutes of technology, and a high relevance of formalised qualifications as market barriers. By contrast, the social status of engineers in the United Kingdom was lower, and dependent more on their performance as practitioners than as formally qualified academics. This makes it easier to integrate engineers within large firms and reduce the incentives of engineers to be 'something special'.

Summing up the last two sections, changing framing conditions have caused geographical proximity to become less important within aeronautical value-chain systems. Regional innovation systems in aeronautical regions need access to knowledge from other aeronautical regions or related techno-logical fields. This transfer, however, is hindered by cultural distances, which restrict mutual communication codes and trust in compliance with coopera-tion agreements between communication partners. On a regional level, regional innovation systems could provide opportunities to develop common communication codes by frequent F2F meetings and common formal and informal rules based on trust and social control. Within inter-regional con-texts, however, these approaches need to be adjusted. In the next section we look at attempts for adjustments in Hamburg as a metropolitan aeronautical region.

Adjustments in the aero-engineering sector in Hamburg

In no other segment within its value chain did Airbus assert its sourcing strat-egy so strictly as in the aero-engineering sector. Within one year the number of direct suppliers was reduced from 700 to seven. In Hamburg the con-sequences are obvious: the number of employees in engineering service firms had never been so high as it was in 2004, and none of the engineering service firms located in Hamburg is still completely independent. Most of them have been integrated within international or national diversified firms, and others initiated formal cooperation with firms in Germany or other European coun-tries. Thus, competitiveness has grown, but independence disappeared. Did openness of learning structures and creativity within the existing regional sys-temic linkages play a role in this process, and did the adjustment in Hamburg reveal any specificities of a metropolitan region?

In the second section the model of Orlikowski (2002) was introduced to show strategic options to promote openness within a single distributed organ-isation. We will use these factors within this section to take a look at possible influences tending to increase openness not only within a single organisation but also within a regional knowledge system. The results are based on a set of interviews executed in spring and summer 2005 with firm representatives and representatives from public authorities and regional research institutes in the region. We define this region as a metropolitan region, as it shows typical characteristics of a metropolitan region: a high population density, a high proportion of employment and sales in service sectors, a diversified set of qualification and research facilities, a huge variety of leisure and cultural ser-vices options, a relatively high proportion of foreign inhabitants, a high employment inflow from surrounding regions, and a dominant role as an administrative and services centre. The region is one of the two main aero-space clusters in Germany, with the other cluster, in the area of Munich, being mainly dominated by power-engine companies. During the past decade, employment has increased sharply, owing to the decision by Airbus to allocate the final assembly and cabin interiors for all single-aisle models to

Hamburg. Meanwhile, more than 30,000 employees work in the aircraft and spacecraft sector in northern Germany, with the City of Hamburg having only 18,000 employees. Including engineering and design services, more than 55,000 employees belong to the aircraft and spacecraft sector in northern Germany (the City of Hamburg has 35,000 such employees). The core companies within the aircraft sector are Airbus SAS (Airbus Germany GmbH), with five production sites within the region (more than 10,000 employees in Hamburg only), and Lufthansa Technik as one of the world market leaders in aeronautics services, specialising in maintenance, repair and overhaul (MRO) services and having its headquarters in Hamburg with more than 7,000 employees. Lufthansa Technik is also an important user of cabin interior services, as it is developing special products for VIP customers. About 350 SMEs with expertise in aerospace products and services are located in northern Germany as a whole. The main expertise in the region is influenced by the agreement within Airbus SAS to allocate responsibilities for development and production of the cabin, fuselage and rudders as well as the final assembly and cabin interiors for the single-aisle models to Hamburg.

A major challenge for the regional innovation system in Hamburg in the context of cabin interior is the relatively weak linkage between the SMEs in this segment so far. The network linkages are dominated by hierarchical relationships between the OEMs and single component suppliers. Research and development is mainly driven by the demand of the OEMs and their facilities. The SMEs are too small and not sufficiently focused to rise within the value chain and to become system suppliers integrating knowledge interactions between supplying firms themselves. Therefore, the foreign engineering service firms are coming into a system that so far has only poor decentralised structures and is mainly dependent on the OEMs. The description of the strategic attempts in the aeronautical region deals therefore also with the interrelationships between these weak – more agglomerative – linkages and the integration of foreign organisations into the system.

The first aspect of the strategic model by Orlikowski refers to the development of a *common identity* within the organisation. In the case of aero-engineering in Hamburg, this means the emergence of identification with the region and its development as a location for the aeronautics industry not only for the firms that are already located in Hamburg but also for those newly attracted by the proximity to Airbus and Lufthansa Technik. A supporting tool for this is a public–private initiative to bundle all relevant players together within a network following the common objective to improve the location condition and the image of the location for customers and new firms. Within the aero-engineering sector in Hamburg, several firms joined together to form an association (the Hanseatic Engineering and Consulting Association – HECAS). Originally, independent engineering firms within the metropolitan region were the members and used the association to present their products jointly to the public. Three of these firms even tried to cooperate to form a bigger service supplier. Within the past three years, however, most of the companies have been taken over by national and international firms. Even subsidiaries of international companies that relocated their engineering offices to Hamburg are now members of the association. This process of forming joint initiatives and visions on an informal and private basis, while remaining open to new members, profits from the historical development of the region. As a typical trade metropolitan region, Hamburg was used to having a sovereign citizenship interested in private (or public–private) partnerships. Furthermore, the ongoing international contacts through transportation, international tourism, and foreign direct investment caused the citizens to acquire the routines and mental outlook needed to deal with new firms and actors within the region. Thus, the metropolitan history and framing conditions support this institutionalising process.

The main driving forces for the common identity, however, are the OEMs. The clear commitment of Airbus and Lufthansa Technik to the production locations in the region – although in large part politically motivated – creates a signal even for newly locating engineering firms that they need to improve the conditions for knowledge sharing and interaction, as they have to build up a long-term profile in the region. Accordingly, the common identity of the organisations is the development of competitive structures against other (and potentially new) Airbus and Boeing locations. Airbus supports this process by its decentralised strategy of specific competences at the CoE according to the location conditions, which means working closely with R&D organisations and regional authorities available in the region, and looking at suitable topics for the region and the CoE. In the case of Hamburg these topics particularly affect the emergence of new qualification schemes and infrastructures to overcome knowledge bottlenecks.

The second aspect refers to frequent *face-to-face contacts*. These forms of communication help to initiate social interaction and trust between the partners. Regional networks in Hamburg support frequent F2F contacts in two ways. First, they organise social events with presentations by regional or foreign actors from the aeronautics segment. These activities include regular meetings with entrepreneurs from other aeronautics locations, in particular Toulouse, and the organisation of meetings with Airbus representatives at other locations. Second, the biggest trade fair for cabin interior systems ('Aircraft Interior') was launched in Hamburg. Again, this temporary event creates opportunities for firm representatives from different countries to interact, and analyse options for cooperation. The attractiveness of a metropolitan region with several cultural highlights and a strong international reputation makes it easier for the organising company to achieve a critical mass of exhibitors and visitors. This intensive interaction also brings incentives for firm or association representatives to visit foreign aeronautics locations. Consequently, actors in Hamburg become more familiar with partners and routines in other countries.

The third aspect refers to *standardisation*. In this context, regional policy actively supports future standardisation processes by joint qualification schemes with the biggest regional aeronautics cluster in Toulouse. These qualification schemes include a joint programme for vocational trainees with mutual stays within firms and exchanges between local universities. These practical experiences help to overcome differences in qualification cultures and cognitive patterns. Similarly, exchange programmes and cooperation with engineering schools in other countries help to overcome skill shortages by the importing of engineers. Again, the attractiveness of this metropolitan region with its urbanisation and cultural diversity increases the incentives for foreigners to immigrate, at least temporarily, as they relate at least an image to the region of destination. Additionally, the OEM support this process by offering training and apprenticeship places for foreign trainees and by participating in recruitment fairs in other regions or countries. Within this sector, regional and company strategies are barely separable.

The fourth and fifth aspects refer to *individual benefits and opportunities to participate*. In this context, Hamburg has relatively low barriers for foreign companies wishing to locate in the region or to take over regional engineering firms. Regional policy did not prohibit the takeovers, and instead promoted the growth of the merged firms via support for new academic qualification schemes and joint recruitment shows in other regions and countries to attract new engineers. Consequently, there is no difference in dealing with regionally bounded or new international firms. Again, long historical experiences with foreign direct investment and intensive international trade within this metropolitan area supported this openness towards new foreign organisations. The clear benefits of these relocations were set out by the OEMs as they announced a restructuring of their value chain and sourcing strategies. As local firms could not offer proof of their capacity for upgrading in the value chain, foreign engineering companies were welcome by Airbus and Lufthansa Technik to fill the gap.

Conclusion

Traditionally, agglomeration advantages of metropolitan areas have been closely connected to the availability of diversified and thick labour markets. Because of these advantages, metropolitan regions are able to adjust to structural changes relatively quickly, as in theory a 'sticky workforce' attracts new firms from growing sectors (Scharmer, 2001; Florida, 2002). In the case of aero-engineering in Hamburg, shortages in the regional labour markets and forces tending to reorganise value-chain systems caused new challenges for the metropolitan region, challenges that could be solved only by adjusting existing systemic links to necessary interactions with partners outside the region or newly locating organisations. The chapter has shown the importance of complementary strategies by dominant OEMs and other regional actors to implement prerequisites for an integration of new partners. The advantages of metropolitan regions should not be restricted to the benefits of an agglomerative labour pool and cultural services but extended to contributions for openness strategies such as historical experiences and mental outlooks concerning foreign exchange, and relatively low barriers for international cooperation due to the reputation and image of the region. These contributions, however, work only in a framework of internationalisation defined by sourcing and innovation strategies of regional OEM. Accordingly, regional as well as organisational issues have to be taken into account to understand why some regions are more successful in opening up regional innovation systems than others.

References

Aghion, P. and Tirole, J. (1994) The management of innovation, *Quarterly Journal of Economics*, 109, pp. 1185–1209.

Amara, N., Landry, R. and Ouimet, M. (2005) Milieux innovateurs: determinants and policy implications, *European Planning Studies*, 13, pp. 939–965.

Argyris, C. and Schön, D. (1996) *Organizational Learning II: Theory, Method, and Practice*, Reading, MA: Addison-Wesley.

Asheim, B. T. (2002) Temporary organisations and spatial embeddedness of learning and knowledge creation, *Geografiska Annaler, Series B: Human Geography*, 84, pp. 111–124.

Bartlett, C. A. and Ghoshal, S. (1989) *Managing across Borders: The Transnational Solution*, Boston, MA: Harvard Business School Press.

Bathelt, H., Malmberg, A. and Maskell, P. (2004) Clusters and knowledge: local buzz, global pipelines and the process of knowledge creation, *Progress in Human Geography*, 28, pp. 31–56.

Blum, U. and Müller, S. (2004) The role of intellectual property rights regimes for R&D cooperation between industry and academia, in Wink, R. (ed.) *Academia–Business Linkages: European Policy Strategies and Lessons Learnt*, Basingstoke, UK: Palgrave Macmillan.

Boschma, R. A. (2005) Proximity and innovation: a critical assessment, *Regional Studies*, 39, pp. 61–73.

Braczyk, H.-J., Cooke, P. and Heidenreich, M. (eds) (2004) *Regional Innovation Systems: The Role of Governances in a Globalized World*, 2nd ed., London: Routledge.

Cantwell, J. (1989) *Technological Innovation and Multinational Corporations*, Oxford: Basil Blackwell.

Cantwell, J. and Piscitello, L. (1999) The emergence of corporate international networks for the accumulation of dispersed technological competences, *Management International Review*, 39, Special Issue 1, pp. 123–147.

Capello, R. and Faggian, A. (2005) Collective learning and relational capital in local innovation processes, *Regional Studies*, 39, pp. 75–87.

Chen, J., Zhu, Z. and Xie, H. Y. (2004) Measuring intellectual capital: a new model and empirical study, *Journal of Intellectual Capital*, 5, pp. 195–212.

Coleman, J. S. (1986) Social structure and the emergence of norms among rational actors, in Diekmann, A. and Mitter, P. (eds) *Paradoxical Effects of Social Behavior*, Heidelberg: Physica-Verlag.

Dahl, M. S. and Pedersen, C. O. R. (2003) Knowledge flows through informal contacts in industrial clusters: myths and realities? DRUID Working Paper 03-01, Copenhagen.

de Vries, M. J. (2003) The nature of technological knowledge: extending empirically informed studies on what engineers know, *Techné*, 6, pp. 1–21.

Denzau, A. T. and North, D. C. (1994) Shared mental models: ideologies and institutions, *Kyklos*, 47, pp. 3–31.

Ferguson, E. S. (1992) *Engineering and the Mind's eye*, Cambridge, MA: MIT Press.

Florida, R. (2002) *The Rise of the Creative Class and how it's transforming work, leisure, community, and everyday life*, New York: Basic Books.

Foss, N. J. (1999) The use of knowledge in firms, *Journal of Institutional and Theoretical Economics*, 155, pp. 458–486.

Gann, D. M. and Salter, A. J. (2000) Innovation in project-based, service-enhanced firms: The construction of complex products and systems, *Research Policy*, 29, pp. 955–972.

Garcia-Vega, M. (2006) Does technological diversification promote innovation? An empirical analysis for European firms, *Research Policy*, 35, pp. 230–246.

Gertler, M., Wolfe, D. and Garkut, D. (2000) No place like home? The embeddedness of innovation in a regional economy, *Review of International Political Economy*, 7, pp. 688–718.

Giuliani, E. and Bell, M. (2005) The micro-determinants of meso-learning and innovation: evidence from a Chilean wine cluster, *Research Policy*, 34, pp. 47–68.

Grabher, G. (2004) Learning in projects, remembering in networks? Communality, sociality, connectivity in project ecologies, *European Urban and Regional Studies*, 11, pp. 103–123.

Hall, A. R. (1974) What did the industrial revolution in Britain owe to science? In McKendrick, N. (ed.) *Historical perspectives: Studies in English Thought and Society*, London: Europa.

Harada, T. (2003) Three steps in knowledge communication: the emergence of knowledge transformers, *Research Policy*, 32, pp. 1737–1751.

Harhoff, D., Henkel, J. and van Hippel, E. (2003) Profiting from voluntary information spillovers: how users benefit by freely revealing their innovations, *Research Policy*, 32, pp. 1753–1769.

Harmaakorpi, V. and Melkas, H. (2005) Knowledge management in regional innovation systems: the case of Lahti, Finland, *European Planning Studies*, 13, pp. 641–659.

Hart, O. and Holmström, B. (1987) The theory of contracts, in Bewley, T. (ed.) *Advances in Economic Theory*, Cambridge: Cambridge University Press.

Harzing, A. W. (2000) An empirical analysis and extension of the Bartlett and Ghosal typology of multinational companies, *Journal of International Business Studies*, 31, pp. 101–120.

Hassink, R. (2005) How to unlock regional economies from path dependencies? From learning region to learning cluster, *European Planning Studies*, 13, pp. 521–535.

Karl, H., Möller, A. and Wink, R. (2004) *Innovation Policy in Germany: An Economic Assessment*, Torino.

König, W. (1993) Technical education and industrial performance in Germany: a triumph of heterogeneity, in Fox, R. and Guagnini, A. (eds) *Education, Technology and Industrial Performance in Europe, 1850–1939*, Cambridge: Cambridge University Press.

Law, J. (1992) The Olympus 320 engine: a case study in design, development, and organization control, *Technology and Culture*, 33, pp. 409–440.

Lublinski, A. E. (2003) Does geographic proximity matter? Evidence from clustered and non-clustered firms in Germany, *Regional Studies*, 37, pp. 453–467.

Machlup, F. (1980) *Knowledge: Its Creation, Distribution, and Economic Significance*, vol. 1: *Knowledge and Knowledge Production*, Princeton, NJ: Princeton University Press.

Maskell, P., Pedersen, T., Petersen, B. and Dick-Nielsen, J. (2005) Learning paths to offshore outsourcing: from cost reduction to knowledge seeking, DRUID Working Paper 05-17, Copenhagen.

Mol, M. J. (2005) Does being R&D intensive still discourage outsourcing? Evidence from Dutch manufacturing, *Research Policy*, 34, pp. 571–582.

Muller, E. and Zenker, A. (2001) Business services as actors of knowledge transformation: the role of KIBS in regional and national innovation systems, *Research Policy*, 30, pp. 1501–1516.

Nonaka, I., Toyama, R. and Nagata, A. (2000) The firm as a knowledge-creating entity: a new perspective on the theory of the firm, *Industrial and Corporate Change*, 9, pp. 1–20.

Nooteboom, B. (2002) *Trust: Forms, Foundations, Functions, Failures and Figures*, Cheltenham, UK: Edward Elgar.

North, D. C. (1990) *Institutions, Institutional Change, and Economic Performance*, Cambridge: Cambridge University Press.

Orlikowski, W. J. (2002) Knowing in practice: enacting a collective capability in distributing organizing, *Organization Science*, 13, pp. 249–273.

Pitt, J. C. (2001) What engineers know, *Techné*, 5, pp. 17–30.

Rizzello, S. (2000) *Cognition and Evolution in Economics*, Jena: Max Planck Institut.

Sanchez, R. and Mahoney, J. T. (1996) Modularity, flexibility, and knowledge management in product and organisation design, *Strategic Management Journal*, 17 (Special Issue Winter), pp. 63–76.

Scharmer, C. O. (2001) Self-transcending knowledge: organizing around emerging realities, in Nonaka, I. and Teece, D. (eds) *Managing Industrial Knowledge: Creation, Transfer and Utilization*, London: Sage.

Simmie, J. (2005) Innovation and space: a critical review of the literature, *Regional Studies*, 39, pp. 789–804.

Staudenmaier, J. M. (1985) *Technology's Storytellers: Reweaving the Human Fabric*, Cambridge, MA: MIT Press.

Thornton, D. W. (1995) *Airbus Industrie: The politics of an international industrial collaboration*, New York: St Martin's Press.

Tirole, J. (1999) Incomplete contracts: where do we stand? *Econometrica*, 67, pp. 741–781.

Tura, T. and Harmaakorpi, V. (2005) Social capital in building regional innovative capability, *Regional Studies*, 39, pp. 1111–1126.

UNCTAD (United Nations Conference on Trade and Development) (2004) *World Investment Report: The shift towards services*, New York: UNCTAD.

Vincenti, W. G. (1990) *What Engineers Know and How They Know It*, Baltimore: Johns Hopkins University Press.

Wengenroth, U. (2000) Science, technology, and industry in the 19th century, Working Paper, Center for the History of Science and Technology, Munich.

Wink, R. (2003) Transregional effects of knowledge management: implications for policy and evaluation design, *International Journal of Technology Management*, 26, pp. 421–438.

Zaheer, A., McEvily, B. and Perrone, V. (1998) Does trust matter? Exploring the effects of interorganizational and interpersonal trust on performance, *Organization Science*, 9, pp. 141–159.

Zuliani, J. M. and Leriche, F. (2004) *Mondialisation et métropolisation: Airbus et les recompositions territoriales à Toulouse et Bristol*, Paris: Editions l'Harmattan.

Zuliani, J. M., Jalabert, G. and Leriche, F. (2002) *Système productif, réseaux internationaux de villes, dynamiques urbaines: les villes européennes de l'aéronautique*, Toulouse: CIEU/CNRS.

14 Boundary spanning and the 'knowledge community'

Hiro Izushi

Introduction

In the literature on innovation, boundary-spanning communication is considered to be a key to successful management of innovation. Communication between individuals or organisational units tends to occur more frequently within certain boundaries than across them. Boundaries of inter-personal or inter-organisational communication exist at different levels both within and outside a business firm. Intra-firm boundaries are found between project teams, functional departments, or product divisions, to name a few. As for boundaries external to firms, they are strengthened when firms are located at different regions or nations as well as when they operate in different industries or sectors. Within a group of individuals flanked by such a boundary, they show a strong tendency to assume similarity in terms of skills, educational backgrounds, and statuses, which in turn brews similarity in their norms and values. This tendency is in part embedded in the evolution of corporate organisations seeking for specialisation and efficiency. For instance, each functional department under the unitary form of enterprises (Chandler, 1977) gathers individuals specialising in similar areas of expertise, such as engineering, marketing, and accounting. Further, iterations of interaction and exchange of information between individuals within a group leads to a convergence of their norms, values, and behaviours through the creation of local languages (Rogers and Bhowmik, 1971; Rogers and Kincaid, 1981). The maturation of an industry or sector represents a case of this. An industry or sector consists of individuals with more similar mindsets towards the end of its life cycle after it has undergone earlier stages of creation of varieties in products, firms, and organisations, their replication, and their selection (Nelson, 1995; Metcalfe, 1998; Malerba, 2002).

There is a tension between efficiency and originality that arises from the social, cultural, and technical homogeneity of a group's members (Conway, 1997).[1] On the one hand, efficient and effective communication takes place most frequently between members who share skills, backgrounds, and statuses. In fact, it is a cumulative, self-reinforcing process. The evolution of local languages and coding schemes through iterative communication within an organisational unit further lubricates exchange of information among its members (Rogers and Bhowmik, 1971). This allows speedier and more accurate exchange of information, thus increasing efficiency in intra-organisational communication.

On the other hand, communication between individuals (or groups of individuals) who are mutually distant in their skills, backgrounds, and statuses is more likely to produce fresh and novel ideas and approaches to problem solving (Rogers and Kincaid, 1981). Outsiders provide a more critical perspective, in part because they are less subject to the convergence pressures within a group of individuals (Boissevain, 1974). A study of US federal R&D projects provides evidence that the level of interaction within a group of scientists and engineers with similar backgrounds shows no relation to the performance of problem solving (Allen, 1977).

Given an inherent drive on the part of an organisation towards efficiency at a steady state of market and technological conditions, organisational boundaries often stifle innovation by deterring coordination, exchange, and combination of different sets of resources (and knowledge in particular), and preventing novel ideas from emerging. As a countermeasure to this, boundary spanning refers to communication and collaboration across individuals or groups of individuals separated by such organisational boundaries. Boundary spanning is important particularly at the state of discontinuous changes, namely when radical new opportunities arise and challenge existing players to reframe what they are doing in the light of new conditions. Under such a state, the existing channels and flows of information may not be appropriate or sufficient to support innovation, and firms need to develop new ones (Tidd *et al.*, 2005, pp. 16–18).

Against that background this chapter highlights a new boundary formed by R&D workers researching into the same area of knowledge: the 'knowledge community'. The idea of the 'knowledge community' derives from a contradiction in the growth performance of advanced economies: in spite of a phenomenal growth in the number of workers devoted to creation of new ideas, advanced economies have exhibited constant mean productivity growth rates during the past fifty years.

A growth puzzle in endogenous growth theory

As stated, during the past 50 years, advanced economies have witnessed phenomenal growths in the number of workers devoted to the creation of new ideas. For instance, the US economy, which is seen as one of the leading economies around the globe, has become more knowledge based in a number of key aspects. The number of R&D scientists and engineers per 10,000 persons in the labour force hit an all-time high of 90 persons in 1999, up from 25 in 1950 and 75 in 1987.[2] Also, those with a bachelor's degree or higher as a percentage of the population aged 25 and over grew from 4.6 per cent in 1940 to 24.4 per cent in 2000.[3] Furthermore, the investment in knowledge creation and utilisation (i.e. the sum of R&D expenditures, software purchases, and public and private spending on higher education) as a percentage of GDP grew from 5.8 per cent in 1991 to 6.8 per cent in 2000.[4] Similar growths in the number of R&D scientists and engineers are found in other advanced economies such as France and Japan (Jones, 1995a).[5] Growths of R&D workers and technology-based entrepreneurs are also observed at a regional level in most of the 15 older members of European Union.[6]

In spite of the growth of R&D workers and technology-based entrepreneurs over the past five decades, productivity growth rates in advanced economies have exhibited rather constant means. For example, labour productivity in the US private business sector grew at an annual average rate of 3.5 per cent between 1948 and 1965. By contrast, the average growth rate from 1990 to 2000 stood only at 2.1 per cent. The rate in the 1990s is even lower than that during the period from 1965 to 1972, when labour productivity grew at an average rate of 2.8 per cent. The trend is the same for total factor productivity (TFP), which measures the ratio of the output of goods and services to a combination of inputs including labour and capital. Even in the period from 1990 to 2000, when the US economy enjoyed a long boom, the TFP of US private businesses grew at an average annual rate of 1 per cent. The growth rate is significantly lower than 2.4 per cent in the period 1948–1965 and 1.6 per cent in the period 1965–1972.[7] Advanced economies in Europe also showed similar long-run trends during the post-Second World War period and experienced declines in productivity growth rates in the 1990s.[8]

The contrasting patterns of productivity growth rates and growths of R&D workers posed a puzzle to researchers of endogenous growth theory (Jones, 1995a, b). Endogenous growth theory emerged in the 1990s, aiming to account for technical progress in the growth process (Malecki, 1997; Armstrong and Taylor, 2000). The traditional Solow–Swan growth model (Solow, 1956; Swan, 1956) leaves technical progress as an exogenous factor outside the economic domain (i.e. the 'manna from heaven' view of technology). In his seminal 1990 article Paul Romer explores this uncharted territory, linking technical progress to production of knowledge by R&D workers at profit-seeking businesses.

The origin of Romer's model can be traced back to the idea of 'research capital' developed by Griliches (1964, 1979, 1980, 1986). When estimating production functions at the firm level, Griliches assumes that technology is a function of two factors: accumulated research capital (i.e. 'knowledge') and other forces affecting technology. Of these, the second factor is considered as disembodied technical change external to the model, which is the same treatment of technical change as in the Solow–Swan model. By contrast, the accumulated research capital is expressed as a linear combination of the number of R&D workers and the level of past research in relation to the current state of knowledge (i.e. a depreciation rate for knowledge created previously) in each year to this date. The model assumes that the amount of new knowledge created in year t is proportional to the stock of R&D

workers (or the real gross investment in research) in that year. Then, if disembodied, external technical change and depreciation of research capital from past research are set aside, the growth of productivity in an infinitesimal period is proportional to the number of R&D workers in that period in this model. Importantly, the Griliches model implies that each piece of knowledge created by R&D workers adds up to a single stock at the firm level, determining the productivity of the firm concerned. In other words, each piece of knowledge has a pervasive effect upon the productivity of the firm (i.e. intra-firm knowledge spillovers).

Romer extends this idea to an economy (assuming intra-economy knowledge spillovers) in addition to the assumption of inter-temporal knowledge spillovers. When examining the role of knowledge in growth, Romer focuses on its informational component, as opposed to skills embodied in individuals, and examines its nature in relation to two aspects of economic goods: whether a good is 'rival' and whether a good is 'excludable'. A pure 'rival' good has the property that its use by one firm or person precludes its use by another. Table as an example any kind of food. If it is purchased and consumed (i.e. eaten), it is not available for any further consumption by another person. On the other hand, a purely 'non-rival' good has the property that its use by one firm or person in no way limits its use by another. Knowledge is normally 'non-rival', as it remains unchanged even after being used by one person. It is available for further use by other individuals, and can be used an infinite number of times. Knowledge is not consumable like raw materials, energy, and food, and is more durable than machinery, appliances, and buildings, unless it is refuted.

'Excludability', the other aspect of economic goods Romer focuses upon, refers to whether the owner of a good can prevent others from using it. For example, when a house is purchased, the house is the owner's property so that the owner can prevent others from living in it. Accordingly, privately owned houses are 'excludable'. On the other hand, air in the open space is normally 'non-excludable', because one cannot prevent others from breathing it. Further, there is a type of economic good called 'public goods', such as public roads, bridges, and parks. Public goods are by definition 'non-excludable', as any member of the public cannot keep others from using them. As for knowledge, it can be either 'excludable' or 'non-excludable'. When a piece of new knowledge is created, the creator can keep it for his or her own use through secrecy, unless another person discovers the same knowledge. Knowledge becomes subject to 'non-excludability' when its creator makes the knowledge known to others or sells a product that embodies the knowledge, thus allowing others to do reverse engineering. One way of preventing 'non-excludability' from coming into effect is to patent knowledge. Legal use of patented knowledge by other persons can only take place when some form of financial compensation is paid to its patentee. Accordingly, knowledge is normally 'non-excludable' when it is known to others, but can be 'excludable' when its copyright is protected through regulations.

As was noted earlier, Griliches's scope of knowledge implicitly focuses on the use of knowledge by its creator alone. While he considers the impact of a firm's research upon its own productivity, he excludes from his analysis the impact the firm receives from research conducted by other firms. In other words, he does not consider the 'non-excludable' aspect of knowledge in his 'research capital' model. In contrast, Romer asserts that knowledge enters into production in both 'non-excludable' and 'excludable' ways. In his model a firm that creates new knowledge and invents a new product (or a new production process) has property rights over the use of the knowledge in the market. Accordingly, no other firm can copy and sell the product. However, the new knowledge that enables the production of the new good also increases the total stock of knowledge in the economy and becomes available to researchers in all firms, stimulating research throughout the economy. In other words, the owner of a new piece of knowledge has property rights over its use in the production of a new product but not over its use in research. This is a departure from Griliches's analysis of research and productivity at the firm level, as Romer takes into account 'spillovers' of knowledge from its creator to other individuals and firms who benefit from it in their research activity.

Romer's model asserts that the rate of productivity growth is proportional to the stock of workers engaged in R&D within the economy. In his model, the economy's productivity is represented by its stock of knowledge that is accessible to all workers engaged in R&D. Each R&D worker is assumed to have the same chance of creating new knowledge (i.e. making innovations) on average. If more workers are engaged in R&D, it is likely that a greater amount of new knowledge is created during each period of time. Accordingly, Romer posits a proportional relationship between them: the amount of new knowledge created within the economy is proportional to its stock of R&D workers in level. Romer further presumes that the amount of new knowledge each R&D worker creates is also influenced by the economy's stock of knowledge. A greater stock of knowledge stimulates research, enhancing productivity of research (i.e. the amount of new knowledge each R&D worker creates each year). He asserts that the amount of knowledge each R&D worker creates each year is proportional to the economy's stock of knowledge. In sum, the total amount of new knowledge created in an economy each year is a function of two factors: the number of workers engaged in R&D, and the level of the economy's stock of knowledge (that is, the economy's productivity level). Thus, Romer reaches his key conclusion that the economy's growth rate of productivity is proportional to the number of workers engaged in R&D.

This positive association between the number of R&D workers and the rate of productivity growth was shared by a number of early R&D-based growth models such as Grossmann and Helpman (1991a, b) and Aghion and Howitt (1992). However, they subsequently came under attack for their failure to account for the diverging trends of R&D workers and productivity growth rates in advanced economies (Jones, 1995a, b).

The 'knowledge community'[9]

There are a number of attempts to modify the Romer model with regard to its failure to account for the diverging trends of R&D workers and productivity growth rates in advanced economies. Those attempts largely fall into two groups. The first approach assumes that innovation becomes harder to come by, referring to a long-run decline in the number of patents registered (Jones, 1995a; Kortum, 1997; Segerstrom, 1998).[10] The second approach asserts product fragmentation over time (Young, 1998; Howitt, 1999). A rising population induces product proliferation and increases the number of intermediate products.[11] Innovation efforts of R&D workers are more thinly spread, keeping productivity from undergoing explosive growth. The model of the 'knowledge community' falls within the second approach but has its own growth implications, distinct from those suggested by the approach.

The model of the 'knowledge community' is based on two phenomena: specialisation of R&D workers, and a widening technological base of industry. First, the model stresses that the learning and understanding of knowledge is a costly process, and each R&D worker understands and uses only a small part of the stock of knowledge at a particular point of time (Griliches, 1994, p. 16). Use of knowledge by R&D workers is not simply to access a stock of knowledge but also to learn and understand it so that they can apply and further develop knowledge to create commercial value in production. Unless learning and understanding of knowledge take place, its application and development for the creation of commercial value does not ensue. Such learning and understanding of technical knowledge requires skills on the part of R&D workers. The ability of an R&D worker to recognise the value of new knowledge, assimilate it, and apply it to commercial ends is critical to his or her capabilities of innovation.

This ability, which Cohen and Levinthal (1990) call 'absorptive capacity', is history dependent. It reflects how much an R&D worker has invested through education, training, and on-the-job learning in an area of expertise he or she specialises in. The history-dependent nature of the capacity forces each R&D worker to focus upon and specialise in a limited area of knowledge. As the theory of human capital demonstrates, skills formation and development incur costs. In particular, R&D workers possess high-cost skills, as their skills are formed through a long period of costly education and training. The costly investment underpinning learning and understanding of knowledge means that the knowledge each R&D worker understands and embodies is limited in its amount and specialised in its area. As the level of technology goes up and each knowledge area accumulates a greater amount of knowledge to be learnt, such specialisation further advances.[12]

The other phenomenon the model is based on is a widening technological base of industry. Granstrand (1998) argues that technological opportunities are generated through the combination and recombination of various technologies, new as well as old. Though not all combinations are technically and/or economically feasible, technological opportunities grow progressively, owing to the exponential growth in the number of possible combinations of technologies. This gives rise to multi-technology products and processes that involve a larger number of technologies at the industry level as well as increasing technology diversification and multidisciplinary R&D at the firm level. Many consumer and business products of today contain a larger number of technologies than before. For example, a study of the automotive industry by Miller (1994) shows that the industry's technology base has expanded from mechanical engineering to take in electromechanical systems, now stretching to include fuel cell technology. A widening technological base has furthered the growth of specialist groups of R&D workers within many industries.

In the model, each R&D worker researches into a specific area to produce state-of-the-art knowledge from which commercial value is created. The area of knowledge in which an average R&D worker stays at a state-of-the-art level and contributes to creation of new knowledge is constrained by his or her limited capacity. Within an industry a number of R&D workers specialise in the same area of knowledge. Call the group of R&D workers researching into the same area of knowledge a 'knowledge community'. It is assumed that, because of his or her limited capacity, each R&D worker belongs to only one 'knowledge community'. An R&D worker in one 'knowledge community' understands knowledge in other 'knowledge communities' at an elementary or intermediate level, which helps to develop knowledge in the area of his or her community. By contrast, an R&D worker does not understand state-of-the-art knowledge outside his or her own community. Also, the R&D worker's creation of state-of-the-art knowledge is confined to the area of his or her own 'knowledge community'. Across 'knowledge communities', spillovers effects of elementary or intermediate knowledge are assumed to be constant, so they can be ignored in model estimation. Within a 'knowledge community' all knowledge is shared by R&D workers to enhance their research. This is an application of Romer's model to a 'knowledge community' rather than to an intermediate product, an industry, or an economy.

A growth in the level of state-of-the-art knowledge within a 'knowledge community' adds commercial value to the industry's product. It also causes further specialisation of a 'knowledge community'. The amount of knowledge an R&D worker needs to master grows as his or her community's knowledge accumulates and the level of state-of-the-art knowledge becomes higher (Jones, 2005). This forces the community's R&D workers to specialise in a narrower area. In the model it is envisaged that the area of knowledge covered by existing 'knowledge communities' becomes narrower, while new 'knowledge communities' emerge by drawing R&D workers from existing communities as well as new entrants into the R&D workforce.

Value added to an industry's product arises from the sum of state-of-the-art knowledge created in the industry's 'knowledge communities'. Commercial value arising from state-of-the-art knowledge is combined across 'knowledge communities' at the industry's product development. Accordingly, an industry's total factor productivity is determined by the average level of state-of-the-art knowledge of its 'knowledge communities' and the total area of state-of-the-art knowledge covered by them. An economy is viewed as consisting of a number of industries, in each of which the above dynamics take place.

In short, the model suggests that a growth of 'knowledge communities' slows down productivity growth by segmenting the R&D workforce into a growing number of groups by specialty and thus deterring sharing of knowledge across the groups. This model is tested against the productivity growth performance of over 30 European regions in the 1990s. In the cross-section growth accounting, the number of 'knowledge communities' in each region is estimated using the data provided by patent applications to European Patent Office (EPO). The EPO divides patent applications by 120 International Patent Classification (IPC) categories. The data provide a proxy for the level of R&D activities by knowledge area, although they have some known limitations.[13] In the estimation it is assumed that the growth rate of the number of 'knowledge communities' is proportional to that of patent applications, while the number of 'knowledge communities' grows at a slower pace than that of patent applications. This introduces variations in the size of 'knowledge communities': the number of R&D workers belonging to a 'knowledge community' is on average greater in an IPC category where more patent applications are filed.

Key results of the model estimation are as follows.[14] First, the model accounts for the productivity growth rates of the regions sufficiently well. The econometric analysis shows that all variables in the model enter the equation significantly. In other words, there is evidence supporting the model that the R&D workforce is divided into a number of groups by specialty, reducing the scale effects of the R&D workforce found in the Romer model.

Second, there are two sources of productivity growth: the growth in the area of state-of-the-art knowledge covered by a region's 'knowledge communities' as a whole; and the creation of new knowledge by R&D workers in each 'knowledge community'. Whereas the first is expressed in part by the growth of the R&D workforce, the second is accounted for by the number of R&D workers in an average 'knowledge community'. Of them, the number of R&D workers in an average 'knowledge community' enters the tested equation less significantly than the growth of the R&D workforce does. This suggests either errors in specifying the number of 'knowledge communities' or potential cross-regional variations in productivity growth rates when the size of 'knowledge community' is controlled for. For the latter, the sources of such cross-regional variations include variations in the degree of boundary spanning across 'knowledge communities' as well as variations in research productivity due to the style of R&D management and the quality of R&D personnel.

Concluding comments

This short chapter describes a model that resolves the contradiction between the significant rate of growth in R&D workers devoted to the creation of new ideas and the constant productivity growth rate exhibited by advanced economies in the past fifty years. It highlights the 'knowledge community' as a group of R&D workers researching into the same area of knowledge. While boundary spanning is normally discussed under the context of organisational units (e.g. teams, functional departments, product divisions, firms), sectoral units (e.g. sectors, industries), or geographical units (e.g. regions, nations), the chapter adds another boundary, defined by the area of knowledge. The model assumes a constant level of spillovers across 'knowledge communities'. However, it is most likely that the level of cross-community knowledge spillovers varies significantly among regions. Some studies, and most notably that of Saxenian (1994), argue that some regions have an industrial system or culture more conducive to boundary-spanning communication than others, thus achieving faster productivity growth. Nonetheless, those studies are largely qualitative, not offering more robust, quantitative evidence. Another account of the variations can be potentially provided by a debate on diversity versus specialisation (e.g. Glaeser *et al.*, 1992; Feldman and Audretsch, 1999; de Lucio *et al.*, 2002). The debate, which has yet to be settled, centres on the question of whether diversity of economic activities in a region, rather than specialisation, facilitates knowledge spillovers and promotes industrial growth. If the diversity thesis holds, then it suggests that the location of diverse yet complementary R&D activities within the same region facilitates boundary spanning better than when they are located apart. There is little doubt that further research is needed in this area.

Another issue is how policy programmes can facilitate knowledge spillovers across 'knowledge communities'. Given the growing numbers of R&D workers and technology-based entrepreneurs, the importance of boundary spanning across 'knowledge communities' to regional competitive advantage and sustainable productivity growth is greater than ever. For innovative businesses, it is imperative to link in a diversity of external sources while managing to connect individuals across disciplines within their organisations. A key role identified in boundary spanning is the gatekeeper who can effectively transfer external ideas and information to his or her project groups (Tushman and Katz, 1980; Katz and Allen, 1982, p. 16). Although gatekeepers are seen to keep relations with experts in a diverse range of fields outside their immediate working environment, their role in cross-disciplinary communication and exploration of ideas needs to be more properly recognised and built into the mechanism of their organisations. At a regional level there may be a role for policy to strengthen such boundary-spanning capacities. Yet it is not clear how this is achieved most effectively. This is in part because studies of technology transfer and learning networks often fall short of paying due attention to boundary-spanning functions and outcomes at a regional level. This calls for a new angle in the research into the role of policy in promoting regional creativity.

Notes

1 Conway (1997) provides a review of the literature on the tension between efficiency and originality. The ensuing discussion about the tension is based on the review.

2 US National Science Foundation (2003) 'National patterns of R&D resources: 2002 data update' (NSF 03-313). Table 8. As the National Science Foundation has revised its definitions of R&D scientists and engineers a number of times, a direct comparison of the 2002 update with earlier definitions needs care. However, numbers taken from National Science Foundation (1989) *Science and Engineering Indicators* and various issues of the *Statistical Abstract of the U.S. Economy* show that the number of R&D scientists and engineers per 10,000 labour force increased from about 25 in 1950 to nearly 80 in 1988, an increase of over three-fold. See Jones (1995a).

3 US Census Bureau, *Educational Attainment in the United States: March 1995*; and *2000 Census of Population and Housing*.

4 OECD, *OECD Science, Technology and Industry Scoreboard*, 2001 edition, p. 146, and 2003 edition, p. 16; also US Department of Commerce, Bureau of Economic Analysis, *National Economic Accounts*.

5 By contrast, Germany and the United Kingdom have experienced either a slow-down of growth or even a decline in the number of R&D workers during the past two decades. Nonetheless, rapid employment growths in high-technology services (including telecommunications, computer and related services, and R&D, where start-ups by technology-based entrepreneurs are often found) have compensated for them since the middle 1990s.

6 Eurostat, http://epp.eurostat.cec.eu.int.

7 US Department of Labor, Bureau of Labor Statistics, 'Major sector multifactor productivity index', www.bls.gov/.

8 Jones (1995a) and Eurostat, http://epp.eurostat.cec.eu.int.

9 This section is based on the author's forthcoming articles.

10 This assumption remains open to question in view of a surge in patenting since the mid-1980s. Whereas there is a view that the surge of patenting reflects an increase in innovation spurred by changes in the management of research, an explosion of new firm formation in high-technology industries, and a growth in venture capital organisations (Kortum and Lerner, 1999), others argue that major changes in patent regimes and new corporate behaviours known as 'patent port-folio races' are more responsible (Hall and Ziedonis, 2001; OECD, 2004).

11 A variant of this approach is Young's (1998) assumption of an increased variety of differentiated solutions to similar products. Young argues that the continued improvement of a greater variety of differentiated solutions requires additional research inputs, thus suppressing productivity growth rates.

12 Drucker (1999) observes increased specialisation of skilled workers in a wide range of knowledge-intensive occupations, which results in their status of 'associates' rather than 'subordinates' within corporate hierarchies, owing to their superior expertise relative to that of people occupying higher positions.

13 For instance, the propensity to patent is known to vary widely across industries (Pavitt, 1982). Also, many patents turn out to be worthless, while a few are extremely valuable. Yet patent statistics are the most widely available data on research outputs (Griliches, 1990). There is some evidence that suggests close

association between patents and other productivity-based measures at the national and regional levels (Acs *et al.*, 2002).

14 Detailed discussions of the model's formula and its econometric test will be found in the author's forthcoming articles.

References

Acs, Z. J., Anselin, L., and Varga, A. (2002) Patents and innovation counts as measures of regional production of new knowledge. *Research Policy* 31 (7), pp. 1069-1085.

Aghion, P. and Howitt, P. (1992) A model of growth through creative destruction. *Econometrica*, 60, pp. 323–351.

Allen, T. (1977) *Managing the Flow of Technology: Technology Transfer and the Dissemination of Technological Information within the R&D Organization.* Cambridge, MA: MIT Press.

Armstrong, H. and Taylor, J. (2000) *Regional Economics and Policy*, 3rd ed. Oxford: Blackwell.

Boissevain, J. (1974) *Friends of Friends: Networks, Manipulations and Coalitions.* Oxford: Basil Blackwell.

Chandler, A. (1977) *The Visible Hand.* Cambridge, MA: Harvard University Press.

Cohen, W. M. and Levinthal, D. A. (1990) Absorptive capacity: a new perspective on learning and innovation. *Administrative Science Quarterly* 35 (1), pp. 128–152.

Conway, S. (1997). Strategic personal links in successful innovation: link-pins, bridges, and liaisons. *Creativity and Innovation Management* 6 (4), pp. 226–233.

de Lucio, J. J., Herce, J. A. and Goicolea, A. (2002) The effects of externalities on productivity growth in Spanish industry. *Regional Science and Urban Economics* 32 (2), pp. 241–258.

Drucker, P. F. 1999. *Management Challenges for the 21st Century.* Oxford: Butterworth-Heinemann.

Feldman, M. P. and Audretsch, D. B. (1999) Innovation in cities: science-based diversity, specialization and localized competition. *European Economic Review* 43 (2), pp. 409–429.

Glaeser, E. L., Kallal, H. D., Scheinkman, J. A. and Shleifer, A. (1992) Growth in cities. *Journal of Political Economy* 100 (6), pp. 1126–1152.

Granstrand, O. (1998) Towards a theory of the technology-based firm. *Research Policy* 27 (5), pp. 465–489.

Griliches, Z. (1964) Research expenditures, education, and the aggregate agricultural production function. *American Economic Review* 54 (6), pp. 961–974.

Griliches, Z. (1979) Issues in assessing the contribution of research and development to productivity growth. *Bell Journal of Economics* 10 (1), pp. 92–116.

Griliches, Z. (1980) Returns to research and development expenditures in the private sector. In J. W. Kendrick and B. N. Vaccara (eds) *New Developments in Productivity Measurement and Analysis.* Chicago: University of Chicago Press.

Griliches, Z. (1986) Productivity, R&D, and basic research at the firm level in the 1970's. *American Economic Review* 76 (1), pp. 141–154.

Griliches, Z. (1990) Patent statistics as economic indicators: a survey. *Journal of Economic Literature* 28 (4), pp. 1661-1707.

Griliches, Z. (1994) Productivity, R&D, and the data constraint. *American Economic Review* 84 (1), pp. 1–23.

Grossman, G. M. and Helpman, E. (1991a) Quality ladders and product cycles. *Quarterly Journal of Economics* 106, pp. 557–586.

Grossman, G. M., and Helpman, E. (1991b) Quality ladders in the theory of growth. *Review of Economic Studies* 58 (1), pp. 43–61.

Hall, B. H. and Ziedonis, R. H. (2001) The patent paradox revisited: an empirical study of patenting in the U.S. semiconductor industry, 1979–1995. *RAND Journal of Economics* 32 (1), pp. 101–128.

Howitt, P. (1999) Steady endogenous growth with population and R&D inputs growing. *Journal of Political Economy* 107 (4), pp. 715–730.

Jones, B. F. (2005) The burden of knowledge and the 'death of the renaissance man': is innovation getting harder? NBER Working Paper no. 11360.

Jones, C. I. (1995a) R&D-based models of economic growth. *Journal of Political Economy* 103 (4), pp. 759–784.

Jones, C. I. (1995b) Time series tests of endogenous growth models. *Quarterly Journal of Economics* 110 (2), pp. 495–525.

Katz, R. and Allen, T. (1982) Investigating the not invented here (NIH) syndrome: a look at the performance, tenure, and communication patterns of 50 R&D projects. *R&D Management* 12 (1), pp. 7–19.

Kortum, S. S. (1997) Research, patenting, and technological change. *Econometrica* 65 (6), pp. 1389–1419.

Kortum, S. and Lerner, J. (1999) What is behind the recent surge in patenting? *Research Policy* 28 (1), pp. 1–22.

Malecki, E. J. (1997) *Technology and Economic Development: The Dynamics of Local, Regional and National Competitiveness*, 2nd ed. Harlow, UK: Longman.

Malerba, F. (2002) Sectoral systems of innovation and production. *Research Policy* 31 (2), pp. 247–264.

Metcalfe, J. S. (1998) *Evolutionary Economics and Creative Destruction*. London: Routledge.

Miller, R. (1994) Global R&D networks and large-scale innovations: the case of the automobile industry. *Research Policy* 23 (1), pp. 27–46.

Nelson, R. R. (1995) Recent evolutionary theorizing about economic change. *Journal of Economic Literature* 33 (1), pp. 48–90.

OECD (2004) *Patents and Innovation: Trends and Policy Challenges*. Paris: OECD.

Pavitt, K. (1982) R&D, patenting and innovative activities: a statistical exploration. *Research Policy* 11 (1), pp. 33–51.

Rogers, E. and Bhowmik, D. (1971) Homophily–heterophily: relational concepts for communication research. *Public Opinion Quarterly* 34, pp. 523–538.

Rogers, E. and Kincaid, D. (1981) *Communication Networks*. New York: Free Press.

Romer, P. M. (1990) Endogenous technological change. *Journal of Political Economy* 98 (5), pp. S71–S102.

Saxenian, A. (1994) *Regional Advantage: Culture and Competition in Silicon Valley and Route 128*. Cambridge, MA: Harvard University Press.

Segerstrom, P. S. (1998) Endogenous growth without scale effects. *American Economic Review* 88 (5), pp. 1290–1310.

Solow, R. M. (1956) A contribution to the theory of economic growth. *Quarterly Journal of Economics* 70, pp. 65–94.

Swan, T. W. (1956) Economic growth and capital accumulation. *Economic Record* 32, pp. 334–361.

Tidd, J., Bessant, J., and Pavitt, K. (2005) *Managing Innovation: Integrating Technological, Market and Organizational Change*, 3rd ed. Chichester, UK: Wiley.

Tushman, M. and Katz, R. (1980) External communication and project performance: an investigation into the role of gatekeepers. *Management Science* 26 (11), pp. 1071–1085.

Young, A. (1998) Growth without scale effects. *Journal of Political Economy* 106 (1), pp. 41–63.

Index